Y0-BSX-048

STUDIES IN APPLIED MECHANICS 15

Computational Methods for Predicting Material Processing Defects

STUDIES IN APPLIED MECHANICS

STUDIES IN APPLIED MECHANICS 15

Computational Methods for Predicting Material Processing Defects

Proceedings of the International Conference on Computational Methods for Predicting Material Processing Defects, September 8–11, 1987, Cachan, France

Edited by

M. Predeleanu

Laboratoire de Mécanique et Technologie, E.N.S. de Cachan, C.N.R.S., Université Paris VI, Cachan, France

ELSEVIER

Amsterdam — Oxford — New York — Tokyo   1987

ELSEVIER SCIENCE PUBLISHERS B.V.
Sara Burgerhartstraat 25
P.O. Box 211, 1000 AE Amsterdam, The Netherlands

Distributors for the United States and Canada:

ELSEVIER SCIENCE PUBLISHING COMPANY INC.
52, Vanderbilt Avenue
New York, NY 10017, U.S.A.

ISBN 0-444-42859-3 (Vol. 15)
ISBN 0-444-41758-3 (Series)

PREFACE

The predicting of defects arising in mechanical forming processes is one of the main objectives of the theoretical and applied research in this field, with obvious economic implications. From these investigations, forming limitations and evaluation of straining effects induced in the material during its processing history are to be determined through an optimization approach.

This book contains the papers presented at the International Conference on Computational Methods for Predicting Material Processing Defects, held at Cachan, France, during the period 8th - 11th September 1987.

Using recent advances in finite strain plasticity and viscoplasticity, damage modelling, bifurcation and instability theory, fracture mechanics and computer numerical techniques, new approaches to mechanical defect analysis are proposed.

Appropriate methods for explaining and avoiding the defects leading to fracture, high porosity, strain localization or undesirable geometrical imperfections are presented. In addition, some papers are devoted to new formulations and new calculation algorithms to be used for solving the forming problems.

Finally, two papers deal with physical description of defects occurring in forming and cutting operations, focusing on the academical and practical interest of these topics.

June 1987   M. Predeleanu

STRUCMAT 87 : International Conference on Computational Methods
for Predicting Material Processing Defects,
September 8-11, Cachan, France.

Organised by : Laboratoire de Mécanique et Technologie
(E.N.S Cachan, C.N.R.S., Université Paris VI)

In collaboration with : G.R.E.C.O. "Grandes déformations et endommagement",
G.I.S. "Mise en forme".

International Advisory Committee :

T.	ALTAN	U.S.A.
B.	AVITZUR	U.S.A.
B.	BAUDELET	France
N.	BAY	Danemark
J.L.	CHENOT	France
M.M.	DENN	U.S.A.
D.	FRANÇOIS	France
S.K.	GHOSH	G.B.
J.W.	HUTCHINSON	U.S.A.
W.	JOHNSON	G.B.
P.	LADEVEZE	France
J.	LEMAITRE	France
H.	LIPPMAN	R.F.A.
A.G.	MAMALIS	Grèce
Z.	MARCINIAK	Pologne
R.M.	Mc MEEKING	U.S.A
G.	OUDIN	France
G.	POMEY	France
M.	PREDELEANU	France, Conference Chairman
O.	RICHMOND	U.S.A.
F.	SIDOROFF	France

CONTENTS

Computational Methods for Predicting Material Processing Defects, edited by M. Predeleanu
Elsevier Science Publishers B.V., Amsterdam, 1987 — Printed in The Netherlands

NUMERICAL SIMULATIONS OF DYNAMICAL METAL FORMING

F. ARNAUDEAU[1] and J. ZARKA[2]

[1] MECALOG, 68 Quai de la Seine, 75019 Paris.
[2] Laboratoire de Mécanique des Solides, Ecole Polytechnique, 91128 Palaiseau, (FRANCE).

SUMMARY

The **Radioss** system enables the solution of 3D **transient dynamics** of highly non-linear problems during dynamical metal forming or cutting and thus allows the full predictions of their eventual defects. The equations are written for large strains in an arbitrary moving frame which gives the use of a wide range of spatial discretisation from full Lagrange to full Euler. The numerical techniques are based on finite element for the spatial discretization and on an explicit integration scheme for the discretization relative to time. Various examples show the power and the utilities of this system.

NOTATIONS

$\vec{b}$	body forces with components b_i
$\vec{\gamma}$	acceleration.
Δt	time step
$\mathbf{D}$	strain rate tensor
$\dfrac{d}{dt}$	material time derivative
μ	shear modulus
ρ	density
$\mathbf{s}$	deviatoric stress tensor
σ	Cauchy stress tensor with components σ_{ij}
S_F	external surface where forces are prescribed
S_U	external surface where displacements are prescribed
$\vec{v}$	velocity with components v_i
$\mathbf{V}$	gradient of velocities with components v^i_j
Ω	rotation tensor

PRINCIPLES OF THE SYSTEM

Numerical formulations and techniques are based on the works of various authors :

i)   (ref. 1) (ref. 2) and (ref. 3) for the large displacement formulation and time integration scheme,

ii) (ref. 4) (ref. 5) for finite element discretization,

iii) (ref. 1) (ref. 5) for the LAG (Lagrangian) interfaces description,

iv) (ref. 6) for the ALE (Arbitrary Lagrangian Eulerian) formulation.

Field equations

The reference frame is taken as the actual configuration in the Lagrangian formulation or as a moving one with an arbitrary velocity w in the ALE formulation.

The continuum is discretized with 8-nodes bricks finite elements, the velocities field is related to the nodal values via the shape functions Φ_I :

$$v_i = [\ \Phi_I\ ] \cdot \begin{bmatrix} v_i^I \end{bmatrix} \quad I=1,8 \quad i=1,3.$$

Then the equations of motion are written in matrix notation :

$$M\ \{\ \frac{dv}{dt}\ \} = M\ \{\ \gamma\ \} = \{\ F^{ext}\ \} + \{\ F^{bod}\ \} + \{\ F^{trm}\ \} - \{\ F^{int}\ \} + \{\ F^{hgr}\ \}$$

where

$$M = \sum_e \int_V \rho\ [\ \Phi_I\ ]\ [\ \Phi_J\ ]\ dV \quad \text{is the mass matrix,}$$

$$\{\ F^{ext}\ \} = \sum_e \int_S \Phi_I\ F_i\ dS \quad \text{is the externally applied loads vector,}$$

$$\{\ F^{bod}\ \} = \sum_e \int_V \Phi_I\ \rho\ b_i\ dV \quad \text{is the body forces vector,}$$

$$\{\ F^{int}\ \} = \sum_e \int_V \sigma_{ij}\ \frac{\partial \Phi_I}{\partial x_j}\ dV \quad \text{is the internal forces vector,}$$

$$\{\ F^{hgr}\ \} = \sum_e f^{hgr} \quad \text{is the hourglass resistant forces vector and is described later,}$$

$$\{\ F^{trm}\ \} = \sum_e f_{iI}^{trm} \quad \text{is the momentum transport forces in an ALE formulation.}$$

Numerical schemes

If one uses a diagonal mass matrix M, the above equations are straightforward to solve.

We use one integration point in order to greatly reduce the calculation cost even by taking into account the need to control zero energy modes or hourglass modes. This implies that the volume integral are simply :

$$\int_V F(x,y,z) \, dV = F_{\text{integration point}} \cdot \text{Volume}$$

i) Time integration

Central difference time integration is used to compute velocity :

$$v^{n+1/2} = v^{n-1/2} + \gamma^n \cdot \Delta t$$

The scheme is of Δt^3 order of accuracy.

The stability condition or Courant condition is :

$$\Delta t \leq k \ \Delta l/c$$

where, k is a safety factor, Δl is a characteristic element length, $c = \sqrt{(\partial P/\partial \rho)_s + 4/3 \ \mu}$

is the sound speed. The characteristic element length Δl is the volume divided by the largest side surface. The sound speed is calculated according to each material law.

For the ALE scheme, the Courant condition is modified as follows : the sound speed is

replaced by an effective velocity $c + v$, where $v = \sqrt{\sum_i \frac{1}{8} \sum_{I=1}^{8} \left(v_i^I - w_i^I\right)^2}$

ii) Constitutive laws

The major task is the computation of internal forces $\{ F^{int} \}$ which are linked to the particular materials of the continuum. Various material laws are in the system :

Elastic plastic or viscoplastic for metals, hydrodynamic law for fluids, high explosive according to Wilkins (ref. 1), special laws for concretes and rocks. Every law is compatible with an ALE formulation.

For example, in the case of an isotropic elastic-plastic material, the procedure is the following (ref. 2) :

$$\sigma \ (t+\Delta t) = \sigma \ (t) + \dot{\sigma} \ \Delta t \ .$$

With

$$\dot{\sigma} = \aleph + \overset{*}{\sigma}$$

$$\overset{*}{\sigma} = V \ \sigma \ (t) + \sigma \ (t) \ V - \sigma \ (t) \ \text{div}(\vec{V})$$

$\aleph$ is the Truesdell objective derivative of the stress tensor

The following steps are executed in order to obtain $\dot{\sigma}$,

$$V = D + \Omega$$

$$D = D^e + D^p$$

$$\dot{\sigma} = L (D - D^p)$$

where D^p the plastic strain rate tensor is such that :

$$s^e = s(t) + \overset{*}{s} \, \Delta t + 2\mu \, dev(D) \, \Delta t$$

$$s^{pa} = s^e - 2\mu \, \Delta t \, D^p$$

we compute the Von Mises equivalent stress and plastic test criterion :

$$\sigma^e_{eq} = \sqrt{\frac{3}{2} s^e_{ij} s^e_{ij}}$$

if $\sigma^e_{eq} \leq \sigma_y \rightarrow$ the increment is elastic and

$$s^{pa}_{ij} = s^e_{ij} \, , \, D^p = 0$$

if $\sigma^e_{eq} > \sigma_y \rightarrow$ the increment is plastic and

$$D^p_{eq} \, \Delta t = \frac{\sigma^e_{eq} - \sigma_y}{3\mu + h} \, , \, h = \frac{d\sigma_y}{d\epsilon^p} \text{ is the hardening modulus}$$

$$\epsilon^p_{eq} (t+\Delta t) = \epsilon^p_{eq} (t) + D^p_{eq} \, \Delta t$$

$$\sigma_y (t+\Delta t) = a + b \, \epsilon^{p \, n}_{eq}$$

and

$$D^p \, \Delta t = \left(\frac{\sigma_y}{\sigma^e_{eq}} -1 \right) s^e$$

iii) Hourglass resistance

Hourglass modes are deformation modes that cannot be resisted with the use of one integration point. There are 4 hourglass modes (α) per direction (i) for the 3D element. The formulation by Koslov and Frasier (ref. 7) is used :

Define hourglass normalized vectors Γ^α :

$$\Gamma^1 = (1 ,-1 , 1 ,-1 , 1 ,-1 , 1 ,-1)$$

$$\Gamma^2 = (1 , 1 ,-1 ,-1 ,-1 ,-1 , 1 , 1)$$

$$\Gamma^3 = (1 ,-1 ,-1 , 1 ,-1 , 1 , 1 ,-1)$$

$$\Gamma^4 = (1 ,-1 , 1 ,-1 ,-1 , 1 ,-1 , 1)$$

Define hourglass velocity rates :

$$\frac{\partial q_i^\alpha}{\partial t} = \Gamma_I^\alpha \; v_i^I$$

The hourglass resisting forces are :

$$f_{iI}^{hgr} = \sum_\alpha \frac{h}{4} \rho \, c \, V^{\frac{2}{3}} \; \frac{\partial q_i^\alpha}{\partial t} \; \Gamma_I^\alpha$$

where i is the direction index, I is the node index and α is the mode index, h is a scaling coefficient given in input.

iv) Grid velocity

In the ALE formulation, it is necessary to define the grid velocity which is computed, according to Donea (ref. 6), in the form :

$$w_I^{t+\Delta t} = \frac{1}{N} \sum_J w_J^t + \frac{\alpha}{\Delta t} \sum_J L_{IJ}^t \sum_J \frac{u_J^t - u_I^t}{N^2 \, L_{IJ}^t}$$

with the limitation : $1 - \chi < \dfrac{w}{v} < 1 + \chi$;

N is the number of nodes connected to node I,

L_{IJ} is the distance between node I and node J,

α , χ are given in input.

v) Material interfaces

Interfaces between materials are of two types.

The first type is an interface between a LAG (lagrangian) material and an ALE (arbitrary lagrangian eulerian) material. Tangential flow is free along the lagrangian surface. The algorithm (ref. 6) is :

      - Compute the velocities of structural (lagrangian) nodes with added masses and forces from the fluid (ALE) nodes

      - Compute the material velocities, in the normal direction, of the fluid nodes from the velocities of the lagrangian nodes.

      - The grid velocities of the fluid nodes are set equal to the velocities of the structural nodes.

Around corners the normal direction is a function of the element side surfaces in order to minimize mass losses.

The second type is an interface between two Lagrangian materials. The basic methodology of (ref. 5) is used and is improved. Lengthless springs with rigidity K are introduced between the materials and whenever a penetration δ occurs a normal force $F = - K \delta$ is introduced to reduce this penetration .

APPLICATIONS

Taylor cylinder

This test is widely used to characterize material constitutive laws in dynamics. It is a good illustration of the differences between a Lagrangian and an ALE calculations. The material is copper, the impact velocity is 227 m/s, the initial diameter is 0.64cm, the initial length is 3.24cm.

The results of both LAG and ALE simulations are the same (Fig. 1 , Fig. 2), however the CPU time in the ALE case is three times less.

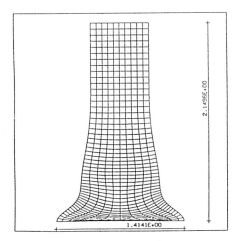

Fig. 1. Final shape at 80 μs, LAG calculation.

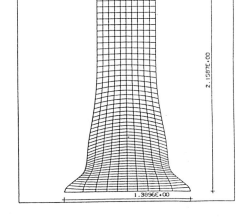

Fig. 2. Final shape at 80 μs, ALE calculation.

Explosively formed projectile

Fig. 3 and Fig. 4 show the forming of a metallic projectile by the detonation of a cylindrical block of explosive.

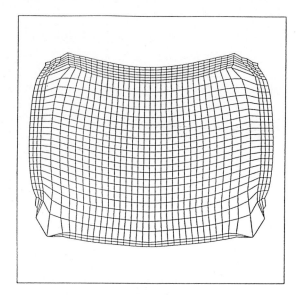

Fig. 3. Deformed mesh during the detonation phase

Fig. 4. Forming of the projectile

Dynamical cutting

The physical phenomena of interaction between the tool and the specimen are extremely complex : friction, adiabatical shear bands, heating. Fig. 5 and Fig. 6 illustrate the stationnary problem in plane strain without friction.

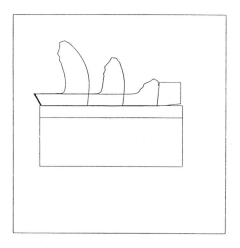

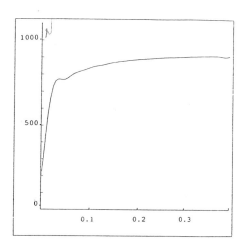

Fig. 5. Formation of the chip

Fig. 6. Stabilized force on the tool.

REFERENCES

1 M. Wilkins, Calculation of elastic plastic flow, LLNL UCRL-7322 (1981)
2 B. Halphen, Sur le champ des vitesses en thermoplasticité finie, Laboratoire de Mécanique des Solides, Ecole Polytechnique, International Journal of Solids and Structures, Vol.11 pp 947-960 (1975)
3 W. Noh, Numerical methods in hydrodynamics calculations, LLNL UCRL-52112 (1976)
4 O.C. Zienkiewicz, The Finite Element Method in Engineering Science
5 J. Hallquist, Theoretical manual for DYNA3D, LLNL (1983)
6 J. Donea and al, An arbitrary lagrangian-eulerian finite element method for transient dynamic fluid-structure interactions, Computer methods in applied mechanics (1982)
7 D. Koslov , G. Frazier, Treatment of hourglass pattern in low order finite element code, International journal for numerical and analytical methods in geomechanics (1978)

Computational Methods for Predicting Material Processing Defects, edited by M. Predeleanu
Elsevier Science Publishers B.V., Amsterdam, 1987 — Printed in The Netherlands

NUMERICAL COMPUTING TECHNIQUE FOR PREDICTING THICKNESS CHANGING IN DEEP
DRAWING OF AUTOBODY SHEET

D. BAUER[1]

[1]Institute of Production Engineering, University Siegen, P.O. Box 101240,
D-5900 Siegen (FRG)

SUMMARY
   Using the numerical computing technique submitted thickness changing can
be predicted for different cases of simulation. In case of high coefficients
of friction and blank holder pressures the sheet thickness decreases more
than with low values. For an ideally plactic material it is below that obtained
when hardening is taken into account. Sheet thickness changing is greater with
isotropy than with anisotropy. But vertical and planar anisotropy have the
maximal influence.

INTRODUCTION

   In deep drawing of autobody sheets nonuniform thickness changing, caused

by the anisotropy and hardening of sheet material, friction and blank holder

pressure, is a well-known problem. Though the distribution of sheet thickness

is of great importance for the press shop to avoid necking and fracture of

drawn parts up to the present it has been impossible to predict it. In this

work first results are presented to solve the problems regarded. By extending

Hill's theory of plastic flow for anisotropic metals a numerical computing

technique is submitted in order to predict flange thickness in a deep drawing

part in 0° and 90° to direction of rolling considering the influences of

vertical and planar anisotropy, material hardness increase, friction and

distribution of blank holder pressure.

BASIC THEORY

   The general theoretical analysis of the yielding and plastic flow of

anisotropic metals presented by R. Hill (ref. 1) is in this case limited to

the deep drawing of cold-rolled autobody sheets. Therefore, choosing Cartesian

co-ordinates (I, II, III) Hill's theory can be written as (ref. 2):

Yield criterion

$$R_{00} \, (\sigma_I - \sigma_{III})^2 + R_{90} \, (\sigma_{III} - \sigma_I)^2 + R_{00} \cdot R_{90} \, (\sigma_I - \sigma_{II})^2 = R_{90} \, (1 + R_{00}) \, Y_{00}^2 \qquad (1)$$

$$R_{90} (1+R_{00}) Y^2_{00} = R_{00} (1+R_{90}) Y^2_{90} \tag{2}$$

Rule of plastic flow

$$\frac{d\ \varepsilon_{III}}{d\ \varepsilon_{I}} = \frac{R_{00}\ (\sigma_{III} - \sigma_{II}) + R_{90}\ (\sigma_{III} - \sigma_{I})}{R_{00}\ R_{90}\ (\sigma_{I} - \sigma_{II}) + R_{90}\ (\sigma_{I} - \sigma_{III})} \tag{3}$$

$$\frac{d\ \varepsilon_{III}}{d\ \varepsilon_{II}} = \frac{R_{00}\ (\sigma_{III} - \sigma_{II}) + R_{90}\ (\sigma_{III} - \sigma_{I})}{R_{00}\ (\sigma_{II} - \sigma_{III}) + R_{00}\ R_{90}\ (\sigma_{II} - \sigma_{I})} \tag{4}$$

Equivalent strain-increment

$$d\ \varepsilon = (\ \frac{2}{3}\ \frac{R_{00}+R_{90}+R_{00} \cdot R_{90}}{1 + R_{00}+R_{90}}\ (R_{90}/d\varepsilon^2_{I} + R_{00}/d\varepsilon^2_{II} + d\varepsilon^2_{III})\ )^{\frac{1}{2}} \tag{5}$$

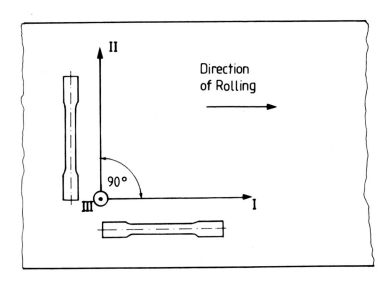

Fig. 1 Principal axes of anisotropy in a cold-rolled autobody sheet, and positions of test specimen taken for tensile testing.

Condition of incompressibility

$$d\ \varepsilon_{I} + d\ \varepsilon_{II} + d\ \varepsilon_{III} = 0 \tag{6}$$

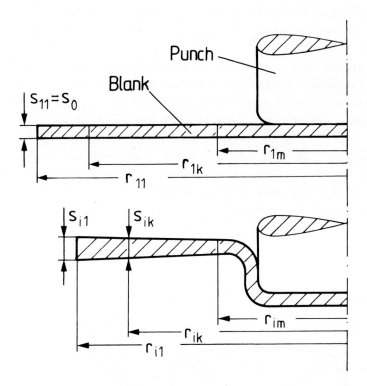

Fig. 2. Thickness changing in the flange of a deep drawing part.

These equations assume that the principal axes of stress and strain coincide with Hill's principal axes of anisotropy. For example in a cold-rolled autobody sheet I, II and III would lie in 0° and 90° to direction of rolling and normal to the plane of sheet, Fig. 1. Where Y_{00} and Y_{90} describe the effective stress-strain curves of plastic flow in 0° and 90° direction of rolling by means of a power law (ref. 3):

$$Y_x = R_{p/x} \qquad 0 \leq \epsilon_x \leq \epsilon_{xg} \tag{7}$$

$$Y_x = C_x \, \epsilon_x^{\,n_x} \qquad \epsilon_{xg} \leq \epsilon_x \leq \epsilon_{xmax} \tag{8}$$

$$\epsilon_{xg} = (R_{p/x}/C_x)^{1/n_x}$$

$$x = 00, \ 90$$

The material constants C_{00} and C_{90}, the work-hardening exponents n_{00} and n_{90}, the moduli for the beginning of yielding $R_{p/00}$ and $R_{p/90}$ and the coefficients of normal anisotropy R_{00} and R_{90} can be measured in uniaxial tensile tests (ref. 3).Fig. 1.

The aim of the present work is to propose a numerical computing technique for predicting the influence of these values and in addition of the coefficient of friction μ and the pressure of the blank holder σ_{zik} on the thickness s_{ik} in the flange of the deep drawing part considered for an arbitrary drawing ratio β_i, Fig. 2. Therefore it is necessary to change the Cartesian co-ordinates (I, II, III) into cylindrical co-ordinates (r, t, z) as shown in Fig. 3. But this changing is only valid for 0° and 90° because for other angles, the principal axes of stress and strain do not coincide with Hill's principal axes of anisotropy.

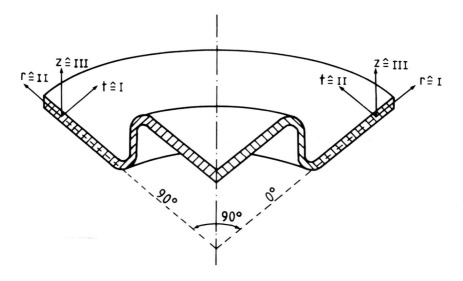

Fig. 3. Cartesian co-ordinates (I, II, III) and cylindrical co-ordinates (r,t, z) in 0° and 90° to direction of rolling of a deep drawn part.

The pressure of the blank holder is measured by using a special pressure sensitive film (ref. 4).

$$\sigma_{zik} = p_N \; (r_{ik}, \; \beta_i) \tag{9}$$

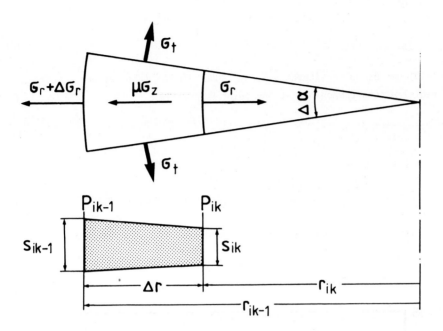

Fig. 4. Equilibrium of principal stresses in radial direction of an arbitrary element of flange.

NUMERICAL COMPUTING TECHNIQUE

Assuming that shear stresses can be neglected at $\Delta\alpha = 0$ the equilibrium equations in 0° and 90° to direction of rolling can be written as, Fig. 4.

$$\frac{\Delta\sigma_r}{\Delta r} = \frac{1}{r} \left(\frac{\sigma_t}{2} \left(1 + \frac{s_{ik}}{s_{ik-1}} - \sigma_r \right) \right)$$

$$+ \frac{\sigma_r}{\Delta r} \left(\frac{s_{ik}}{s_{ik-1}} - 1 \right) - 2\mu|\sigma_z|\frac{1}{s_{ik-1}}$$

(10)

$i = 1$, j and k = 2, m

14

Regarding the boundary conditions

$$\sigma_{ri1} = 0 \tag{11}$$
and

$$s_{11} = s_o \tag{12}$$

eqn. (10) can be solved iteratively by "RUNGE-KUTTA" (ref. 2) and we obtain the radial stress components at any point P_{ik}

$$\sigma_{rik} = \sigma_{rik-1} + \Delta\sigma_r \tag{13}$$

$$r_{ik} = r_{ik-1} + \Delta r \tag{14}$$

where the transverse stress component σ_{zik} is delivered by eqn.(9). Combining eqn. (13) and eqn. (14) with eqn. (1) and eqn. (2) the circum-ferential stress component σ_{tik} can be found if $\Delta\,\epsilon_{ik}$ is known (ref. 2).

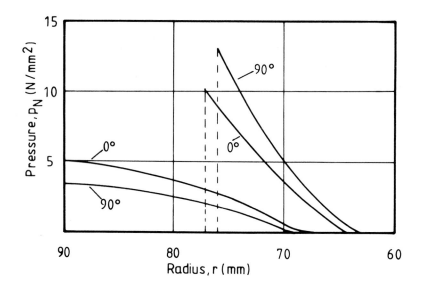

Fig. 5. Distribution of blank holder pressure at the beginning and end of deep drawing process.

Using the definitions

$$\Delta \varepsilon_{rik} = \ln (r_{i-1k} s_{i-1k})/r_{ik} \tag{15}$$

$$\Delta \varepsilon_{tik} = \ln (r_{ik}/r_{i-1k}) \tag{16}$$

$$\Delta \varepsilon_{zik} = \ln (s_{ik}/s_{i-1k}) \tag{17}$$

and rearranging eqn. (16) and (17) s_{ik} is found to be

$$s_{ik} = s_{i-1k} \left(\frac{r_{ik}}{r_{i-1k}} \right)^{\Delta \varepsilon_{zik}/\Delta \varepsilon_{tik}} \tag{18}$$

Substituting the values of eqn. (15), (16) and (17) into eqn. (5) we obtain $\Delta \varepsilon_{ik}$. Hence, the equivalent strain at P_{ik} can be written

$$\varepsilon_{ik} = \varepsilon_{i-1} - \Delta \varepsilon_{ik} \tag{19}$$

Placing the value of ε_{ik} into eqn. (7) and (8) and substituting the values of these equations into eqn. (3) and (4) we obtain finally $\Delta \varepsilon_{zik}/\Delta \varepsilon_{tik}$. Thus, the thickness s_{ik} can be calculated at any radius r_{ik} by use of eqn. (18)

The complete description of the developed numerical computing technique has been given by H. Schmidt, (ref. 5).

TABLE 1

Cases of simulation for different values of work hardening, anisotropy, friction and blank holder pressure

case	n_{00}	n_{90}	C_{00} N/mm^2	C_{90} N/mm^2	$Y_{p/00}$ N/mm^2	$Y_{p/90}$ N/mm^2	R_{00}	R_{90}	μ -	P_N N/mm^2
1	0,232	0,223	548	532	163	166	1,75	1,91	0,0	> 0
2	0,232	0,223	548	532	163	166	1,75	1,91	0,2	> 0
3	0,232	0,223	548	532	163	166	1,75	1,91	0,4	> 0
4	0,232	0,223	548	532	163	166	1,75	1,91	0,2	= 0
5	0,000	0,000	548	532	163	166	1,75	1,91	0,2	> 0
6	0,232	0,223	548	532	163	166	1,00	1,00	0,2	> 0

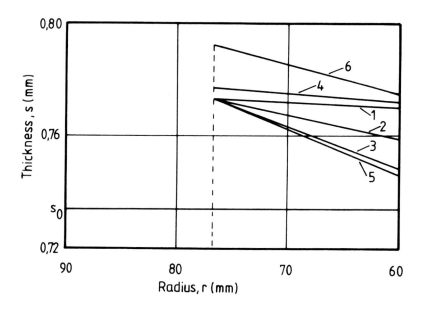

Fig. 6. Sheet thickness calculated in 0° to direction of rolling for different cases of simulation listed in Table 1.

RESULTS

In the following first results of the numerical computing technique submitted are plotted in Fig. 6 and Fig. 7, respectively. The different cases of simulation calculated are listed in Table 1. Sheet material used has been autobody steel St 1405. The initial dimensions of blank investigated have been r_{11} = 90 mm and s_{11} = s_o = 0,734 mm. The drawing ratio has been β_i = 1,18, the punch diameter 100 mm and r_{im} = 60 mm. The corresponding distribution of blank holder pressure measured has been plotted in Fig. 5.

The results are obtained from Fig. 6 and Fig. 7 are demonstrating that with high coefficient of friction (case 3) the sheet thickness decreases more pronouncedly than with low values (case 2). It can be seen, too, that the sheet thickness changing is more affected by friction (case 1) than by blank holder pressure (case 4). For an ideally plastic material (case 5) the thickness predicted is below that obtained when work hardening (case 2) is taken into account. Sheet thickness calculated is greater with isotropy (case 6) than with anisotropy (case 2).

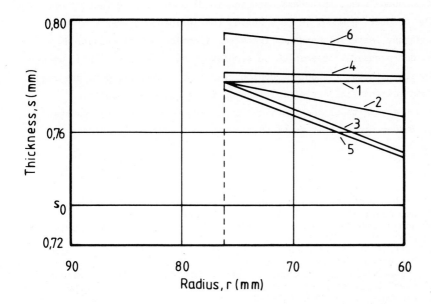

Fig. 7. Sheet thickness calculated in 90° direction of rolling for different cases of simulation listed in Table 1.

Comparing Fig. 6 to Fig. 7 it can be also seen that for all cases of simulation the sheet thickness calculated is generally higher in 90° than in 0° to direction of rolling. Form Table 1 it is clear, that the amount of R_{90} is greater than the amount of R_{00}. But on the other hand by investigations of H. Schmidt (ref. 5) it could be shown that the sheet thickness changing in 0° to direction of rolling is more sensitively affected by the anisotropy R_{90} in 90° to direction of rolling than by the anisotropy R_{00} in 0° to direction of rolling, and vice versa (ref. 8).

Comparing these results of simulation with results obtained experimentally (ref. 6) it is found that for case 1 the best agreement could be ascertained between calculated an measured values. The maximum deviation occuring has been lower than 10 μm.

CONCLUSION

Measuring blank holder pressure by a special sensitive film and extending Hill's theory of plastic flow for anisotropic metals with work hardening a numerical computing technique using the method of Runge-Kutta is developed to predict autobody sheet thickness changing in 0° and 90° to direction of rolling. The predicting model is considering the influence of vertical and planar anisotropy, material hardness increase, friction and distribution of blank holder pressure. It enables to simulate some other cases, too, as ideally plastic material without hardening, isotropic material, deep drawing with high and low coefficients of friction or blank holder pressures.

Comparing the results of the numerical computing technique with results obtained by measurements it it found that sufficient agreement could be ascertained.

REFERENCES

1 R. Hill, A theory of the yielding and plastic flow of anisotropic metals, Proceedings, Royal Society, 193 A (1948) 281-297.
2 D. Bauer and H. Schmidt, Planare Anisotropie beim Tiefziehen von Stahl- blech, Bänder, Bleche, Rohre 25 (1984) 259-262.
3 D. Bauer, Werkstoffverfestigung und Anisotropie beim Tiefziehen von Fein- blechen, Maschinenmarkt Würzburg 86 (1982), 156-159.
4 D. Bauer, Measurement of blank holder contact pressure in deep drawing of autobody sheet by pressure sensitive film, Proceedings of 14 th IDDRG biennial congress, Köln 1986, 442-443.
5 H. Schmidt, Blechdickenbestimmung beim Tiefziehen unter Berücksichtigung der planaren Anisotropie und der Niederhalterdruckverteilung, Dr.-Ing. Diss. Universität Siegen 1983.
6 D. Bauer and H. Schmidt, Experimentelle Untersuchungen der Flanschdicke eines Ziehteils, Bänder, Bleche, Rohre 25 (1984), 304-305.
7 D. Bauer and H. Schmidt, Blechdickenverteilung beim Karosserieblech- Tiefziehen, Bänder, Bleche, Rohre 26 (1985), 296-299.
8 S.P. Jakovlev and V.V. Sevelev, Über die Möglichkeit der Beseitigung von Zipfelbildung beim Tiefziehen anisotroper Werkstoffe, Umformtechnik, 3 (1969), 43-47.

Computational Methods for Predicting Material Processing Defects, edited by M. Predeleanu
Elsevier Science Publishers B.V., Amsterdam, 1987 — Printed in The Netherlands

AN EULERIAN FINITE ELEMENT METHOD FOR THE THERMAL AND VISCO PLASTIC DEFORMATION
OF METALS IN INDUSTRIAL HOT ROLLING

J.H. BEYNON[1], P.R. BROWN[1], S.I. MIZBAN[1], A.R.S. PONTER[1] and C.M. SELLARS[2]

[1]Department of Engineering, The University, Leicester LE1 7RH (Great Britain)
[2]Department of Metallurgy, The University, Sheffield S1 3JD (Great Britain)

SUMMARY
    The combined thermal and deformation problem of steady state hot rolling can
be set up as the solution of three coupled sets of continuum problems, giving
rise to a second order partial differential equation, which is solved using the
Petrov-Galerkin method. The relationship between effective strain and
instantaneous strain rate may be found as the solution of a first order
hyperbolic differential equation, also solved using the Petrov-Galerkin method.
The thermo-mechanical solution for the hot rolling of aluminium is described,
with examples for different reductions, together with the potential consequences
for the break down of casting defects.

INTRODUCTION
    Much attention has been given to the application of numerical techniques for
metal working processes. The first numerical attempts to compute the
temperature distribution were made by Bishop (ref. 1) and later by Altan and
Kobayashi (ref. 2), both were concerned with metal extrusion and applied a
finite difference method to the problem. Since the 1970's a finite difference
solution has also been developed by Sellars and co-workers (refs. 3-4) for the
temperature fields in multistand hot flat rolling of metals and extended to
include metallurgical changes throughout the rolling process. All these methods
above were essentially uncoupled thermo-plastic ones. The first attempts to use
the flow formulation for solving coupled thermal flow problems were made by Jain
(ref. 5), Dawson and Thompson (ref. 6) and Zienkiewicz et al (ref. 7) who used a
finite element iterative procedure. Along with the finite element development
to solve conduction-convection problems at moderate to high Graetz number (or
Reynolds or Peclet numbers) has come an awareness that the Standard Galerkin
procedures, which yield "best approximations", tend to give mixed results in the
presence of convection terms. Non-physical oscillations occur in the
numerically computed solution which may be avoided by mesh refinement, but that
can prove to be prohibitively expensive in terms of computer time and storage.
The finite difference methods may perform slightly better than finite element in
producing wiggle-free solutions, but in most cases the side effect is loss of
accuracy.
    Various schemes of so called 'up-wind' differencing have been formulated
using different trial and test functions (that is, a Petrov-Galerkin or weighted
residual finite element method), such as those by Christie et al (ref. 8),
Heinrich et al (ref. 9), Christie and Mitchell (ref. 10), Hughes (ref. 11), and
Hughes and Atkinson (ref. 12). These methods yield satisfactory results in
one-dimensional problems yet they overly diffuse solutions in multi-dimensional
cases. More acceptable Petrov-Galerkin schemes have been demonstrated to
overcome such difficulties as in Griffiths and Mitchell (ref. 13) and Hughes and
Brooks (refs. 14-15). Although appropriate forms of the test functions have
been used in these formulations, perturbation terms should be added in such a

way that it is effective only along the direction of flow. For this case,
methods to calculate the flow lines should be used to validate such forms. In
this paper, we consider a more general form of the Petrov-Galerkin scheme in the
absence of the exact (or computed or experimental) flow lines where the test
functions are obtained by correcting the trial functions with velocity dependent
additive terms. This removes the requirement of prior knowledge of the flow
lines, particularly important for modelling practical forming processes where
experimental and computing techniques for determining the flow lines are
inexact.

THERMAL PROBLEM
   Hot rolling generally takes place at temperatures larger than about 0.6 of
the absolute melting temperature of the metal. At such temperatures, the flow
stress is sensitive to the strain rate and temperature of deformation. The heat
balance during the rolling of flat products comprises: (a) heat loss to the
environment before and after each rolling pass, (b) heat loss to the rolls
during the rolling pass and (c) heat gain due to the deformational work during
the rolling pass. The temperature changes during each rolling pass are more
rapid than between the rolling stands because of the considerable chilling by
the cool rolls and the temperature rise due to deformational heating.
Furthermore, contact times during rolling are short compared with the interstand
times. As a consequence, the finite element modelling of thermal changes during
the rolling pass are on a different scale to the changes between passes. Two
separate programs are being developed; one for the roll gap and its immediate
vicinity which is discussed in this paper, and another for the interstand
changes which will not be presented here.
   The problem being modelled is the hot rolling of wide flat products (ref. 16)
and the process is treated as steady state in two dimensions for any given roll
stand. The governing equation of conductive-convective heat transfer in a
flowing material is written as

$$\frac{\partial}{\partial x}\left(k\,\frac{\partial T}{\partial x}\right) + \frac{\partial}{\partial y}\left(k\,\frac{\partial T}{\partial y}\right) = \rho c\left[\nu_x\,\frac{\partial T}{\partial x} + \nu_y\,\frac{\partial T}{\partial y}\right] - Q \tag{1}$$

where T is the temperature of the rolling stock, k is the thermal conductivity
coefficient, ρ is density, c is the specific heat at constant pressure, Q is the
heat production rate due to the energy dissipated by plastic deformation and ν_x
and ν_y are the x and y components of the velocity vector $\underline{\nu}$. The problem is
further simplified by assuming that the top and bottom halves of the slab are
symmetrical.
   Equation (1) is taken with the boundary conditions, see Fig.1,

$$\frac{\partial T}{\partial n} = 0 \quad \text{on FE, AB, CD and AF} \tag{2a}$$

$$T = T_0 \quad \text{on DE} \tag{2b}$$

and

$$\frac{\partial T}{\partial n} = H'(T-T_R) \text{ on BC} \tag{2c}$$

where T_0 is the specified temperature on DE, T_R is the roll temperature, $\partial/\partial n$
refers to differentiation along the normal to the boundary, and H' is the
temperature dependent constant which can be evaluated by

$$H' = -H/k$$

where H is the coefficient for heat transfer to the rolls. The boundary-value
problem is to find T = T(x,y) which satisfies (2a), (2b) and (2c).

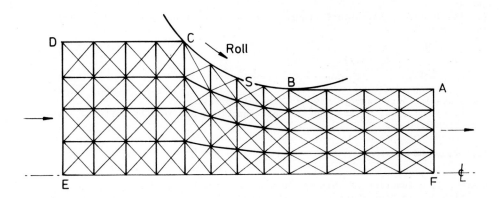

Fig. 1. Finite element mesh

PLASTIC FLOW PROBLEM
In most hot working processes the elastic deformation can be neglected when compared to the large plastic strain. The equation of state defines the deformation as a plastic strain rate function of effective stress ($\overline{\sigma}$), effective plastic strain ($\overline{\epsilon}^p$) and temperature (T);ie,

$$\dot{\epsilon}^p = f \quad (\overline{\sigma}, \overline{\epsilon}^p, T) \tag{3}$$

The formulation of equation (3) is obtained from Sellars and Tegart (ref. 17) where the von-Mises yield criterion for plane strain is considered.
The effective (and total) strain is formulated by setting up a first order hyperbolic differential equation in terms of the velocity field $\underline{v} = (v_x, v_y)$, which can be written as

$$\left. \begin{array}{l} v_x \dfrac{\partial \overline{\epsilon}^p}{\partial x} + v_y \dfrac{\partial \overline{\epsilon}^p}{\partial y} = \dot{\overline{\epsilon}}^p \\[2em] \dot{\overline{\epsilon}}^p = 0 \text{ on DE} \end{array} \right\} \tag{4}$$

where $\dot{\overline{\epsilon}}^p$ is the effective plastic strain rate.

FINITE ELEMENT FORMULATIONS
A new Petrov-Galerkin formulation is proposed for the heat conduction-convection problem, equation (1). The weak form of (1), employing the boundary conditions (2a), (2b) and (2c), for a domain of volume V, is given by

$$\int_V k\left(\frac{\partial \phi_i}{\partial x} \frac{\partial \Psi_j}{\partial x} + \frac{\partial \phi_i}{\partial y} \frac{\partial \Psi_j}{\partial y}\right) T_i dV + \int_V \rho c \left(v_x \frac{\partial \phi_i}{\partial x} + v_y \frac{\partial \phi_i}{\partial y}\right) \Psi_j T_i dV -$$

$$\int_S k\alpha \, \phi_i \Psi_j T_i dS = \int_V Q \, \Psi_j dv + \int_S k\alpha T_R \, \Psi_j dS \tag{5}$$

where:
(1) The dependent variable T (temperature as a function of spatial position x,y) is approximated by

$$T^*(x,y) = \sum_{i=1}^{N} T_i \, \phi_i \, (x,y) \tag{6}$$

for suitable shape functions ϕ_i at N nodal points.
(2) Ψ_j (j = 1,2,...,N) are the new test functions obtained by perturbing the shape functions as

$$\Psi_j = \phi_j \, + \eta g \tag{7}$$

where g is a function of velocity and shape functions gradient and η is a function of a perturbation parameter dependent on the Graetz number and the roll speed. When the shape functions are unperturbed, a Galerkin approximation will present (ie $\eta = 0$).
(3) S is that part of the boundary where the condition (2c) holds.
(4) The velocity components $\underline{\nu}$ are approximated by

$$\nu_x = \sum_{i=1}^{N} \nu_{x_i} \, \phi_i \text{ and } \nu_y = \sum_{i=1}^{N} \nu_{y_i} \, \phi_i \tag{8}$$

and are formulated for each element. In our case, as in most situations of practical interest, $\underline{\nu}_i$ will be given the same finite element approximation as that used for T_i. However, it is not necessary for the order of approximation employed for both $\underline{\nu}_i$ and T_i to be the same.
    The Petrov-Galerkin method is also applied to the effective strain, equation (4), giving,

$$\left[\int_V \Psi_j \left(\nu_x \frac{\partial \phi_i}{\partial x} + \nu_y \frac{\partial \phi_i}{\partial y} \right) dV \right] \bar{\varepsilon}_i^p = \int_V \Psi_j \, \dot{\bar{\varepsilon}}^p \, dV \quad ; \; j=1,2,\ldots,n \tag{9}$$

where $\bar{\varepsilon}^p$ is approximated by

$$\bar{\varepsilon}^p = \sum_{i=1}^{N} \bar{\varepsilon}_i^p \, \phi_i \tag{10}$$

    It is clear from the above that the plastic deformation process represented by (3) and (4) is highly coupled with the thermal equation (1). Hence, a direct coupled formulation can be presented through the heat production rate

$$Q = \underline{\sigma}^T \, \dot{\varepsilon}^p \tag{11}$$

(the superscript T in $\underline{\sigma}^T$ refers to transpose). This emphasises that the temperature development depends not only on the heat loss to the environment and rolls, but also on the heat gained from the energy dissipated in the deformation process.

RESULTS AND DISCUSSION
    A triangular finite element mesh has been used for the approximations, where the variables vary linearly; this could be adapted for different approximations. Initially, the velocity distribution is chosen using the specified roll speed, and with Q=0 in equation (5) the temperature distribution is obtained which is

necessary for the solution of equation (3). Also, for equation (3) the initial effective stress $\bar{\sigma}$ is calculated from the elastic problem where the finite element approximation has been applied to the displacement components. ε^p appears in the function of strain $f(\varepsilon^p)$ in equation (3), which here is taken to be $f(\varepsilon^p)=1$ by assuming ideal plastic behaviour ($f(\varepsilon^p)$ could be specified by an alternative relationship). Two iterative procedures in the program have been developed, the one for the plastic flow problem being contained within the one for the thermal problem. A non-dimensionalised procedure has been adopted for all the numerical calculations.

Results for temperature and effective strain are illustrated in figures 2-9. They are from direct-coupled thermo-plastic solutions for aluminium where the slab entry temperature = 819 K (546 °C), roll temperature = 353 K (80 °C), roll radius = 0.406 m, roll speed = 1 m/s, and slab entry thickness = 0.2425 m. Different reductions have been computed: 10, 23 and 46 % reduction in thickness, for which the lengths of slab considered are 0.413, 0.599 and 0.835 m, respectively.

The temperature profiles demonstrate the strong cooling effect of the rolls. Figure 2 compares the temperature profiles for the three different reductions. The larger the reduction, the greater is the drop in slab surface temperature. This is due to the longer time of contact with the roll, which is the same size for all the reductions. The contact times are 0.14, 0.21 and 0.31 s for 10, 23 and 46 % reduction, respectively. The nearly linear rate of temperature rise just below the surface is affected by the finite element mesh size. This also gives rise to a knee in the curve for the 46 % reduction example.

Below the surface, the temperature rise due to the heat of deformation can be clearly seen as peaks in the temperature profiles above the entry temperature level of 819 K, Figure 2. At the centre of each slab the deformation heating has produced a near uniform rise in temperature, corresponding to the steady value of effective strain, Figure 3. As the effective strain increases on approaching the surface, the temperature of the stock also rises. The temperature peaks and falls again due to the chilling effect of the roll.

The levels of strain at the slab centre, Figure 3, correspond closely to the values of nominal true strain for each reduction: 0.12, 0.30 and 0.71 for 10, 23 and 46 %, respectively. The strain in excess of these values represents the redundant shearing brought about by constraining the metal to flow through the roll gap. Figure 3 also shows how the larger reductions bring a deeper penetration of the deformation. This is important in the rolling of as-cast slabs when the cast microstructure needs to be broken down to produce stronger and tougher microstructures. Furthermore, defects in the form of porosity must be welded up; this also requires sufficient depth of penetration of the deformation (ref. 18).

Figures 4-9 show the full patterns of temperature and strain for the whole region. Use of the Petrov-Galerkin technique has resulted in very smooth profiles, virtually completely free of the non-physical wiggles or oscillations which result from application of the Standard Galerkin method (ref. 16).

CONCLUSIONS
     A coupled thermo-plastic program has been developed using a new modified Petrov-Galerkin method. A finite element solution has been obtained for the hot rolling of aluminium where the temperature and mechanics of the process can be determined. The model has the ability to solve problems with large reductions and for any material for which basic material properties are available. The program is continually being enhanced, particularly with the inclusion of microstructural development during rolling and with more complex and realistic friction conditions.

ACKNOWLEDGEMENT
     The support of the Science and Engineering Research Council is gratefully acknowledged.

24

REFERENCES

1 J.F.N. Bishop, An approximation method for determining the temperature
   reached in steady-state motion problem of plane strain, Q. Mech. Appl. Math.,
   9 (1956) 236.
2 T. Altan and S. Kobayashi, A numerical method for estimating the
   temperature distributions in extrusion through conical dies, J. Eng. Ind., 90
   (1968) 107-118.
3 C.M. Sellars, The physical metallurgy of hot working, in: C.M. Sellars and
   G.J. Davies (Eds.), Hot working and forming processes, Metals Society,
   London, 1980, pp.3-15.
4 C.M. Sellars, Computer modelling of hot-working processes, Mat. Sci. Engg. 1
   (1985) 325-332.
5 P.C. Jain, Plastic flow in solids (static, quasistatic, and dynamic
   situations including temperature effects), University College of Swansea,
   Wales, 1976, Ph.D. thesis.
6 P.R. Dawson and E.G. Thompson, Steady-state thermomechanic finite element
   analysis of elastoviscoplastic metal forming processes, Numerical
   Modelling of Manufacturing Processes, ASME, PVP-PB-025, 1977, pp.167-182.
7 O.C. Zienkiewicz, P.C. Jain and E. Onate, Flow of solids during forming and
   extrusion: some aspects of numerical solutions, Int. J. Solids Struct., 14
   (1978) 15-38.
8 I. Christie, D.F. Griffiths, A.R. Mitchell and O.C. Zienkiewicz, Finite
   element methods for second order differential equations with significant
   first derivatives, Int. J. Num. Engg., 10 (1976) 1389-1396.
9 J.C. Heinrich, P.S. Huyakorn, A.R. Mitchell and O.C. Zienkiewicz, An upwind
   finite element scheme for two dimensional convective transport equations,
   Int. J. Num. Engg., 11 (1977) 131-143.
10 I. Christie and A.R. Mitchell, Upwinding of high order Galerkin methods in
   conduction-convection problems, Int. J. Num. Engg. 12 (1978) 1746-1771.
11 T.J.R. Hughes, A simple scheme for developing upwind finite elements, Int. J.
   Num. Engg., 12 (1978) 1359-1365.
12 T.J.R. Hughes and J. Atkinson, A variational basis for upwind finite
   elements, IUTAM Symposium on Variational Methods in the Mechanics of Solids,
   Northwestern Univ., Evanston, U.S.A., 1978, pp. 387-391.
13 D.F. Griffiths and A.R. Mitchell, On generating upwind finite element
   methods, in: T.J.R.Hughes (Ed.) ASME Meeting on Finite Element Methods for
   Convection Dominated Flows, New York, 1979, AMD-34, pp. 91-104.
14 T.J.R. Hughes and A. Brooks, A multidimensional upwind scheme with no
   crosswind diffusion, in: T.J.R.Hughes (Ed.) ASME Meeting on Finite Element
   Methods for Convection Dominated Flows, New York, 1979, AMD-34, pp. 19-35.
15 T.J.R. Hughes and A. Brooks, A theoretical framework for Petrov-Galerkin
   methods with discontinuous weighting function: application to the streamline
   upwind procedure, in: R.H. Gallagher et al (Eds.), Finite elements in fluids,
   John Wiley and Sons Ltd., London, 1982, 4: pp. 47-65.
16 J.H. Beynon, P.R. Brown, S.I. Mizban, A.R.S. Ponter and C.M. Sellars,
   Inclusion of metallurgical development in the modelling of industrial
   hot rolling of metals, in: K. Mattiasson et al (Eds.), NUMIFORM 86: Numerical
   methods in industrial forming processes, Gothenburgh, Sweden, Aug. 25-29,
   1986, A.A. Balkema, Rotterdam, 1986, pp. 213-218.
17 C.M. Sellars and W.J.McG. Tegart, La relation entre la resistance et la
   structure dans la deformation a chaud, Mem. Sci. Rev. Metall., 63 (1966)
   731-746.
18 L. Leduc, T. Nadarajah and C.M. Sellars, Density changes during hot rolling
   of cast steel slabs, Metals Tech., 7 (1980) 269-273.

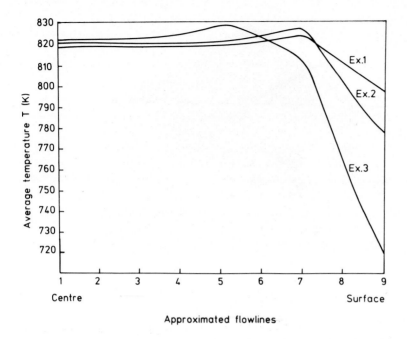

Fig. 2. Temperature profiles from centre to surface of slabs; EX1 = 10 % reduction, EX2 = 23 % reduction, EX3 = 46 % reduction.

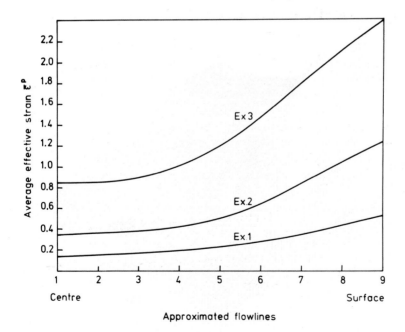

Fig. 3. Effective strain profiles from centre to surface of slabs at the exit plane; EX1 = 10 % reduction, EX2 = 23 % reduction, EX3 = 46 % reduction.

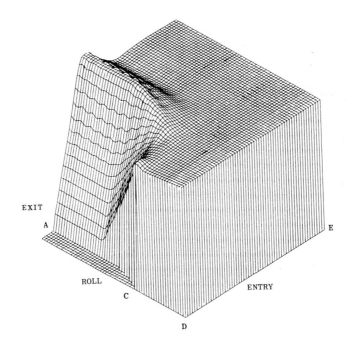

Fig. 4. Temperature distribution for EX1: 10 % reduction.

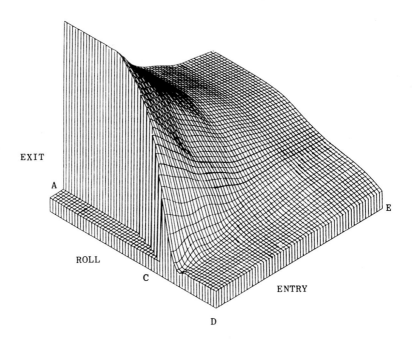

Fig. 5. Effective strain distribution for EX1: 10 % reduction

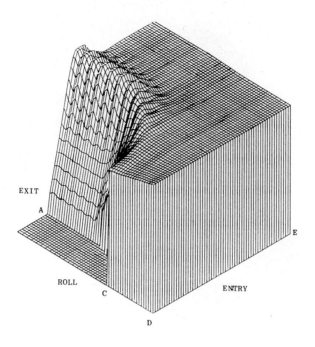

Fig. 6. Temperature distribution for EX2: 23 % reduction.

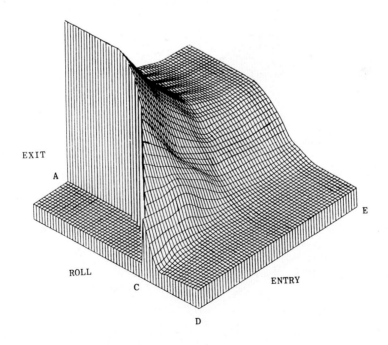

Fig. 7. Effective strain distribution for EX2: 23 % reduction.

28

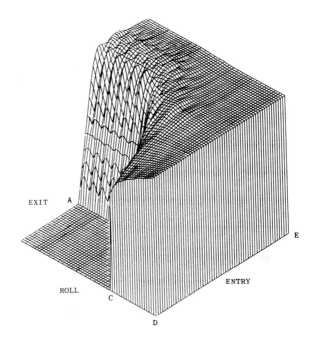

Fig. 8. Temperature distribution for EX3: 46 % reduction.

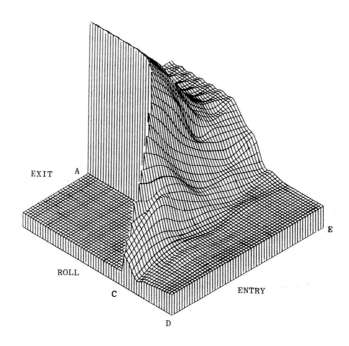

Fig. 9. Effective strain distribution for EX3: 46 % reduction.

Computational Methods for Predicting Material Processing Defects, edited by M. Predeleanu
Elsevier Science Publishers B.V., Amsterdam, 1987 — Printed in The Netherlands

SURFACE DEFECTS IN HOT ROLLING OF COPPER BASED ALLOYS HEAVY INGOTS:
THREE DIMENSIONAL SIMULATION INCLUDING FAILURE CRITERION

O. BRANSWYCK[1], C. DAVID[1], C. LEVAILLANT[1], J.L. CHENOT[1],
J.P. BILLARD[2], D. WEBER[2], J.P. GUERLET[2]

[1]CEMEF, Ecole des Mines de Paris
UA CNRS n° 852 et GRECO Grandes Déformations et Endommagement
SOPHIA-ANTIPOLIS - 06560 VALBONNE

[2]C.L.A.L., 13, rue Montmorency, 75003 PARIS CEDEX

SUMMARY
    Using 3D finite element method (F.E.M.), hot rolling of heavy ingots is
modeled using viscoplatic description of the material through a NORTON-HOFF
law. In order to predict surface defects the material damage is estimated by
the OYANE criterion.

INTRODUCTION

    Heavy ingots (200 x 400 x 2500 mm) of copper based alloys are hot rolled
(at 925 °C) to a final thickness of 11 mm for further cold rolling. Sometimes,
defects appear on the side or the surface of ingots during the first steps of
rolling in the form of cracks perpendicular to the rolling direction. These
defects are detrimental to the process cost because of machining which is
required to eliminate the surface cracks, and recycling of high values scraps.

    A 3D numerical model LAMEF 3 (ref. 1) is used to test the effects of rol-
ling parameters (temperature, reduction ratio, ...) on damage occurence.

PRESENTATION OF THREE DIMENSIONAL ROLLING PROGRAM

    During hot rolling of copper based alloys slabs, the material is assumed to
be homogeneous, incompressible, isotropic and obeys a viscoplastic law. The
deviatoric stress tensor S is related to the strain rate $\dot{\varepsilon}$ through the equivalent strain rate $\dot{\bar{\varepsilon}}$, the strain rate sensitivity m, and K the material consistency, with the NORTON-HOFF relation.

$$S = 2K \, (\sqrt{3}\dot{\bar{\varepsilon}})^{m-1} \, \dot{\varepsilon} \quad , \quad \text{tr} \, \dot{\varepsilon} = 0 \tag{1}$$

coefficients K and m depend on thermo mechanical conditions occuring during
the deformation.

    The domain Ω with Γ as boundary is shown on figure 1. Using the penalty me-
thod to impose the incompressibility condition, the solution of the stationary
flow in the domain Ω, is obtained by solving the dual variational problem where the first order variation of the functional Φ vanishes :

$$\Phi \ (\ V \ ) = \int_{\Omega} \frac{K}{m+1} \ (\sqrt{3} \ \dot{\bar{\varepsilon}})^{m+1} \ d\Omega + \int_{\Omega} \frac{\rho K}{2} \ (\mathrm{div} \ v \ )^2 \ d\Omega \tag{2}$$

ρ is the penalty coefficient $(\rho \gg 1)$

Figure 1. System geometry

Moreover, the boundary condition at the interface Γ_c between roll and slab is introduced with a friction shear stress τ defined by the following relation, consistent with the Norton-Hoff behaviour :

$$\tau = - \alpha K \ \| \ \Delta V \ \|^{p-1} \ \Delta V \tag{3}$$

ΔV is the sliding velocity at the interface Γ_c, α and p are parameters depending on contact conditions. Then, the exit velocity is determined with the value of α.

Equation (3) can be introduced into the variational principle with the following additional functional :

$$\Phi_\tau \ (\ V \ ) = \int_{\Gamma_c} \frac{\alpha \ K}{p+1} \ \| \ \Delta V \ \|^{p+1} \ ds$$

Without external forces, the solution V of the initial problem with friction corresponds to the minimum of functional :

$$\phi = \phi_\Omega + \phi_\tau \tag{4}$$

The domain Ω is subdivided into a number of hexahedral elements. The three components of the unknown velocity field V are discretized using standard finite element procedure within each isoparametric element (ref. 2).

The minimization of the functional (4) leads to a set of non linear algebraic equations with nodal velocity components as the unknowns. These equations are solved by the iterative Newton Raphson method.

The stress field is computed on the basis of the velocity field and of the equilibrium equation.

$$S_{ij,j} - P_{,i} = 0$$

P is the pressure field in the domain Ω.

In order to be able to predict the work hardening distribution inside the domain Ω, the equivalent strain $\bar{\varepsilon}$, defined by equation (5) is computed as a post-processing result of the velocity field determination.

$$\bar{\varepsilon} = \int_{\text{deformation path}} \dot{\bar{\varepsilon}} \ dt \tag{5}$$

Equation (5) is solved in the domain Ω, using a finite difference technique along the streamlines of each point. Then, the equivalent strain is added step by step along the deformation path, for M_1 and M_2, two points on the deformation path :

$$\Delta\bar{\varepsilon}(M_1 M_2) = \int_{\widehat{M_1 M_2}} \dot{\bar{\varepsilon}}\ dt = 1/2\ (\dot{\bar{\varepsilon}}(M_1) + \dot{\bar{\varepsilon}}(M_2))\Delta t,$$

Δt is the time step between M_1 and M_2.

Then, assuming that an initial two dimensional equivalent strain map $\varepsilon_0(x_0,y,z)$ is given at entry (for $x = x_0$), the equivalent strain distribution is evaluated in the domain Ω as shown on figure 2 :

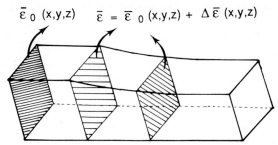

Figure 2 : Initial strain and integration scheme

After the numerical calculation of the stress field and equivalent strain field, the OYANE's criterion may be introduced in our model to obtain a prediction of damage. Based on a mechanical macroscopic approach and a critical volume of voids this criterion (ref. 3) is defined by :

$$\int_0^{\bar{\varepsilon}f} (1 + A\ \frac{\sigma m}{\sigma_{eq}})\ d\bar{\varepsilon} = B \tag{6}$$

where A and B are material characteristics, depending on thermal conditions, σ_m et σ_{eq} are respectively the mean stress and equivalent stress values.

Then, at each studied temperature, for given material, equation (6) is integrated and a distribution of B_s inside the rolled product is determined.

APPLICATION TO COPPER BASED INGOTS :

Material properties

(i) Rheological data. Torsion tests of specimens ($\emptyset$ 6 mm, L 35 mm) cut in the ingot length direction, in basaltic area, are run to determine the rheological behaviour.

We show on figure 3 $\bar{\sigma},\bar{\varepsilon}$ curves for various test temperatures. Up to 800 °C, the stress-strain curves present only a hardening stage leading to a "brittle" failure. From 800 °C to 900 °C the hardening stage is followed by a softening stage : fail strain is more important and nearly constant. At higher temperatures, a steady state appears for an equivalent strain which decreases with temperature.

Thus, the rheological characteristics of this alloy vary markedly with temperature. Material ductility increases when recristallization occurs (at higher temperature).

Hot forming material behaviour can be generally described by a viscoplastic with strain hardening double power law relationship. From 700 °C the rheological behaviour is found to be properly fitted to a Norton-Hoff incompressible law :

$$\sigma_0 = K_0 \dot{\bar{\varepsilon}}^m \exp((mQ)/(RT)) \quad \text{with } m = 0,14 \quad Q = 266 \text{ KJ.mol}^{-1}$$

and the consistency K of the equation (1) is :

$$K = \frac{\sigma_0 (\dot{\bar{\varepsilon}}, T)}{(\sqrt{3}^{m+1} \dot{\bar{\varepsilon}}^m)} = \frac{K_0 \exp[(mQ)/RT]}{\sqrt{3}^{m+1}}$$

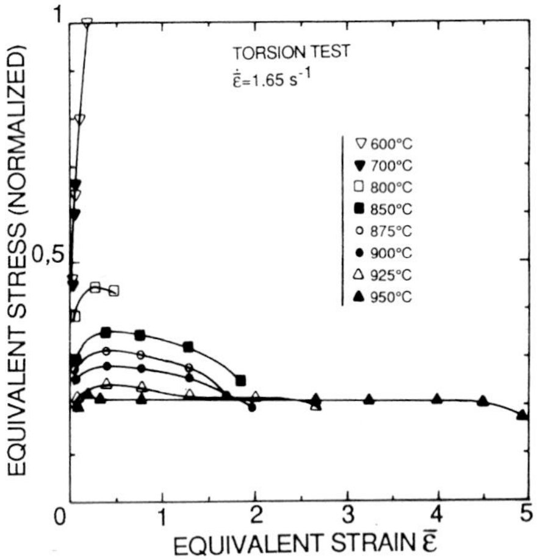

Figure 3 : Rheological data deduced from torsion tests

(ii) Material Damage : experimental OYANE criterion

The torsion test gives experimental B value (B exp) since $\sigma_m = 0$. A value is then determined through tension test results. The curves B exp and A exp as functions of temperature are plotted respectively on figure 4 and 5.

The A exp curve is a straight line with an equation fitted by

$A(T) = 0,096 \ T - 57,6$ where T is in °C.

Numerical simulation of hot rolling

The F.E.M. computations give all components of the stress tensor and equivalent strain. At first, in order to predict the surface defects, we concentrate only on the axial component σ_{xx} (of stress tensor) and the equivalent strain $\bar{\varepsilon}$; this choice will be justified in the discussion. Then we discuss a failure risk estimated by the OYANE criterion.

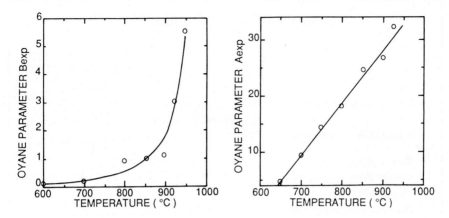

Figure 4 and 5 : Determination of material characteristic B exp and A exp.

Maps of σ_{xx} iso-values and $\bar{\varepsilon}$ iso-values are plotted respectively on figure 6 and 7. The simulation values of OYANE criterion, B_s, are shown on figure 8.

On these figures, we had written out the rolling conditions (temperature, reduction ratio, ...) and we represent only (xOz) planes labelled from 1 to 7 (see scheme on figure 9). The maximum values of σ_{xx} and $\bar{\varepsilon}$ and mesh geometry (280 nodes) are shown on the scheme of figure 9.

DISCUSSION OF FAILURE RISK PREDICTION.

At first, to predict the failure risks we have chosen to compare only the maximum computed values of σ_{xx} and $\bar{\varepsilon}$ to the critical values of mechanical tests because cracks are thought to be due to tensile stress perpendicular to experimentally observed defects (mode I cracking assumption) caused by a too low material ductility.

On Fig. 9, we note a compressive σ_{xx} area (A) located under the rolls (what can be easily expected) and two tension areas ($\sigma_{xx} > 0$) near the same locations where the surface defects are experimentally observed. One is centered on the slab outer sides, under the hold (C) and the other is located on the great faces after the hold (D).

In these tension areas, in particular in C area, the stress values obtained by numerical simulation are relatively near the experimental critical values but the strains are markedly lower than experimental failure strains. Therefore we cannot predict failure without taking both strains and stresses values into account . this can be done using a damage criterion.

MAPS OF σ_{xx} ISO-VALUES

- MAPS IN (XOZ) PLANES

- SIMULATION CONDITIONS:

  Reduction rate:7,5%
  Cylinder diameters:692mm
  Ω_{cyl} : 94 Tr.mm^{-1}
  Homogeneous temperature:925°C
  $\alpha = 0.195$

- VALUES OF ISOS IN MPa :

1 : σ_{xx}=-45	11 : σ_{xx}=5
2 : σ_{xx}=-40	12 : σ_{xx}=10
3 : σ_{xx}=-35	13 : σ_{xx}=15
4 : σ_{xx}=-30	14 : σ_{xx}=20
5 : σ_{xx}=-25	15 : σ_{xx}=25
6 : σ_{xx}=-20	16 : σ_{xx}=30
7 : σ_{xx}=-15	17 : σ_{xx}=35
8 : σ_{xx}=-10	18 : σ_{xx}=40
9 : σ_{xx}=-5	19 : σ_{xx}=45
10 : σ_{xx}=0	

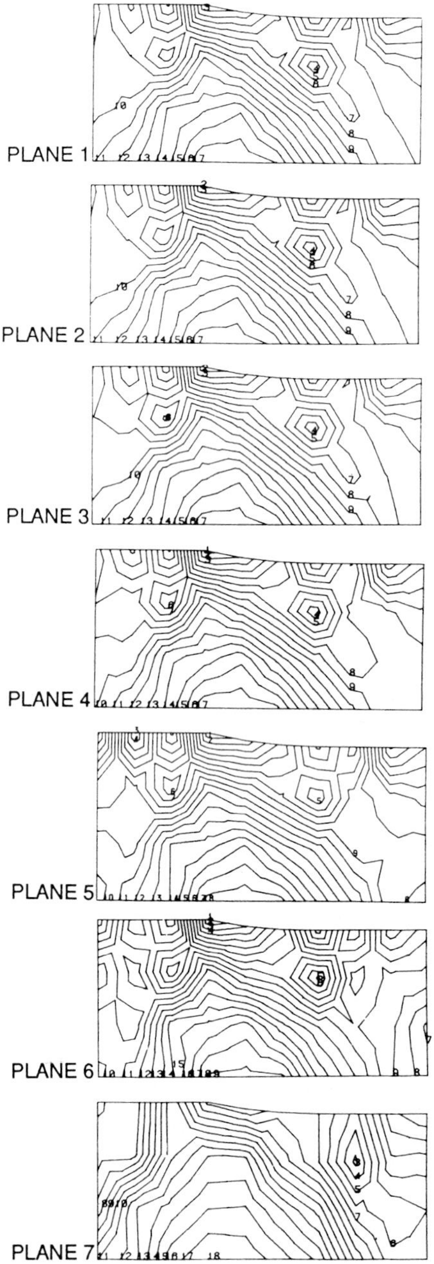

PLANE 1

PLANE 2

PLANE 3

PLANE 4

PLANE 5

PLANE 6

PLANE 7

Figure 6 : Maps of σ_{xx} iso-values

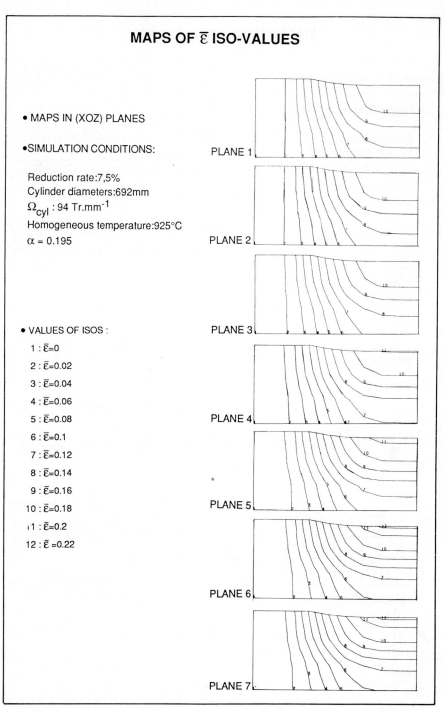

Figure 7 : Maps of ε̄ iso-values

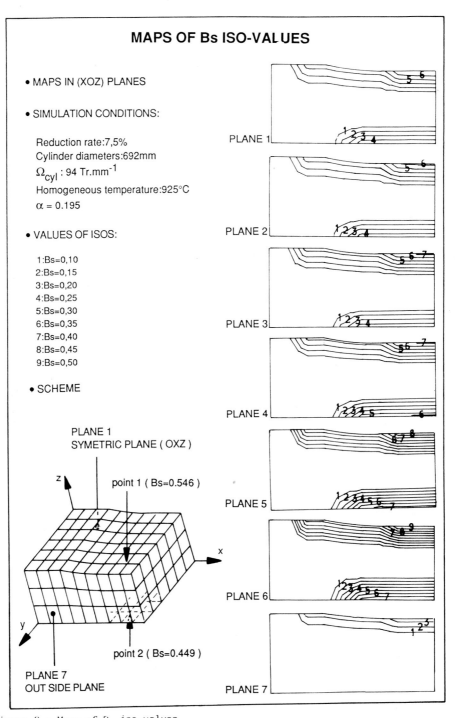

Figure 8 : Maps of B_S iso-values

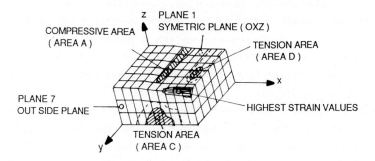

Figure 9 : scheme

Classical criteria are proposed in the litterature (réf 4.5). In our investigation, the criterion must involve mechanical parameters . the Latham and Cockroft criterion was found to be unable to fit with both torsion and tension results : so we take the two parameters OYANE's criterion.

The experimental and simulation values (B exp and B_S) of B parameter are plotted on figure 10 as a function of slab temperature, for the two critical points where the B_S values are maxima (cf figure 8).

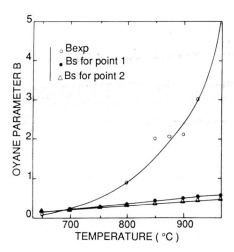

Figure 10 : Variation of B in function of slab temperature

It appears that for an homogeneous slab temperature of 925 °C, material should not crack. Conversely a temperature of 700 °C would be a critical homogeneous temperature of rolling.

Then it is clear that temperature has a dramatic influence on failure risk. But, in practice, ingot temperature is not homogeneous and surfaces cool more rapidly than slab bulk due to radiation, convection, and water cooling of cylinders. Thermal characteristics of the present material being not yet accurately known (some slabs cooling recordings are in progress), it is not possible to run anisothermal computations. Nevertheless, Figure 10 can be used for a quantitative assessment of failure risk if it assumed that the slab surface strains are imposed by the slab core and are close to the isothermal results obtained for a rolling temperature equal to the core temperature. As the σ_m/σ_{eq} ratio does not depend on temperature (from the constitutive equation), the surface region failure risk would be independent on the surface temperature. Following this assumption, the surface failure risk, estimated through B_s, has to be compared to the B_{exp} parameter at the surface temperature. For example, taking a nominal temperature of 925 $^{\circ}$C, surface cracks would appear if surface temperature is lower than 750 $^{\circ}$C (cf figure 10).

This procedure for surface failure prediction from isothermal computations is rather simple and avoids more expensive thermomechanical calculations. The benefical influence of material purity, which is related to the scraps recycling ratio, to ductility (which is sensible in the 700 $^{\circ}$C - 750 $^{\circ}$C range) may thus explain the absence of rolling defects in the case of no recycled scraps containing materials. The validity of anisothermal failure predictions from isothermal computations will be checked in near future by anisothermal computation with various supposed thermal gradients.

CONCLUSION

This study enlightens the efficiency of 3D rolling simulation program LAMEF3 for estimating local stresses and strains and to deal with defects risks prediction by indicating the influence of rolling parameters such as temperature and rolling steps scheme. In the present case, neither temperature can be raised to lower the failure risk because of alloy stability nor the first step reduction ratio can be lowered because of strain heterogeneities. Material purity related to recycling material was found to be the only adjustable (although with some difficulty) parameter.

REFERENCES
1 J.L. CHENOT, Etude de l'élargissement en laminage à chaud. Calcul tridimensionnel par éléments finis, Rapport GRECO N° 63/1982.
2 O.C. ZIENKIEWCIZ, the Finite Element Method.
3rd, London, MAC GRAW HILL, 1977.
3 M. OYANE, T. SATO, K. OKIMOTO and S. SHIMA. Criteria for ductile fracture and their applications, J. Mech. Working Technol. 4 (1980) 65. 81.
4 M. PREDELEANU, J.P. CORDEBOIS and L. BELKINI, Failure analysis of cold upseting by computer and experimental simulation, in : Proc. of the Numiform' 86 Conference, Gothenburg, 25-29 August, 1986, 277-282.
5 N.L. DUNG, Fracture intiation in upsetting Tests, in : Proc. of the Numiform' 86 Conference, Gothenburg, 25-29 August, 1986, 261-269.

ANALYSIS OF THE PROFILE OF LOCAL NECKING IN SHEET METAL DEFORMATION PROCESSES

J.D. BRESSAN

Departamento de Engenharia Mecanica, Faculdade de Engenharia de Joinville, 89200 Joinville, SC (Brazil)

SUMMARY

In this paper an analytical model to predict the strain distri- bution in the neck during plastic deformation of sheet metal sub- jected to biaxial and uniaxial tension is presented. The model is based on the development of strain gradient from the initial thick ness imperfections. The present theoretical analysis incorporates a rate-dependent anisotropic plasticity theory and is compared with experimental results. A numerical method is presented to solve the partial differential equation which describe neck growth. The neck profile is determined as a function of the initial thickness gradient,strain and strain-rate coefficients, normal anisotropy and stress state. Predicted strain distribution for AK steel is reasonably good for uniaxial test. The limit strain is defined as an acceptable strain gradient in the neck.

INTRODUCTION

The limit of performance of a sheet metal during stretch forming is known as the Forming Limit Diagram(FLD). Predicting such a limit in the sheet metal forming processes is an important task to improve formability. However, the definition of limit strain in a pressed product depends on the local strain gradient or thickness imperfection considered acceptable. From the practical point of view a diffuse neck (low strain gradient) is acceptable whereas a local neck (high strain gradient) would not. The onset of local necking during stretch forming originates at preexisting thickness imperfections or material inhomogeneities. The neck has a conti- nuous development resulting in a very localised defect which can be interpreted as failure. Important advances have been made in recent years in the development of the theorectical limit strain mainly the work by Marciniak (1), Storen and Rice (2) and Williams and Bressan (3). The purpose of the present work is to present a numerical method which can be used to construct the entire profile of the neck and to provide a detailed description of neck growth. The analytical model of strain gradient development (3) has been employed in the present work.

THEORETICAL ANALYSIS

The development of nonuniform plastic flow is analyzed for a thin sheet loaded under various in-plane biaxial loading conditions. The specimen is assumed to have an initial geometric non-uniformity $\mu=(1/t_0).(dt_0/dx)$ where t_0 is the initial thickness. Considering an anisotropic and strain and strain-rate dependent material under general state of stress, the equivalent stress $\bar{\sigma}$ may be expressed in terms of the equivalent strain $\bar{\varepsilon}$ as:

$$\bar{\sigma} = K (\varepsilon_0 + \varepsilon)^n \dot{\bar{\varepsilon}}^M \tag{1}$$

where n is the work hardening coefficient and M is the strain-rate sensitivity coefficient. On the other hand, Hill (4) recently proposed an anisotropic yield function which lead to the relationships for equivalent stress and strain increment,

$$\bar{\sigma} = \frac{1}{2(1+R)} [(1+2R)(\sigma_1-\sigma_2)^m + (\sigma_1+\sigma_2)^m]^{1/m} \tag{2}$$

and

$$d\bar{\varepsilon} = \frac{[2(1+R)]^{1/m}}{2} \left\{ \frac{1}{(1+2R)^{1/(m-1)}} |d\varepsilon_1-d\varepsilon_2|^{\frac{m}{m-1}} + |d\varepsilon_1+d\varepsilon_2|^{\frac{m}{m-1}} \right\}^{\frac{m-1}{m}} \tag{3}$$

where R and m are parameters that describe the degree of normal plastic anisotropy exhibited by the sheet and subscripts 1 and 2 refer to the directions of the principal axes in the plane of the sheet.

The equation derived by the author (3) which describe the development of strain gradient in sheet metal during biaxial stretching is given by:

$$\partial\lambda/\partial\bar{\varepsilon} = \mu/M + (1/M)[\alpha/((1+\alpha)Z) - n/(\varepsilon_0+\bar{\varepsilon})]\lambda \tag{4}$$

where $\lambda=\partial\bar{\varepsilon}/\partial x$ is the strain gradient, $\alpha=d\varepsilon_1/d\varepsilon_2$ is the strain ratio and, the critical subtangent Z,

$$Z = \frac{BD^{(m-1)/m}}{2(1+\alpha)} \tag{5}$$

$$B = [2(1+R)]^{1/m} \quad ; \quad D = \frac{1}{(1+2R)^{1/(m-1)}} |\alpha-1|^{m/(m-1)} + |\alpha+1|^{m/(m-1)}$$

The entire profile of the neck could be investigated by applying equation (4) to a number of material points along the normal neck direction 1 or x-axis. The numerical method is outlined below for the entire neck profile.

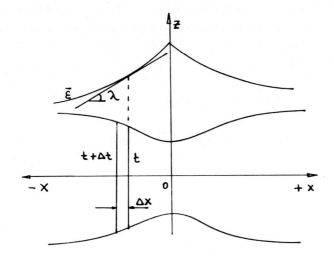

Fig. 1. Neck Profile.

Considering Fig.1 the thickness variation along the neck was obtained using a computer program to solve numerically equation (4) as,

$$\Delta\bar{\varepsilon} = \mu.(\lambda/\mu).\Delta x \tag{6}$$

$$\bar{\varepsilon} = \bar{\varepsilon} + \Delta\bar{\varepsilon} \tag{7}$$

$$t = t_0.\exp(-\bar{\varepsilon}/Z) \tag{8}$$

$$\Delta(\lambda/\mu) = \frac{1}{M}.(1+\frac{\lambda}{\mu}.(\frac{\alpha}{(1+\alpha).Z} - \frac{n}{(\varepsilon_0+\varepsilon)})).\Delta\bar{\varepsilon} \tag{9}$$

$$\lambda/\mu = \lambda/\mu + \Delta(\lambda/\mu) \tag{10}$$

$$x = x + \Delta x \tag{11}$$

where Δ is the increment of the variable. It was adopted $\Delta x = -.001$ mm for $t_0 = 1$ mm. For $\Delta x = -.0001$ mm the results were essentially the same. The adopted boundary conditions were: at $x=0$, inside the neck, $t=t_{min}$ and

$$\bar{\varepsilon} = Z.\log(t_0/t_{min})$$

$$\lambda/\mu = \frac{1}{M} . \int_0^{\bar{\varepsilon}} \exp[-\alpha.(\bar{\varepsilon}-\xi)/((1+\alpha).Z)]/[(\varepsilon_0+\bar{\varepsilon})/(\varepsilon_0+\xi)]^{n/M}.d\xi \qquad (12)$$

where t_{min} is the minimum thickness and equation (12) is the analytical solution of equation (4). Within the drawing or negative
quadrant of principal-strain space the development of local neck
is described by (3),

$$\partial\lambda/\partial\bar{\varepsilon} = \mu/M + (1/M).[1/Z - n/(\varepsilon_0+\bar{\varepsilon})].\lambda \qquad (13)$$

For a material showing no rate-dependence, M=0 , eqution (13)
becomes:

$$\lambda/\mu = [n/(\varepsilon_0+\bar{\varepsilon}) - 1/Z]^{-1} \qquad (14)$$

RESULTS AND DISCUSSION

One possible means of investigating the validity of the present
approach could be to examine the experimental data available in
the literature and compare it with the present theory. A reasonable
value of the thickness nonuniformity μ is .01 to .07/mm as reported
by D. Lee and F. Zaveri (5) for AK steel.

In order to test the theory presented above, comparisons have
been made between the predictions of the neck profile and the experimental data for uniaxial and plane strain test. The correlation
is reasonably good for AK steel for μ=.01 to .02/mm. For brass the
discrepancies may be due to the theoretical yield loci and constitutive equation. Besides, the coefficients n and M may exhibit
variations between uniaxial and biaxial tests and thereby neck
profile be influenced.

A relevant aspect in the analysis of the profile of local
necking is to provide an alternative method to determine M-value
and the influence of the initial thickness inhomogeneity on the
morphology of the neck. In addition, the approach provides a
method to investigate the neck profile in order to correlate the
local limit strain with the corresponding strain gradient λ/μ. It
is clear that the definition of local neck depends on the value of
λ/μ adopted as acceptable.

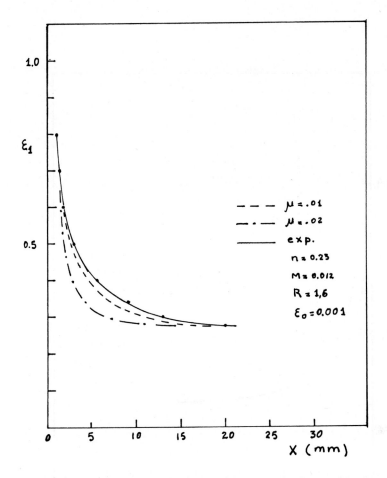

Fig. 2.Neck profile for AK steel for uniaxial tension test. Theoretical strain distribution for μ=0.01/mm fits reasonable well the experimental data (6).

ACKNOWLEDGEMENTS

Financial support provided by CNPq and Faculdade de Engenharia de Joinville, Santa Catarina, Brazil, is gratefully acknowledged.

44

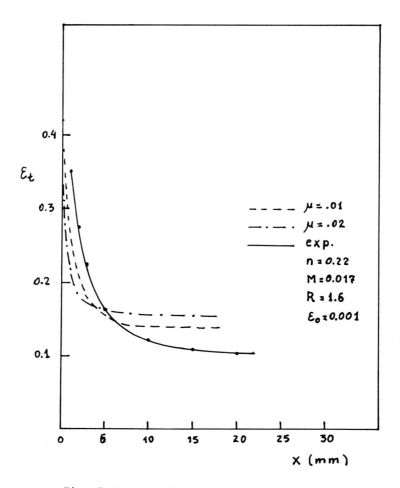

Fig. 3.Neck profile for AK steel for uniaxial tension test. ε_t is the thickness strain. Experimental data taken from reference (7).

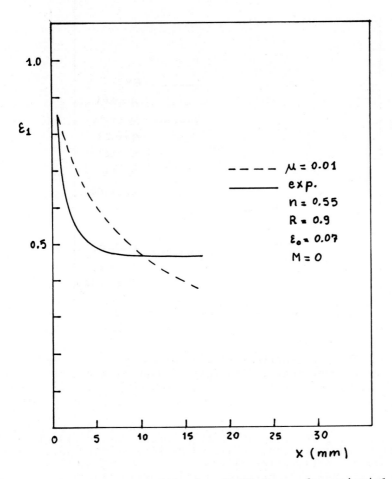

Fig. 4.Neck profile for 70/30 Brass for uniaxial
tension test. Experimental data taken from
reference (6).

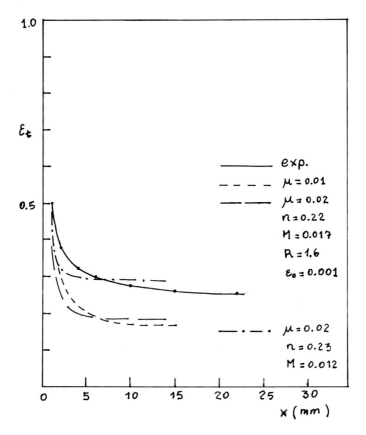

Fig. 5. Neck profile for AK steel for plane strain
test. Experimental data (exp.) taken from reference (7).

REFERENCES

1 Z. Marciniak and K. Kuczynski, Limit strains in the processes of
  stretch-forming sheet metal, Int. J. Mech. Sci., 9(1967)609-620.
2 S. Storen and J.R. Rice, Localized necking in thin sheet, J.Mech.
  Phys. Solids, 23(1975)421-441.
3 J.D. Bressan and J.A. Williams, Limit strains in the sheet forming of strain and strain-rate sensitive materials, J. Mech. Work.
  Tech., 11(1985)291-317.
4 R. Hill, Theoretical plasticity of textured aggregates, Math.
  Proc. Camb. Phil. Soc., 85(1979)179-191.
5 D. Lee and F. Zaveri, Neck growth and forming limits in sheet
  metals, Int. J. Mech. Sci., 26(1981)680.
6 A. K. Gosh, The influence of strain hardening and strain-rate
  sensitivity on sheet metal forming, J. Eng. Mat. Tech.,99(1977)
  264-274.
7 H.J. Kleemola and J.O. Kumpulainen, Factors influencing the
  forming limit diagram, J. Mech. Work. Tech., 3(1980)289-302.

Computational Methods for Predicting Material Processing Defects, edited by M. Predeleanu
Elsevier Science Publishers B.V., Amsterdam, 1987 — Printed in The Netherlands

A FINITE ELEMENT ANALYSIS OF SPRINGBACK IN PLANE STRAIN FOLDING WITH BINDERS OF HIGH STRENGTH STEEL SHEETS

M. BRUNET

Laboratoire de Mécanique des Solides
INSA - 304 - 69621 VILLEURBANNE - FRANCE

SUMMARY
    For the purpose of studying the folding and springback of high strength steel sheets, a plane strain finite element program is developed. The program takes account of the geometrical and material nonlinearities due to the large displacements, finite rotations and plastic deformation, which occur during the metal forming process. The friction forces between the punch, the upper and lower binder and the sheet modelled by Coulomb's friction law, are also taken into account. An iterative procedure is carried out during each increment in order to maintain equilibrium overall the process and to permit unloading during the springback. The numerical results obtained are in good agreement with experimental values of springback.

INTRODUCTION

    This paper deals with the numerical prediction of the tension developed in sheet metal during a plane strain folding operation, and the resultant springback after removing the upper and lower binders.

    The motivation for studying springback is its effect on shape control, while the final yield strengh of the material determines, in part, its structural strength and dent resistance.

    Comprehensive works concerning both experimental, theoretical and analytical analysis of the springback has been carried out by BOYER J.C. and BOIVIN M. (réf. 1 and 2), RONDE-OUSTEAU F. (ref. 3), WENNER M.L. (ref. 4), LIU Y.C. (ref. 5), DAVIES R.G. and LIU Y.C. (ref. 6). In these papers, analytical methods are developed with some assumptions such as pure bending effects without friction or pure constant tensile load under the punch with some friction conditions for instance. Then, it is possible to quantify the roles of various parameters such as the die and punch radius, the restraining force and the material properties of the sheet metal.

    The aim of this work is to simulate the overall process of folding and springback of a high strength steel sheet in order to take account of the geometrical aspects and frictional boundary conditions due to the friction arising from the pressure exerted on the sheet metal by the flat binder areas. Though relatively simple to put into practice, the folding process is complicated when seen from a theoretical point of view , because the sheet metal is exposed to large displacements, finite rotations and plastic flow during the process.

The governing equations of such a process contain therefore nonlinearities of both material and geometrical kinds. Furthermore, the sheet material is often encumbered with anisotropy caused by previous treatments, and friction between the sheet and the binders additionnally complicates a theoretical investigation of the process.

Also the present work makes use of the finite element method, based on a suitable updated Lagrangian formulation of the governing equations incorporating all the latest developments in combining large displacements, finite rotations with elastic-plastic behaviour and unilateral contact with friction.

GOVERNING EQUATIONS

In the folding process, the surface traction acting on the sheet, are dependent on the deformation. Consequently, the work done by the external loads is path-dependent. Furthermore, using a flow theory of plasticity, the current stress-strain relation is dependent on the past stress history of the material points. To account for these non-conservative effects it is necessary to employ an incremental solution procedure of some kind.

Also the present work makes use of the finite element method based on an updated Lagrangian formulation.

The principle of virtual work at time $t + \Delta t$ referred to t can be written in the following way :

$$\int_V \{\Delta S_{ij} \delta u_{i,j} + \sigma_{ij} \Delta u_{k,i} \delta u_{k,j}\} \, dV = \int_{S_t} (t_i + \Delta t_i) \delta u_i \, dS - \int_V \sigma_{ij} \delta u_{i,j} \, dV \quad (1)$$

This permits us to calculate a certain state of equilibrium based on the knowledge of the previous equilibrium state together with the actual increments in the external loads or the prescribed displacements. To avoid error accumulation and to permit unloading, the new equilibrium state is calculated by an iterative technique where the term :

$$R = \int_{S_t} t_i \delta u_i \, dS - \int_V \sigma_{ij} \delta u_{i,j} \, dV \quad (2)$$

is often called the residual which is expected to vanish.

For history-dependent elasto-plastic materials the constitutive relations are readily incorporated in the updated Lagrangian formulation since these equations are functions of variables referred to the current state.

By introducing the corotational (JAUMANN) increment to the KIRCHHOFF stress tensor $\Delta \overset{\triangledown}{S}_{ij}$ such that :

$$\Delta S_{ij} = \Delta \overset{\triangledown}{S}_{ij} - \sigma_{mj} \Delta \varepsilon_{im} - \sigma_{im} \Delta \varepsilon_{jm} \quad (3)$$

where the strain increment are defined by :

$$\Delta\varepsilon_{ij} = 1/2\,(\Delta u_{i,j} + \Delta u_{j,i}) \qquad (4)$$

the constitutive equation of an isotropic material can be written as :

$$\Delta\overset{\triangledown}{S}_{ij} = C_{ijkl}^{e-p}\,\Delta\varepsilon_{kl} \qquad (5)$$

where C_{ijkl}^{e-p} is the classical constitutive law corresponding to the J_2 flow theo-ry of plasticity. Use of the KIRCHHOFF stress tensor leads to a symmetrical stiffness matrices in the finite element approximation to the principle of vir-tual work (1).

BOUNDARY CONDITIONS

In the present formulation of the folding problem, it is assumed that the tool parts are completely rigid bodies, whose relative movements are forced on a deformable elastic-plastic solid. Then we can use parametric functions to describe the rigid surface of the punch and of the die with its upper and lower binders.

In an incremental approach, we determine an appropriate prescribed displace-ment to just bring a or some material points in the surface Sc to be in contact with the rigid surface. Thus, for any two in-contact points we have :

$$\Delta u_n - \Delta g_n = 0 \qquad \text{on Sc} \qquad (6)$$

where $\Delta u_n = \Delta u.n$ is the increment of the normal displacement constrained by the given rigid surface motion Δg_n.

To resolve this boundary constraint, we apply the exterior penalty method. Physically the constraint (6) can be wiewed as to replace the rigid surface by a set of very stiff springs along the contact surfaces. Thus, we define :

$$\Delta f_n = -\,k_n\,(\Delta u_n - \Delta g_n) \quad \text{on Sc} \qquad (7)$$

with a very large penalty parameter k_n (>0). The point will stay in contact as long as the resultant contact pressure f_n updated by Δf_n remains negative.

We assume that there exists a Coulomb's friction law with the basic assump-tion of a constant coefficient of friction μ. No relative motion is observed if:

$$|f_t| < \mu|f_n| \qquad (8)$$

so that the same relation (7) holds between the tangential increment of the friction force and the tangential increments of displacements.

$$\Delta f_t = -\,k_t\,(\Delta u_t - \Delta g_t) \quad \text{on Sc} \qquad (9)$$

when the state of stress of a point on Sc is such that :

$$|f_t| = \mu|f_n| \quad \text{and} \quad |df_t| = \mu|df_n| \qquad (10)$$

the point is said to be sliding.

Frictional effects play an important role in most of the practical applications. Although a tremendous anmount of effort has been put into the investigation of this subject, a precise incremental friction law suitable for large deformation analysis with sufficient and appropriate experimental data is still lacking. In general, an attempt is made toward constructing an incremental friction law in analogy with the elastoplasticity theory.

FINITE ELEMENT APPROXIMATION

Using constitutive equations (5) with the relation (3), the principle of virtual work at time $t + \Delta t$ referred to t including the virtual work done by the pressure contact forces can be rewritten as :

$$\int_V \{L^*_{ijkl}\Delta u_{k,l} + \sigma_{jk}\Delta u_{i,k}\}\delta u_{i,j}\ dV + \int_{Sc}\{k_n\Delta u_n\delta u_n + k_{ti}\Delta u_{ti}\delta u_{ti}\}\ dSc$$

$$= \int_{S_t}(t_i+\Delta t_i)\delta u_i dS_t + \int_{Sc}\{(f_n+k_n\Delta g_n)\delta u_n + (f_{ti}+_tk\Delta g_{ti})\delta u_i\}dS_c - \int_V \sigma_{ij}\delta u_{i,j}\ dV$$

(11)

$\forall\ \delta u_i$ such that $\delta u_i = 0$ on S_u and $S = S_c\ U\ S_t\ U\ S_u$ where :

$$L^*_{ijkl} = C^{e-p*}_{ijkl} - 1/2\ (\delta_{ik}\sigma_{jl} + \delta_{il}\sigma_{jk} +\delta_{jk}\sigma_{il} + \delta_{jl}\sigma_{ik})$$

(12)

and the form of the numerical elasto-plastic moduli C^{e-p*}_{ijkl} is discussed later. The spatial discretization of the integral form (11) follows the standard finite element procedures. Within each element, interpolations are made to the variables Δu, Δg, Δt and δu by properly chosen shape functions $N_\alpha(x)$ which lead to and algebraic matrix equation :

$$\{\sum_e K^{Ve}_{ij\alpha\beta} + \sum_e K^{Sce}_{ij\alpha\beta}\}\Delta u_{j\alpha} = \sum_e (f^{Ste}_{i\beta} + \Delta f^{Ste}_{i\beta}) + \sum_e (f^{Sce}_{i\beta} + \Delta f^{Sce}_{i\beta}) -$$

$$\sum_e \int_{ve}\sigma_{ij}\ N_{\beta,j}\ dVe$$

(13)

where :

$$K^{Ve}_{ij\alpha\beta} = \int_{ve}(L^*_{ijkl} + \delta_{ij}\sigma_{kl})N_{\alpha,l}\ N_{\beta,k}\ dVe$$

(14)

$$K^{Sce}_{ij\alpha\beta} = \int_{Sce}\{k_n n_j n_i + k_t(\delta_{ij} - n_i n_j)\}\ N_\alpha N_\beta\ dSce$$

(15)

are the element stiffness matrices, and :

$$\Delta f^{Ste}_{i\beta} = \int_{Ste}\Delta t_i\ N_\beta\ dSte$$

(16)

$$\Delta f^{Sce}_{i\beta} = \int_{Sce}\{k_n\Delta g_n n_i + k_t\Delta g_{tj}\ (\delta_{ij} - n_i n_j)\}\ N_\beta\ dS$$

(17)

are the element load vectors.

Since the result of an incremental calculation is not known in advance, it is necessary to do a few iterations to assure consistent boundary conditions, equilibrium and path independency as it will be shown in the next section.

SOLUTION PROCEDURE

In the procedure presented here, it is attempted to satisfy equilibrium in the current state of the body. Morever, during iteration, the nodal imbalance is formulated with respect to the state obtained as a result of the previous iteration. Hence upon calculation of the latest estimate for the displacement increment, the geometry of the element will be updated corresponding to this latest estimate.

Using the classical linear expression for deformation rate and spin, but taken in the state at the middle of the total displacement increment, the incremental rotation tensor is calculated first as :

$$\Delta \underline{R} = e^{\Delta \underline{\omega}(t+\Delta t/2)} \underline{I} \tag{18}$$

where $\Delta \underline{\omega}(t+\Delta t/2)$ is the incremental spin tensor evaluated at the middle of the increment.

The stress at the end of the increment then follows from :

$$\underline{\sigma}(t+\Delta t) = \Delta \underline{R} . \underline{\sigma}(t) . \Delta \underline{R}^T + \Delta \underline{R}^{1/2} . \Delta \overset{\triangledown}{\underline{S}} . \Delta \underline{R}^{1/2T} \tag{19}$$

The evaluation of (19) requires an integration of the constitutive rate equations in the middle of the increment in order to obtain the corotational stress increment $\Delta \overset{\triangledown}{\underline{S}}$. For VON-MISES plasticity we have found that the elastic predictor – radial corrector one order implicit method is both simple and accurate.

This approach assure path independency and incremental objectivity.

The next aspect demands that in an efficient solution procedure the elasto plastic stiffness matrix must be consistent with the implicit integration procedure of the constitutive rate equations in order to preserve the quadratic convergence of the quasi NEWTON-RAPHSON method. A derivation of this consistent moduli $\underline{C}^{e-p*}$ can be found in (ref. 7). For VON-MISES plasticity and based on the elastic predictor-radial corrector method, it can be derived that :

$$C^{ep*}_{ijkl} = G^{*}(\delta_{ik}\delta_{jl} + \delta_{il}\delta_{jk}) + \{2(1+\nu)G/(3(1-2\nu)) - 2G^{*}/3\} \delta_{ij}\delta_{kl}$$

$$- 3 G^{*}/((1+H^{*}/3G)\bar{\sigma}^{-2}) s_{ij}s_{kl} \tag{20}$$

In this equation, G is the elastic shear modulus, ν the Poisson's ratio and G and H are the modified shear modulus and plastic hardening coefficient related to the actual G and H' by :

$$G^{*} = G \bar{\sigma}/\bar{\sigma}_E \tag{21}$$

$$H^{\textit{*}} = H'/(\bar{\sigma}(H'/3G) + 1)/\bar{\sigma}_E - H'/3G) \tag{22}$$

In these relations, $\bar{\sigma}_E$ is the equivalent elastic stress of the elastic predictor-radial corrector method. It is seen that the shear modulus and the hardening coefficient are modified accordingly such that the effective shear modulus $G^{\textit{*}}$ is lower than the actual shear modulus G. Thus, use of the moduli $\underline{C}^{ep\textit{*}}$ improve considerably the rate of convergence.

Prior to the solution of practical problems, a number of one-element tests were carried out to check the correctness of the implementation. The calculation were carried out with different increment sizes and the attempted independence of increment size was obtained.

NUMERICAL RESULTS

An experimental and theoretical analysis of the folding process has been carried out by RONDE-OUSTEAU (ref. 3) and BOYER and BOIVIN (ref. 1 and 2) for the theoretical part.

RONDE-OUSTEAU have accomplished a large series of measurements of the springback in high strength steel sheets of various materials. The theoretical investigation is essentially concentrated about the part of the sheet which is under the punch with the assumption of pure bending and constant axial tension. A very good agreement is obtained between theoretical analysis and experimental investigations (ref. 3) for materials with the following mechanical properties :

Material	Yield strength (MPa)	Plastic hardening modulus (MPa)
Typical E.S.	140	4000
SOLPRESS	112	4000
SOLPHOR A400	244	3600
OP HR 40	187	4880
IF HR 45/50	209	3600

But the predicted springback of the theoretical analysis (17.4 degrees) and the average experimental measure (8.3 degrees) are quite different for a (dispersoïd) steel (SOLDUR 40) with a yield strength of 383 MPa and a plastic hardening modulus of 1333 MPa. The geometrical characteristics of the folding are as follow :

- Radius punch : 25 mm
- Thickness of the sheet : 0.69 mm
- Size of the sheet : 180 mm
- Size of the binder area : 80 mm
- Force of the binder on the sheet : 11 400 daN
- Friction coefficient : 0.15
- Folding angle : 90°

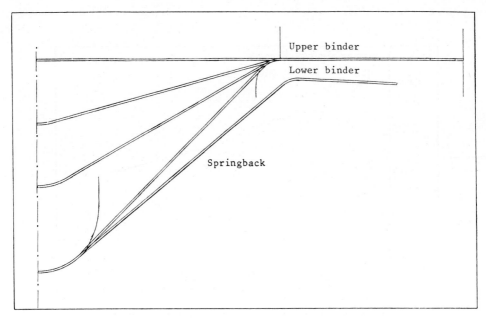

Figure 1 : Folding and springback of the sheet

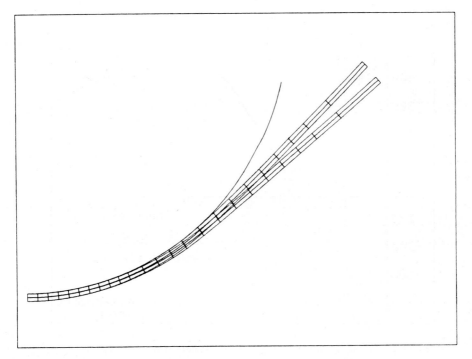

Figure 2 : Mesh under the punch at the end of folding and springback

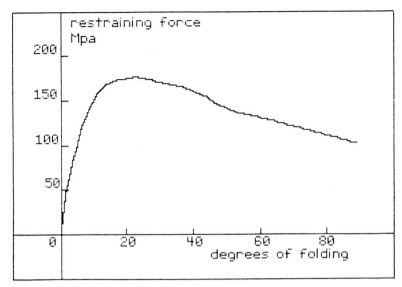

Figure 3 : Restraining force in the free part of the sheet

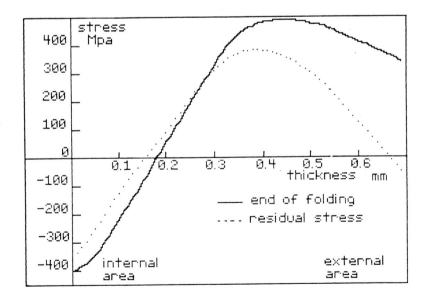

Figure 4 : Bending axial stress in the sheet under the punch

Also, the finite element numerical analysis concentrate on this case. The element used in this bidimensional analysis is a plane strain quadrilateral and isoparametric element with 24 degrees of freedom (see fig. 2 for details under the punch). The area integration are performed using a nine point Gaussian scheme and the contact pressure is supposed to very linearly between two nodes. The folding process has required 784 increments with an average of 3 iterations per increment. In order to avoid numerical difficulties, the upper binder is first removed and then the lower binder until no-contact occurs between the sheet and the die, this analysis required 158 increments and the equilibrium is maintained up to a nodal force imbalance of 4.8 %.

The springback is shown in figure 1 and 2 with a half-angle of 5.5 degrees which is in a better agreement with the experimental result (8.3 degrees) than the theoretical one (17.4 degrees).

This can be explain by the curve of the raistraining force from the binder area (fig.3) wich passes through a maximum of 175 MPa and decreases linearly thereafter up to 120 MPa at the end of folding.

Figure 4 shows the variation of the axial stress through the sheet thickness just under the punch. An important feature observed is that the sheet tends to leave the punch surface near the axe of symetry during the folding, which explains the unloading observed in the stress curve near the external area at the end of the folding. For this material, a plastic articulation seems to develop along the punch due to the much lower level of plastic hardening modulus ($H' = 1333$ MPa) than those of the other materials. This may be an explanation of the drawback of the theoretical investigation which doesnot take account of this special behaviour.

CONCLUSION

A finite element method has been used to analyze the plane strain folding and springback problem. Comparison of predicted versus measured springback in a very high strength steel sheet shows reasonable agreement if we keep in mind that the numerical solution strongly depends on the estimate frictional coefficient. We close this paper by emphasizing that it has presented a bidimensional analysis which leads to a somewhat lenghty calculation but permits to observe specific behaviour such as gap and unloading just under the punch at the end of folding.

REFERENCES

1 J.C. BOYER, M. BOIVIN, Calcul du retour élastique et des contraintes rési-
   duelles lors du pliage des tôles à haute limite d'élasticité, Mécanique Matériaux Electricité n° 385 - Janvier 1982 p. 26
2 J.C. BOYER, M. BOIVIN, Calcul du retour élastique et des contraintes rési-
   duelles lors du pliage des tôles sous tension - Rapport GRECO Grandes défor-

mations et endommagement n° 83/1983

3 F. RONDE-OUSTEAU, Etude du retour élastique en pliage - Rapport GIS - Mise en Forme - Mars 1987

4 M.L. WENNER, On Work Hardening and Springback in Plane Strain Draw Forming J. Appl. Metalworking, V.2 - N° 4 - (1983) 277-287

5 Y.C. LIU, Springback Reduction in U-Channels -"Double Bend" Technique, J. Appl. Metalworking, V.3 - n° 2 - (1984) 148-156

6 R.G. DAVIES, Y.C. LIU, Control of Springback in a Flanging Operation, J. Appl. Metalworking, v.3 - n° 2 - 142-147

7 J.C. NAGTEGAAL, On the implementation of Inelastic Constitutive Equations with special Reference to large deformation Problems. Comp. Meth. Appl. Mech. Eng., V.3, pp. 469-484.

Computational Methods for Predicting Material Processing Defects, edited by M. Predeleanu
Elsevier Science Publishers B.V., Amsterdam, 1987 — Printed in The Netherlands

PREVISION DES DEFAUTS DE GEOMETRIE AU COURS DU LAMINAGE A CHAUD DES ACIERS

S. CESCOTTO[1], H. GROBER[2], R. CHARLIER[1], A.M. HABRAKEN[1]

[1]Services M.S.M., Institut du Génie Civil, Université de Liège (Belgique)

[2]Arbed-Recherches, Service Procédés,(Grand-Duché de Luxembourg)

RESUME
    Cet article présente, dans les grandes lignes, les points importants de la
simulation par éléments finis du laminage à chaud des aciers en vue de prévoir
numériquement la forme du produit après chaque passe de laminage.

INTRODUCTION

    Les études sur la simulation numérique des grandes déformations des métaux
trouvent un champ d'application important dans le domaine du laminage à chaud
des aciers. Le souci de mieux maîtriser ce procédé de formage conduit à déve-
lopper des codes de calcul tridimensionnels capables de prendre en compte les
non linéarités fondamentales liées à ces procédés : loi de comportement inélastique, très grandes déformations et contact unilatéral avec les cylindres.

    Un tel code, bien dominé et bien utilisé, peut s'avérer un outil précieux
d'aide à la mise au point d'une séquence de laminage. En particulier, il peut
permettre de prévoir des défauts de géométrie et de corriger en conséquence la
forme des cannelures des cylindres.

    Dans cet article, nous résumons les points importants dont il faut tenir
compte sur le plan de la simulation numérique du laminage à chaud et nous don-
nons quelques exemples d'applications.

LOI CONSTITUTIVE DES ACIERS DE LAMINAGE

    Les aciers de laminage peuvent différer les uns des autres par leur composi-
tion, le diamètre initial de grain,...
Leurs propriétés mécaniques sont donc influencées par ces facteurs, ainsi que
la température de mise en oeuvre et la vitesse de déformation.

    Des modèles constitutifs capables de représenter les aspects essentiels du
comportement des métaux à haute température existent depuis de nombreuses an-
nées et peuvent, aujourd'hui, être considérés comme classiques (voir [1] par
exemple).
La difficulté se situe donc plutôt au niveau du choix d'un modèle et du calibrage de ses paramètres.
Dans le cas du laminage à chaud, le matériau n'est pas soumis à un chargement

cyclique. Certes, des déformations répétées sont appliquées au cours des passes successives mais, entre celles-ci, la structure du métal se modifie par recristallisation et croissance de grain. De ce fait, on peut se contenter, pour étudier une passe donnée, d'un modèle à écrouissage isotrope aussi simple que possible, mais qui soit néanmoins capable de représenter correctement l'influence prépondérante de la vitesse de déformation et de la température.

Dans ce travail, nous avons opté pour une loi constitutive élasto-viscoplastique représentée par les équations suivantes :

$$\overset{\triangledown}{\sigma}_{ij} = C^e_{ijkl} \, (D_{kl} - D^n_{kl})$$

$$D^n_{ij} = \Phi \, (F) \, \frac{\hat{\sigma}_{ij}}{\sqrt{J_2}}$$

$$\dot{\kappa} = \frac{2}{\sqrt{3}} \, \Phi \, (F) \, h_\kappa \, (\kappa)$$

$$J_2 = \frac{1}{2} \, \hat{\sigma}_{ij} \, \hat{\sigma}_{ij} \; ; \quad F = \frac{\sqrt{J_2}}{\kappa}$$

σ_{ij} est la contrainte de Cauchy et $\hat{\sigma}_{ij}$ son déviateur

κ est une variable d'état scalaire responsable de l'écrouissage isotrope

D_{ij} est la vitesse de déformation dans la configuration courante du solide : c'est la partie symétrique du gradient de vitesse

D^n_{ij} est la vitesse de déformation inélastique

C^e_{ijkl} est le tenseur élastique isotrope de Hooke.

Les fonctions $\Phi \, (F)$ et $h_\kappa \, (\kappa)$ sont les seuls paramètres du modèle.
En se basant sur des essais de torsions à chaud réalisés par l'IRSID, on a pu montrer [1] que les fonctions suivantes :

$$\Phi \, (F) \; = \; B \, (F)^n \; ; \quad h_\kappa \, (\kappa) \; = \; \frac{H_1}{(\kappa)^m}$$

sont satisfaisantes pour le but poursuivi. Les paramètres B, H_1, n et m doivent être calibrés pour chaque nuance d'acier et chaque température. A titre d'illustration, la figure 1 compare ce modèle (EVP) et les résultats de l'IRSID pour un acier de nuance donnée (C = 191 10^{-3} % ; M_n = 826 10^{-3} % ; Al = 410 10^{-3} % ; N_b = 0 %, diamètre initial de grain d_o = 130 μ_m) à 1100° C. On constate que la concordance est satisfaisante dans un domaine de vitesses de déformations variant dans un rapport de 1 à 10000.

MODELISATION DU SOLIDE

Le solide en cours de déformation est discrétisé en éléments finis et son équilibre est exprimé à l'aide de l'équation des puissances virtuelles exprimée dans la configuration déformée ;

$$\int_v \sigma_{ij}\, \delta D_{ij}\, dv = \int_v F_i\, \delta v_i\, dv + \int_a T_i\, \delta v_i\, da$$

où δv_i est un champ virtuel de vitesse et δD_{ij} est le champ correspondant de vitesses de déformations virtuelles

$$\delta D_{ij} = \frac{1}{2}\left(\frac{\partial \delta v_i}{\partial x_j} + \frac{\partial \delta v_j}{\partial x_i}\right) \; ;$$

v et a sont le volume et la surface actuels du solide déformé ;
F_i et T_i sont les forces volumiques et surfaciques qui le sollicitent.
Cette équation fortement non linéaire (à cause des grandes déformations, de la loi constitutive et des problèmes de contact unilatéral aux frontières du solide) est évidemment résolue pas à pas selon les techniques numériques bien connues. Seuls les points particuliers aux grandes déformations seront évoqués rapidement.

Trajet entre deux configurations successives.

Soient γ_A et γ_B deux configurations successives du solide aux instants t_A et t_B (avec $t_B = t_A + \Delta t$). Connaissant la géométrie ($\underline{x}_A$) et les contraintes ($\underline{\sigma}_A$) dans γ_A et connaissant seulement la géométrie ($\underline{x}_B$) dans γ_B, on cherche à y déterminer les contraintes ($\underline{\sigma}_B$) (figure 2). Le premier problème qui se pose est de choisir un trajet du solide entre γ_A et γ_B. Plusieurs possibilités existent.

a. Trajet à vitesse constante.

On admet que chaque particule suit un trajet rectiligne à vitesse constante entre γ_A et γ_B

$$\underline{v}_{AB} = \frac{1}{\Delta t}(\underline{x}_B - \underline{x}_A)$$

$$\underline{x}(t) = \underline{x}_A + \underline{v}_{AB}(t - t_A)$$

Sous cette hypothèse, le gradient de vitesse $\underline{L} = \partial \underline{v}/\partial \underline{x}$ et la vitesse de déformation $\underline{D} = \mathrm{sym}(\underline{L})$ ne sont pas constants. En particulier, si γ_A et γ_B ne diffèrent que par une rotation de corps rigide, le trajet rectiligne induit des déformations dans les configurations intermédiaires. Comme la loi constitutive est inélastique, il en résulte des contraintes indésirables dans γ_B. Le trajet à vitesse constante n'est donc pas incrémentalement objectif.
Toutefois, on peut montrer [3], que si la rotation entre γ_A et γ_B reste modérée, les contraintes parasites sont négligeables.

b. Trajet à gradient de vitesse constant.

Le passage de γ_A à γ_B peut se faire selon un trajet non rectiligne des particules, de telle sorte que le gradient de vitesse $\underline{L}$ soit constant [4]. Par exemple, dans le cas bi-dimensionnel, on trouve :

$$\underline{L} = C_1 \begin{bmatrix} \dfrac{J_{11} - J_{22}}{2} & J_{12} \\ J_{21} & \dfrac{J_{22} - J_{11}}{2} \end{bmatrix} + C_2 \begin{bmatrix} 1 & 0 \\ 0 & 1 \end{bmatrix}$$

où

$$A = \sqrt{\left(\dfrac{J_{11} - J_{22}}{2}\right)^2 + J_{12}J_{21}} \quad ; \quad C_1 = \dfrac{\rho_1 - \rho_2}{2A} \quad ; \quad C_2 = \dfrac{\rho_1 + \rho_2}{2}$$

$$\rho_1 = \dfrac{J_{11} + J_{22}}{2} + A \quad ; \quad \rho_2 = \dfrac{J_{11} + J_{22}}{2} - A$$

Dans ces formules, ρ_1 et ρ_2 sont les valeurs propres du jacobien de la transformation $\gamma_A \to \gamma_B$: $\underline{J} = \partial \underline{x}_B / \partial \underline{x}_A$.

Les valeurs propres ρ_1 et ρ_2 peuvent être imaginaires, mais les coefficients C_1 et C_2 sont toujours réels.

Dans le cas particulier où le passage de γ_A à γ_B se fait par une rotation de corps rigide, le trajet à gradient de vitesse constant n'induit aucune déformarion intermédiaire. Il est donc incrémentalement objectif.

c. Trajet avec rotation finale.

Si l'on effectue la décomposition polaire du jacobien $\underline{J} = \partial \underline{x}_B / \partial \underline{x}_A$

$$\underline{J} = \underline{R} \, \underline{U} \, ,$$

on peut choisir le trajet suivant [5].

Pour $t_A \leqslant t < t_B$, seule se produit la déformation $\underline{U}$, tandis que la rotation $\underline{R}$ se fait instantanément au temps $t = t_B$.

Dans le cas d'une rotation de corps rigide, aucune déformation ne se produit de t_A à t_B, et la rotation se fait en bloc à la fin de l'intervalle de temps. Ce trajet est donc aussi incrémentalement objectif.

Dans le code de calcul utilisé pour ce travail, seules les deux premières méthodes sont implémentées. Diverses comparaisons ont montré que, pour des cas pratiques, elles donnent toutes deux satisfaction, mais que la méthode à gradient de vitesse constant est supérieure lorsque les incréments de déformations sont grands et s'accompagnent de rotations importantes.

Intégration de la loi constitutive le long d'un trajet.

Lorsque le trajet est choisi, on calcule σ_B en intégrant la loi constitutive le long de celui-ci. Les équations constitutives se présentent sous la forme d'un système d'équations différentielles du premier ordre qui peut s'écrire symboliquement :

$$\dot{\underline{y}} = \underline{f}(\underline{y}, t)$$

avec

$$\underline{y} = \underline{y}_A \quad \text{en} \quad t = t_A$$

De très nombreuses techniques d'intégration numérique peuvent être utilisées.

Nous n'en citerons que trois.

a. Schéma explicite.

$$\underline{y}_B = \underline{y}_A + \dot{\underline{y}}_A \, \Delta t$$

C'est le schéma le plus facile à implémenter, mais il n'est stable que pour des pas de temps Δt très petits.

b. Schéma implicite.

$$\underline{y}_B = \underline{y}_A + [(1 - \theta) \, \dot{\underline{y}}_A + \theta \, \dot{\underline{y}}_B] \, . \, \Delta t$$

Ce schéma est stable si $\theta \geqslant \frac{1}{2}$ mais est itératif puisque $\dot{\underline{y}}_B = f \, (\underline{y}_B, \, t_B)$ intervient dans le membre de droite.

c. Schéma à sous-intervalle.

L'intervalle de temps Δt est divisé en sous-intervalles.
Dans notre étude, nous avons choisi des sous-intervalles égaux dt, mais il est possible également [6] de choisir automatiquement la taille de ce sous-intervalle de manière à atteindre une précision désirée. Le passage du sous-intervalle (I) au suivant (I + 1) se fait selon le schéma suivant :

$$\underline{y}_{I+1} = \underline{y}_I + [(1 - \theta) \, \dot{\underline{y}}_I + \theta \, \dot{\underline{y}}_{I+1}] \, dt$$

mais pour éviter les itérations, $\dot{\underline{y}}_{I+1}$ est calculé approximativement par :

$$\dot{\underline{y}}_{I+1} = \dot{\underline{y}}_I + [\frac{\partial f}{\partial \underline{y}_I}] \, (\underline{y}_{I+1} - \underline{y}_I)$$

Il vient donc :

$$[\underline{I} - \theta \, dt \, \frac{\partial f}{\partial \underline{y}_I}] \, (\underline{y}_{I+1} - \underline{y}_I) = \dot{\underline{y}}_I \, dt$$

Cette technique donne une bonne précision et une bonne stabilité numérique. Dans des applications courantes, des pas de temps correspondant à des incréments de déformation de l'ordre de 2 % peuvent être intégrés de manière satisfaisante avec 10 à 20 sous-intervalles. Toutefois, lorsque le comportement du matériau change très fort (passage élastique → élastoplastique) il faut augmenter le nombre de sous-intervalles ou réduire le pas de temps.

Tenseur d'incidence.

Les contraintes σ_B dépendent du trajet choisi entre γ_A et γ_B ainsi que du schéma d'intégration. Elles ne sont généralement pas en équilibre avec les forces appliquées au solide déformé. Il convient donc de rechercher une configuration γ_B^x plus proche de l'équilibre, c'est-à-dire de déterminer un champ de déplacements correctifs $d\underline{u}$ (figure 3) tel que :

$$\underline{x}_B^* = \underline{x}_B + d\underline{u} = \underline{x}_A + (\Delta\underline{u} + d\underline{u}) = \underline{x}_A + \Delta\underline{u}^*$$

Les contraintes $\underline{\sigma}_B^*$ dans γ_B^* doivent se calculer directement le long de AB^*

$$\underline{\sigma}_B^* = \underline{\sigma}_A + \Delta\underline{\sigma}^*$$

et non pas le long du chemin ABB^*

$$\underline{\sigma}_B^* = \underline{\sigma}_B + d\underline{\sigma} = \underline{\sigma}_A + \Delta\underline{\sigma} + d\underline{\sigma}$$

car, non seulement la configuration γ_B n'a aucune existence physique, mais en outre, le trajet fictif BB^* devrait se parcourir en un temps infiniment court, ce qui n'est pas admissible avec un modèle constitutif élasto-visco-plastique où la vitesse de déformation joue un rôle fondamental.

La perturbation $d\underline{\sigma}$ des contraintes induite par la perturbation $d\underline{u}$ des coordonnées ne peut donc être calculée que par la formule

$$d\underline{\sigma} = \underline{\sigma}_B^* - \underline{\sigma}_B = \Delta\underline{\sigma}^* - \Delta\underline{\sigma}$$

Elle dépend donc, elle aussi, du trajet choisi entre γ_A et γ_B ainsi que du schéma d'intégration. Si l'on note $d\underline{L}$ la perturbation des gradients de vitesse induite par $d\underline{u}$, on définit le tenseur d'incidence $\underline{C}$ par la relation :

$$d\underline{\sigma} = \underline{C} \; d\underline{L} \longleftrightarrow d\sigma_{ij} = C_{ijkl} \; dL_{kl}$$

Il dépend lui aussi du trajet et du schéma d'intégration choisis entre γ_A et γ_B. Son calcul analytique est généralement impossible, mais on peut l'évaluer numériquement en perturbant une à une les composantes de $\underline{L}$. Le tenseur d'incidence, sauf cas particulier, est non symétrique [7].

Matrice tangente.

Les forces hors d'équilibre $\{R\}$ dans γ_B résultent de la discrétisation de l'équation des puissances virtuelles. En les dérivant par rapport aux paramètres de discrétisation (les déplacements nodaux en général), on obtient sans peine la matrice tangente qui permet de déterminer les déplacements nodaux correctifs $\{du^N\}$

$$[K_T] \; \{du^N\} = - \{R\}$$

On montre dans [7] que la matrice tangente se calcule le plus facilement dans la configuration γ_B où elle prend la forme :

$$[K_T] = \int_{V_B} [B]^T \; [[\sigma] + [C]] \; [B] \; dv_B$$

où $[B]$ est la matrice qui apparaît dans la relation

$$\{dL\} = [B] \; \{du^N\} \; ;$$

$[\sigma]$ est une matrice non symétrique de contraintes initiales dans γ_B

$[C]$ est la matrice d'incidence introduite précédemment.

En général, $[K_T]$ est non symétrique.

Eléments solides.

Les éléments finis développés sont tous isoparamétriques : quadrilatères à
8 noeuds ou triangles à 6 noeuds pour l'état plan ou axisymétrique, briques à
nombre de noeuds variable (de 8 à 32) pour les solides tridimensionnels.

MODELISATION DU CONTACT

La modélisation du contact unilatéral entre le solide qui se déforme et les
cylindres du laminoir se fait à l'aide d'éléments finis de contact basés sur
une méthode de pénalisation (figure 4) : on permet au solide de pénétrer légère-
ment à l'intérieur de la fondation (les cylindres), et la profondeur de pénétration est contrôlée par un coefficient de pénalisation K.

Loi constitutive de contact.

On utilise une loi du type Mohr-Coulomb avec coefficient de frottement ϕ
(figure 5). Par analogie avec les lois élasto-plastiques des solides, on définit une surface limite d'équation

$$f = p - \phi \, |\tau|$$

Pour f < 0, on est en domaine "élastique" et la loi constitutive est donnée par:

$$\begin{Bmatrix} \dot{p} \\ \dot{\tau} \end{Bmatrix} = \begin{bmatrix} K_p & 0 \\ 0 & K_\tau \end{bmatrix} \begin{Bmatrix} \dot{u} \\ \dot{v} \end{Bmatrix}$$

où u et v sont les composantes normale et tangentielle de la vitesse relative du
solide par rapport à la fondation. On permet donc, dans l'esprit de la méthode
de pénalisation, un léger déplacement relatif entre le solide et la fondation.
Pour f = 0, il y a glissement. La vitesse relative de déplacement se décompose
en une partie "élastique" calculée comme ci-avant, et une partie "plastique",
supposée normale à une surface de glissement non associée $g = |\tau| - C$

$$\dot{u} = \dot{u}^e + \dot{u}^p \; ; \; \dot{v} = \dot{v}^e + \dot{v}^p$$

$$\dot{u}^p = \dot{\lambda} \frac{\partial g}{\partial p} = 0 \; ; \; \dot{v}^p = \dot{\lambda} \frac{\partial g}{\partial \tau} = \dot{\lambda} \, \tau$$

En exprimant, de manière classique, la condition de cohérence, on trouve [8]

$$\begin{Bmatrix} \dot{p} \\ \dot{\tau} \end{Bmatrix} = \begin{bmatrix} K_p & 0 \\ \phi K_p \; \text{sgn} \, \tau & 0 \end{bmatrix} \begin{Bmatrix} \dot{u} \\ \dot{v} \end{Bmatrix}$$

Le tenseur constitutif "élasto-plastique" n'est donc pas symétrique.
Cette loi constitutive est objective puisqu'elle est exprimée dans un repère
local qui tourne avec la frontière du solide (formulation corotationnelle).

Objectivité incrémentale et intégration numérique.

Supposons que les coordonnées du solide et de la fondation varient linéaire-
ment entre γ_A et γ_B, c'est-à-dire que ce trajet est à vitesse constante.

Dans le cas où γ_A et γ_B ne diffèrent que par un mouvement de corps rigide,
la transformation à vitesse constante est conforme : les longueurs sont modi-
fiées mais pas les angles (figure 5). Il s'ensuit que la vitesse tangentielle
$\dot{v}$ du solide par rapport à la fondation est nulle pendant ce trajet. Par con-
tre, la vitesse normale relative $\dot{u}$ n'est pas nulle. Le trajet à vitesse constante est donc objectif pour la composante tangentielle, mais pas pour la composante normale. Pour avoir un schéma totalement objectif, il suffit de calculer
les pressions de contact dans γ_A et γ_B par :

$$p_A = K_p \, d_A \quad ; \quad p_B = K_p \, d_B$$

où d_A et d_B sont les profondeurs de pénétrations dans γ_A et γ_B, sans essayer
d'intégrer la loi incrémentale $\dot{p} = K_p \, \dot{u}$.
Cela revient à écrire que la vitesse instantanée de pénétration vaut :

$$\dot{u} = (d_B - d_A)/\Delta t$$

Pour intégrer les contraintes tangentielles de frottement, on utilise un
schéma numérique à sous-intervalles, semblable à celui déjà mentionné précédemment.

Eléments finis de contact.

On a développé des éléments de contact pour l'analyse bi-dimensionnelle et
tri-dimensionnelle.

Dans le cas bi-dimensionnel, ce sont des arcs de parabole à 3 noeuds, compatibles avec les éléments solides plans du second degré. Dans le cas tri-dimensionnel, ce sont des surfaces gauches compatibles avec les briques solides.
Une particularité intéressante réside dans le fait que le contact avec la fondation est assuré aux points d'intégration.

DEFAUTS DE GEOMETRIE AU COURS DU LAMINAGE

Les défauts de géométrie se présentent sous forme d'une déviation hors tolé-
rances des dimensions du produit laminé. Ils peuvent avoir comme origine un
mauvais positionnement des cylindres ou des dispositifs qui guident le produit
à l'entrée de la passe de laminage. Très souvent, ils résultent du calibrage
des cylindres, c'est-à-dire de la forme géométrique des cannelures successives.
Le remplissage incorrect des cannelures peut conduire à des défauts typiques
tels qu'ils sont montrés à la figure 7.
Le calibrage des cylindres est basé sur l'expérience accumulée en pratique in-
dustrielle. Dans le cas de nouveaux profils, plusieurs essais de laminage peuvent être nécessaires avant d'arriver à un résultat satisfaisant.
La simulation numérique doit permettre de réaliser cette mise au point des cali-
brages sans engager les frais importants occasionnés par les essais de laminage.

La calibration du modèle à éléments finis a été d'abord effectuée dans le cas du laminage bi-dimensionnel, par comparaison avec des résultats d'essais de laminage [9].

Pour valider le modèle dans le cas tri-dimensionnel, nous avons choisi le laminage d'une billette de section 100 x 100 mm^2, en acier à 0.1 % C et à une température de 1050°C, subissant une réduction de 20 % sur cylindres plats. Dans ce cas, la validation porte sur la forme des surfaces du produit laminé qui n'entrent pas en contact avec le cylindre.

La figure 8 montre le maillage avant et après déformation, la discrétisation ayant été effectuée pour un quart de la structure.
La figure 9 montre une section transversale du produit laminé après sa sortie de l'emprise.

Aux hautes températures, l'acier a un comportement proche de celui de la plasticine, qui est utilisée fréquemment comme matériau de simulation en laminage. La figure 9 montre également l'élargissement obtenu lors d'un tel laminage sur plasticine. L'accord avec les prévisions du modèle est excellent.
Nous achevons actuellement la validation du modèle à éléments finis dans le cas de cannelures à forme complexe.

REFERENCES

1. J. LEMAITRE, J-L. CHABOCHE, Mécanique des matériaux solides, Dunod, 1985.

2. S. CESCOTTO, H. GROBER, Calibration and Application of an Elastic Visco-plastic Constitutive Equation for Steels in Hot-Rolling Condition, Engineering Computations, Vol. 2, n° 2, June 1985.

3. R. CHARLIER, Approche unifiée de quelques problèmes non linéaires de mécanique des milieux continus par la méthode des éléments finis (grandes déformations des métaux et des sols, contact unilatéral des solides, conduction thermique et écoulement en milieu poreux), Thèse de Doctorat, mars 1987, Université de Liège.

4. A. GODINAS, S. CESCOTTO, Calcul des gradients de vitesse constants au cours d'un pas, Rapport interne n° 159, Services M.S.M., Université de Liège, 1984.

5. BRAUDEL, An implicit and incrementally objective formulation for solving elastoplastic problems at finite strains by the F.E.M. Application to cold forming. Proc. of NUMIFORM'86, Göteborg, pp. 255-260 (1986).

6. O. DESBORDES, M. EL MOUATASSIM, G. TOUZOT, Local numerical integration of large strain elasto-plastic constitutive laws. Proc. Int. Conf. Tucson, on Constitutive Laws for Engineering Matérials : Theory and Applications, Jan 5-10, 1987.

66

7. S. CESCOTTO, R. CHARLIER, Numerical Simulation of elasto-visco-plastic large
   strains of metals at high temperature, Proc. 8th Conf. on SMIRT, Brussels,
   Aug. 19 - 23, 1985.

8. R. CHARLIER, A. GODINAS, S. CESCOTTO, On the modelling of contact problems
   with friction by the finite element method, Proc. 8th Conf. on SMIRT,
   Brussels, Aug. 19 - 23, 1985.

9. H. GROBER, Finite Element Simulation of Hot Flat Rolling of Steel. Proc.
   NUMIFORM'86 Conference (Göteborg) 25-29 August 1986.

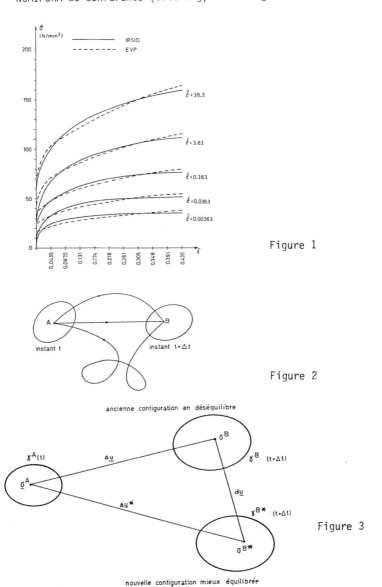

Figure 1

Figure 2

Figure 3

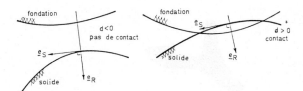

Figure 4

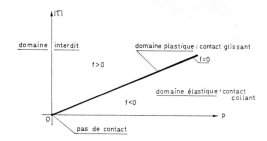

Figure 5

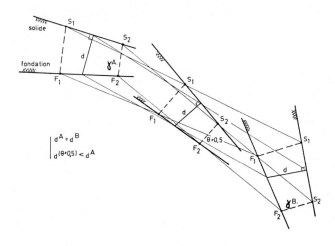

Figure 6

68

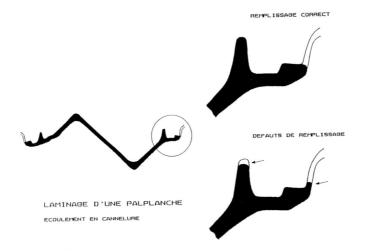

REMPLISSAGE CORRECT

DEFAUTS DE REMPLISSAGE

LAMINAGE D'UNE PALPLANCHE

ECOULEMENT EN CANNELURE

Figure 7

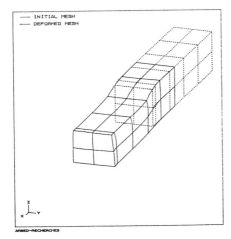

THREE-DIMENSIONAL ANALYSIS OF HOT ROLLING

billet 100 mm × mild steel at 1050°C × r = 20%

AXONOMETRIC VIEW

Figure 8

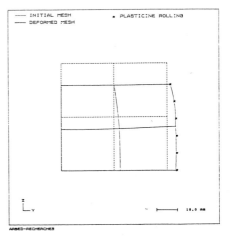

THREE-DIMENSIONAL ANALYSIS OF HOT ROLLING

billet 100 mm × mild steel at 1050°C × r = 20%

FRONT VIEW - SECTION AT x = 20 mm

Figure 9

Computational Methods for Predicting Material Processing Defects, edited by M. Predeleanu   69
Elsevier Science Publishers B.V., Amsterdam, 1987 — Printed in The Netherlands

INFLUENCE OF STRAIN GRADIENTS IN SHEET METAL FORMING

J.P. CORDEBOIS and Ph. QUAEGEBEUR

Laboratoire de Mécanique et de Technologie

E.N.S de Cachan - C.N.R.S. - Université Paris 6 - France.

Abstract . - An overall theory modelizing the onset of necking is presented. We showed that the onset of necking is situated between two extreme modes : diffuse and localized necking modes. Necking criteria in the uniform stress fields on the structure has been developed. But these criteria do not take into account the influence of strain gradients, which are important in sheet metal forming. Thus, we generalize this theory to non-uniform fields. To valid this approach, a comparison between numerical simulations and relevant experiments are presented.

Résumé. - *Influence des gradients de déformation sur l'emboutissabilité des tôles minces.* - Une théorie générale schématisant l'apparition de la striction est présentée. On montre que le début de la striction se situe toujours entre deux modes extrêmes. Dans un premier temps, on a développé des critères de striction dans le cas où le champ de contrainte était uniforme sur la structure. Mais cela ne permettrait pas de prendre en compte l'influence des gradients de déformation. La même théorie a donc été reprise en introduisant la non-uniformité des champs. Pour valider la démarche adoptée, on présente une comparaison entre les résultats numériques et expérimentaux.

1. INTRODUCTION

In sheet metal forming necking arises where strains are the largest. This phenomenon of localization is an important problem. The various criteria predicting the onset of necking are usually expressed by the nullity of a stress function. These criteria are local and are only function of stress evaluated at the considered point. Moreover, the influence of the stress gradient is neglected although they certainly have an effect. The aim of this study is to give a local necking criterion which uses the values of the stress and its gradient.

2. NECKING CRITERIA

2.1. *GENERALITIES*

The necking problem is treated here as a structural instability problem. Consequently, boundary conditions will have an important effect. In this approach, each type of boundary considerations is associated with one necking mode. All modes encountered are situated between two extreme modes.

The less favorable necking mode (in metal forming) named diffuse necking corresponds to the case where no velocities are prescribed on the boundary of the considered domain. In the opposite, when a "maximum velocity conditions" is prescribed on the boundary, we have the most favorable mode, named localized necking. The problem of instability is treated for large strain elasto-plasticity for orthotropic material without evolutive anisotropy, i.-e. with isotropic hardening.

Following many works and as outlined previously, the onset of necking corresponds to a modification of the elasto-plastic behaviour which becomes instable.

During a test, if we maintain constant the external forces the instability is characterized by a velocity field on the domain which is different from that of a rigid body ; in other words, the deformation is localized in some points of the structure. The behaviour is stable if velocity fields are that of a rigid body. Therefore, we may give the following definition : the onset of necking corresponds to a bifurcation of the velocity field when the external loading remains constant.

The rate boundary problem is formulated from the time derivative of the vitual work rate equation which expresses the equilibrium of the structure. The velocity of the stress tensor is eliminated by the elasto-plastic law. It has been shown in <1> that this problem can be reduced to the search of solutions $\vec{V}(M)$ of an equation

$$\mathcal{A}\,(\vec{V},\vec{V}*/\overline{\sigma}) = 0 \qquad \forall\ \vec{V}*(M) \in \mathcal{V} \qquad (1)$$

where $\mathcal{V}$ is the space of virtual admissible velocity fields.

In formula (1), stress $\overline{\sigma}$ is a parameter and function $\mathcal{A}$ has an integral form over the considered volume. When $\overline{\sigma}$ is "small" enough (or rather the stress intensity), it can be shown that the solutions of equation (1), $\vec{V}_0(M)$, correspond to zero strain rate only.

Therefore, there is instability if, for a given external loading, equation (1) admits other solutions than $\vec{V}_0(M)$. For these solutions $\vec{V}(M)$, the associated strain rate field $\mathbb{D}(M)$ is non-zero.

When the stress intensity is small enough, the $\mathcal{A}$ function has the folowing properties :
- it is positive for any kinematically admissible velocity field $\vec{V}(M)$ different from $\vec{V}_0(M)$;
- for any stress tensor $\overline{\sigma}$, the $\mathcal{A}$ function is only reduced to zero by $\vec{V}_0(M)$;
- it is strictly convex in the neighbourhood of $\vec{V}_0(M)$.

With these properties the necking condition may be reformulated in the following way : for a given loading, there will be instability for the values of σ for which it is no longer strictly convex in the neighbourhood of $\vec{V}_0(M)$. An important remark has to be made : for a given loading the tensor σ for which there is instability depends on the space $\mathcal{V}$, therefore it depends on the boundary conditions for velocities.

Consequently, it is sufficient to search for the stress σ which renders zero the minimum of $\mathcal{A}$ for fields $V(M)$ other than $\vec{V}_0(M)$. For a given external loading, these stresses are given by the mathematical expression :

$$\underset{\vec{V}(M)\,\in\,\{\,\mathcal{V}\,-\,\{\vec{V}_0(M)\}\,\}}{Min}\ \mathcal{A}\,(\vec{V},\vec{V}*/\overline{\sigma}) = 0 \qquad (2)$$

2.2. *EXTERNAL NECKING CRITERIA*

Now, we consider two situations :

$\nu = \nu_L$ if non boundary condition is imposed on velocities

$\nu = \nu_0$ if a maximum of velocity conditions is prescribed on the boundary of the considered domain (compatible with plastic strains).

In the first situation, the stress field verifying (2) corresponds to diffuse necking. In the second one, the stress corresponds to localized necking.

The localized mode is more favorable than the diffuse mode because ν_0 is included in ν_L (stress intensity is higher in localized mode than in diffuse mode).

Therefore, for a given loading the localized necking mode needs more important deformation than that characterizing the diffuse one. At the beginning of the study <1>, we try to obtain simplified necking criteria. Thus we solve equation (2) by considering uniform stress field. It is clear that such hypothesis can not take into account the influence of stress gradients.

We have shown that (2) leads to the following necking criteria :

$D(\sigma) = 0$ for a diffuse necking

$D(\sigma) + L(\sigma) = 0$ for a localized necking   (3)

For the chosen behaviour modelization $D(\sigma)$ is a scalar function, depending only on stress. $L(\sigma)$ corresponds to the contribution of locking.

Two particular expressions may be obtained : $L(\sigma) = L_R(\sigma)$ in the case of shrinkage and $L(\sigma) = L_E(\sigma)$ in the case of expansion. We do not give in this paper the explicit form of the function (3), which can be found in <1>.

Figure 1 shows the limit forming diagram given by expression (3) and experimental results in aluminium alloy <3>. Numerical simulations have shown that the necking criteria (3) render account of the influence of predeformation <4>, and of the nature of strain path (figures 2 and 3). This modelization also allows to show the influence of the material rheology.

3. INFLUENCE OF STRAIN GRADIENTS

In the continuation of this work, we studied the strain gradients influence on the solution of equation (2). For that purpose, we considered again the minimisation problem but by looking for non uniform solutions over the considered domain.

Let us consider a particular point of the studied structure for example the point satisfying one of uniform criteria (3). Around this point, we use taylor expansions for all fields involved in (3), up to sufficient order. For instance, if the first stress gradient is reduced to zero at the considered point, we use the second gradient.

The problem is therefore to search for the condition, deduished from (2) which will be satisfied by stress values and the significant stress gradients at the considered point. The solution method is more complex but similar to that used in obtaining relations (3). Briefly the solutions have the following form :

$$D(\sigma_M) + G_D(\nabla_M\sigma,...) = 0 \qquad \text{for a diffuse necking}$$
$$D(\sigma_M) + L(\sigma_M) + G_D(\nabla_M\sigma,...) + G_L(\nabla_M\sigma,...) = 0 \qquad \text{for localized necking.} \qquad (4)$$

The functions D and L depend on the stress values at the considered point. The functions G_D and G_L introduce the first significant stress gradient at the point M. Moreover, these functions depend on inertial element of studied domain.

In the case of diffuse necking, figure 4 shows a numerical simulation of the gradient influence for a strain path. We see that this influence lowers the necking bound. Relations (4) are a generalization of relations (3) obtained with the assumption of uniform stress field.

4. EXPERIMENTAL STUDY

To test the numerical simulation, the method and the results presented in section 3, we proceeded to some experiments. We had to define the specimen types which allow to give some strain gradients during a common test. We decided to make some tensile tests and we chose slotted sheet specimen as seen in figure 5.

The strain measurements were obtained by a visio-plasticity method. Thus, we could test the finite elements package ABAQUS, which has been used for numerical simulations. As we can see in figure 6, the agreement between experimental measurements and numerical results by finite elements is rather good.

By using two sheets with a very different plastic behaviour we displayed the influence of rheological parameters on strain gradients.

Finally, the test confirmed that the necking began at the symmetry central point of the specimen. The TAYLOR expansions were made around this point. As strains and stresses are maxima at that point, their first gradients have zero value. Consequently, it is necessary to introduce the second gradients. The second strain gradients are directly obtained from measurements. The second stress gradients are computed by using the constitutive law.

By relation (4), it is possible to predict along the strain path, the region of the plastic state corresponding to the onset of necking. A comparison is made with measured experimental point.

5. CONCLUSION

In this study, we have shown the influence of the strain gradients on thin sheet metal forming. In this process, it will be useful to minimize the strain gradients. To this end, we propose some solutions. The easiest is to modify the sheet metal forming conditions. Clearly, the main technical parameter will be the pressure of the blank holder. At this stage of the study, we have no indication on that point. The second possibility is to change the form of the metal piece. Numerically, we showed that the form of the specimen, for a given material, influenced the gradients intensity involved in traction. Another possibility is to change the material, as we have seen rheological parameters also influence the strains.

REFERENCES

<1> J.P. CORDEBOIS , P. LADEVEZE
"Sur la prévision des courbes limites d'emboutissage", J.M.T.A., Vol. 5, n° 3, pp. 341-370, 1986.

<2> P. LADEVEZE
"Sur la théorie de la plasticité en grandes déformations", L.M.T., E.N.S. de Cachan - PARIS 6, Rapport interne N° 9. 1981.

<3> M. GRUMBACH, G. SANZ
"Influence des trajectoires de déformation sur les courbes limites d'emboutissage à striction et à rupture". Les Mémoires scientifiques de la revue de métallurgie, n° 4, 1980.

<4> R. ARRIEUX
" Contribution à la détermination des courbes limites de formages du Titane et de l'Aluminium. Proposition d'un critère intrinsèque". Thèse de Docteur-Ingénieur, Université Lyon I, 1981.

<5> G. ROMANO, D. RAULT et M. ENTRINGER
"Courbes limites en trajectoires complexes et répartition des déformations". I.D.D.R.G., G.T. III, Goteborg, 1974.

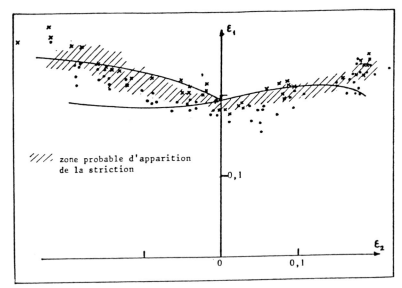

Fig.1 - Comparison between experiments and numerical simulations.

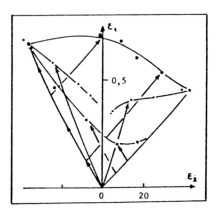

Influence of deformation paths,
according to <5> .

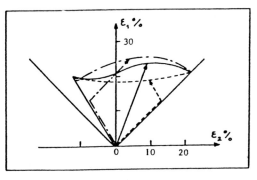

Influence of deformation paths,
according to the proposed model.

Fig. 2

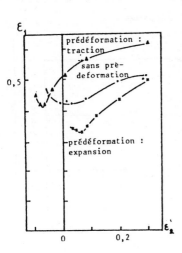

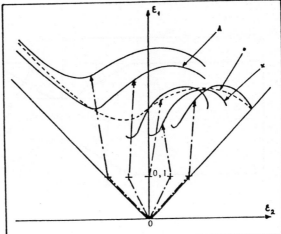

Influence of predeformation,
according to <4> .

Influence of predeformation,
according to the proposed model.

Fig.3

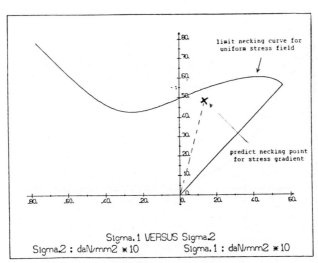

Fig. 4 - Influence of stress gradients on the
onset of diffuse necking.

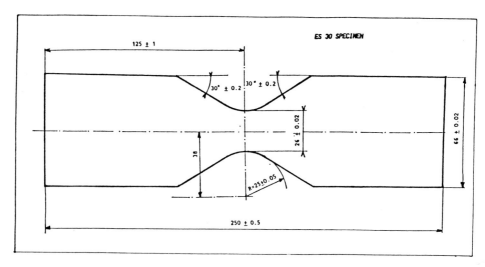

Fig. 5 - ES 30 specimen.

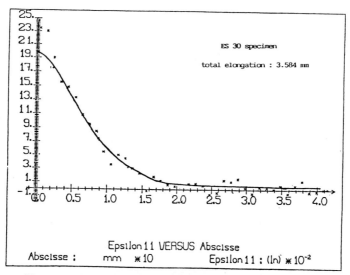

Fig.6 - Comparison between calculated and measured
strains on ES 30 specimen.

Computational Methods for Predicting Material Processing Defects, edited by M. Predeleanu
Elsevier Science Publishers B.V., Amsterdam, 1987 — Printed in The Netherlands

SURFACE DEFECTS AND FAILURE OF ADHESION IN POLYMER MELT EXTRUSION

MORTON M. DENN

Department of Chemical Engineering, University of California, and Center for Advanced Materials, Lawrence Berkeley Laboratory, Berkeley, California 94720 USA

SUMMARY

Surface distortions commonly known as *melt fracture* appear on extrudates of molten polymers when the wall stress exceeds a critical value. These distortions appear to be a consequence of failure of adhesion at the polymer/metal interface, demonstrating the inappropriateness of the classical *no-slip* boundary condition in regions of high stress. The wall boundary condition is likely to be an underlying cause of the lack of convergence of computational algorithms for rheologically-complex fluids at high stress levels. This lack of convergence is a major limitation in processing calculations.

INTRODUCTION

The extrusion of molten polymers to form shaped objects is usually limited by the occurrence of a surface instability known as *melt fracture*. The first occurrence of melt fracture in linear polyethylene is in the form of high-frequency surface variations known as *sharkskin*. At higher throughput there is a transition to a *slip-stick* regime, which consists of alternating sharkskin and smooth sections; a typical extrudate of linear low-density polyethylene (*LLDPE*) in the slip-stick regime is shown in Fig. 1. At still higher throughputs the sharkskin regions disappear and there is a transition to *wavy fracture*; this is a gross distortion of the extrudate, as shown in Fig. 2 for LLDPE. The sharkskin and slip-stick regions are often unobservable in other polymers, and the wavy (or *gross*) melt fracture may be the only form of the instability that is observed. Experimental observations and characterizations of melt fracture have been reviewed by Tordella (ref. 1), Petrie and Denn (ref. 2), and a IUPAC Working Party (ref. 3); our own experimental data on LLDPE have been described in detail in Kalika and Denn (ref. 4).

Ramamurthy (ref. 5) has demonstrated that the onset of sharkskin melt fracture can be moved to higher throughput rates by changing the materials of construction of the extrusion die, and hence that melt fracture is a consequence of a failure of melt/metal adhesion. This observation is inconsistent with the no-slip boundary condition that is conventionally used

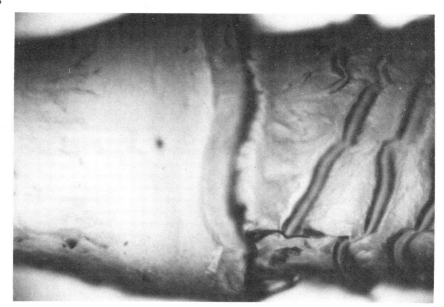

Fig. 1. Extrudate of linear low-density polyethylene in the slip-stick region, showing sharkskin (right) and smooth (left) sections; magnification 75X. (D.S. Kalika)

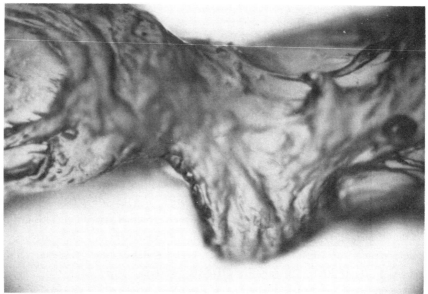

Fig. 2. Extrudate of linear low-density polyethylene in the wavy (gross) fracture region; magnification 75X. (D.S. Kalika.)

in fluid mechanics, but slip velocities can be observed by indirect means; our own experiments on LLDPE indicate an onset of wall slip at a stress of about 0.2 MPa, with a relative velocity between the polymer and the wall given by the linear equation

$$V_{slip}/<V> \approx \tau_w - 0.2, \qquad 0.2 \leq \tau_w \leq 0.4 \qquad (1)$$

V_{slip} is the apparent relative velocity between the polymer melt and the wall, $<V>$ is the average velocity, and τ_w is the measured wall shear stress; the equation (which is a rough fit to data from a number of capillaries) is dimensional, with stress measured in MPa. There is a discontinuity in the slip velocity in the slip-stick region, and slip is nearly complete in the wavy region.

Attempts to predict the onset of melt fracture from linear stability theory have generally been based on the use of viscoelastic constitutive equations together with the no-slip condition (e.g., refs. 6, 7, and 8); all have been unsuccessful. A theory by Pearson and Petrie (ref. 9) that assumes the existence of wall slip is in remarkable agreement with the data correlated by equation (1), although the predicted waveform is not consistent with the surface distortions that are observed experimentally.

NUMERICAL COMPUTATION

Numerical computation of the flow and stress development in rheologically-complex liquids cannot be carried to high values of the *recoverable shear*, which is the ratio of the wall stress to the elastic shear modulus of the liquid, in geometries containing corners and lips. The state of the art of numerical computation for viscoelastic liquids is given by Keunings (ref. 10), but the field is changing rapidly and new developments are reported monthly in *J. Non-Newtonian Fluid Mechanics*. Lipscomb and coworkers (ref. 11) have shown that the stresses computed from an approximate viscoelastic model will exceed the strength of a covalent bond over a finite volume if the no-slip condition is required in regions in which there is a stress singularity for the corresponding flow of a Newtonian fluid. They conclude that the no-slip condition should be relaxed in regions of high stress, and suggest that the use of improper boundary conditions is a major factor in the computational problems. This observation is consistent with the experiments described in the preceeding section.

CONCLUSION

The no-slip boundary condition has traditionally been used in fluid mechanics, but it is clear that it must be relaxed in considering the flow of polymeric liquids under conditions of high stress. The slip boundary condition in a straight channel can be found empirically, as in equation (1), but the underlying physics of adhesion and entanglement between adsorbed and bulk polymer is not understood. This is a major limitation in the current ability to carry out processing calculations and to predict processing instabilities.

ACKNOWLEDGMENTS

This work was supported by the Director, Office of Energy Research, Office of Basic Energy Sciences, Materials Science Division of the U.S. Department of Energy under contract No. DE-AC03-76SF00098. Figures 1 and 2 are from experiments by D.S. Kalika.

REFERENCES

1 J. P. Tordella, Unstable Flow of Molten Polymers, in: F. R. Eirich, ed., Rheology, Vol. 5, Academic Press, New York, 1969.
2 C. J. S. Petrie and M. M. Denn, Instabilities in Polymer Processing, AIChE Journal, 22 (1976) 209.
3 J. L. White and H. Yamane, A Collaborative Study of the Stability of Extrusion, Melt Spinning and Tubular Film Extrusion of Some High-, Low- and Linear-Low Density Polyethylene Samples, Pure & Appl. Chem., 59 (1987) 193.
4 D. S. Kalika and M. M. Denn, Surface Defects and Failure of Adhesion in Polymer Melt Extrusion, J. Rheology, submitted for publication.
5 A. V. Ramamurthy, Wall Slip in Viscous Liquids and Influence of Materials of Construction, J. Rheology, 30 (1986) 337.
6 T.-C. Ho and M. M. Denn, Stability of Plane Poiseuille Flow of a Highly Elastic Liquid, J. Non-Newtonian Fluid Mech., 3 (1977/78) 179
7 K. C. Lee and B. A. Finlayson, Stability of Plane Poiseuille and Couette Flow of a Maxwell Fluid, J. Non-Newtonian Fluid Mech., 21 (1986) 65
8 M. Renardy and Y. Renardy, Linear Stability of Plane Couette Flow of an Upper Convected Maxwell Fluid, J. Non-Newtonian Fluid Mech., 22 (1986) 23
9 J. R. A. Pearson and C. J. S. Petrie, On Melt-Flow Instability of Extruded Polymers, in: R. E. Wetton and R. W. Whorlow, eds., Polymer Systems, Deformation, and Flow, MacMillan, London, 1968.
10 R. Keunings, Simulation of Viscoelastic Fluid Flow, in: C. L. Tucker,III, ed., Fundamentals of Computer Modeling for Polymer Processing, Carl Hanser Verlag, Munich, in press.
11 G. G. Lipscomb, R. Keunings, and M. M. Denn, Implications of Boundary Singularities in Complex Geometries, J. Non-Newtonian Fluid Mech., in press.

Computational Methods for Predicting Material Processing Defects, edited by M. Predeleanu
Elsevier Science Publishers B.V., Amsterdam, 1987 — Printed in The Netherlands

LARGE STRAIN FORMULATION OF ANISOTROPIC ELASTO-PLASTICITY FOR METAL FORMING

A. DOGUI and F. SIDOROFF

Laboratoire de Mécanique des Solides, Ecole Centrale de Lyon,
36, rue Guy de Collongue, 69130 ECULLY, France;
and G.R.E.C.O. Grandes Déformations et Endommagement.

SUMMARY
    Recent progresses in the formulation of anisotropic elasto-plasticity are
reviewed from the point of view of application to metal forming and numerical
computation. After recalling the basic framework of isotropic plasticity, the
anisotropic case is investigated in its two aspects: a) Description of plasticity and hardening in a rotating frame formulation; b) Taking elasticity into
account for the incremental formulation. As a conclusion, practical rules are
presented for the formulation to be implemented in finite element codes.

1. INTRODUCTION

    All metal forming processes are based on large plastic deformations. Their
mechanical analysis, and, in particular, the prediction of defects, therefore,
requires the development of structural analysis in this context, and many finite element codes along this line have been developed in the last ten years. The
modelization and correct formulation of large plastic or elastic-plastic deformations however is not so easy, and quite a large number of theoretical papers
have been devoted to this subject. The purpose of the present work is to provide a guideline and some practical rules for a consistent formulation of large
strain plasticity for computational purpose.

    In the isotropic case (initial isotropy + isotropic hardening), general
agreement has been reached on a generalized Prandt-Reuss relation which will be
briefly recalled together with the basic mechanical, kinematical and thermodynamical background. However, as soon as there is some anisotropy - whether induced or initial - the situation becomes much more complicated (and highly controversial too...). After a discussion of the encountered difficulties and of
the fundamental requirements to be satisfied, a general framework will be presented, including the choice of a rotating frame as one of its basic constitutive component. The most usual choices for this rotating frame will then be
presented, discussed, and their results compared in some typical situations.

    The complete resulting elastic-plastic constitutive equations will then be
derived with particular attention on the hypoelastic equations for unloading
and on the differential formulation which is the most appropriate for numerical

implementation. As a conclusion, practical rules for the numerical formulation of large strain plasticity will be presented.

2. THE ISOTROPIC CASE

2.1 Generalized Prandtl-Reuss relations

It is now generally agreed that the correct formulation of isotropic elasto-plasticity at large strain must be based on an appropriate objective large strain extension of the classical Prandtl-Reuss relations

$$\dot{\varepsilon} = \dot{\varepsilon}^e + \dot{\varepsilon}^p = \Lambda\dot{\sigma} + \lambda(\partial f/\partial\sigma) \tag{1}$$

where Λ is the elastic compliance (Hooke's law) and f is the yield function usually taken as von Mises'. The large strain extension is obtained by repla-cing the strain rate $\dot{\varepsilon}$ by the rate of deformation tensor D, the stress σ by the Kirchhoff's stress tensor τ and the stress rate $\dot{\sigma}$ by the Jaumann rate τ^J of τ. This leads to

$$D = D^e + D^p = \Lambda\tau^J + \lambda(\partial f/\partial\tau) \tag{2}$$

$$\tau = J\,\sigma \qquad\qquad \tau^J = \dot{\tau} - W\tau + \tau W \tag{3}$$

$$D = L^S = \tfrac{1}{2}(L + L^T) \qquad W = L^A = \tfrac{1}{2}(L - L^T) \qquad L_{ij} = \partial v_i/\partial x_j \tag{4}$$

More precisely, using Hooke's law and von Mises yield condition, equation (3) reads

$$D = [(1+\nu)/E]\,\tau^J - (\nu/E)(tr\,\tau^J)1 + (3/2)\lambda\,\tau^D/\bar{\sigma} \tag{5}$$

$$f(\tau) = \bar{\sigma} - \sigma_0(\bar{\varepsilon}) \qquad\qquad \bar{\sigma} = \sqrt{3/2}\,|\tau^D|$$

$$\bar{\varepsilon} = \sqrt{2/3}\,|D^p| = \sqrt{2/3}\,\sqrt{D^p_{ij}D^p_{ij}} \tag{6}$$

where $\bar{\sigma}$ and $\bar{\varepsilon}$ respectively denote von Mises equivalent stress and strain, and τ^D is the deviatoric part of τ.

The plastic multiplier λ in (2) and (5) is to be determined from the yield condition or its time derivative; for instance, an usual form is

$$[(1+\nu)/E]\,\tau^J = D + [\nu/(1-2\nu)]\,(tr\,D)\,1 - (3/2h)[<\tau^D{:}D>/\bar{\sigma}^2]\,\tau^D$$

$$h = 1 + [2(1+\nu)/3E]\,d\sigma_0/d\bar{\varepsilon} \tag{7}$$

From a numerical point of view, these relations are used as a starting point for the evaluation of the stress increment $\Delta\sigma$ resulting from an incremental displacement [1]. Alternatively, this incremental constitutive equation may be directly derived from (5) in a more implicit way [2], [3].

2.2 Plasticity and isotropic hardening

The classical constitutive model presented above can be used directly, but its extension to anisotropic cases requires a deeper analysis of its physical content and meaning. Their complete derivation relies on two basic components: a kinematical description of elastic-plastic deformations which will be presented later, and a mechanical model for plasticity and isotropic hardening, which will be discussed now in a rigid plastic context.

Generally speaking, a plastic material is a highly non linear viscous fluid, and its constitutive equation relates, in the incompressible case, the stress deviator σ^D to the rate of deformation tensor D with a positive dissipation

$$\sigma^D = \sigma^D(D) \qquad \Phi = \sigma^D{:}D \geq 0 \qquad (8)$$

Rate independent plasticity is obtained when the stress σ^D [resp. the dissipation function Φ] is a positively homogeneous function of degree zero [resp. one]

$$\sigma^D(\lambda D) = \sigma^D(D) \qquad \Phi(\lambda D) = \lambda\Phi(D) \qquad (9)$$

for all positive λ. Inversion of this relation easily shows that there exists a relation on σ^D which is the yield function, and that the direction of D only, and not its length $|D|$, depends on σ^D [4].

Standard plasticity is usually assumed, allowing the use of $\Phi(D)$ as a dissipation potential or equivalently of the yield function $f(\sigma)$ as a plastic potential [5]

$$\sigma^D = \partial\Phi/\partial D \qquad D = \lambda\ \partial f/\partial\sigma^D \qquad (\lambda \geq 0) \qquad (10)$$

This is also equivalent to Hill's maximum work. The special case of von Mises plasticity is obtained from

$$\Phi = \sqrt{2/3}\ \sigma_o\ |D| = \sigma_o\ \dot{\bar\varepsilon} \qquad f = \sqrt{3/2}\ |\sigma^D| - \sigma_o \leq 0 \qquad (11)$$

Isotropic hardening is obtained from a uniform dilatation of the yield surface, and is most simply developed in the framework of generalized standard models [6]. The mechanical hardening energy w only depends on the scalar hardening variable p. The dissipation Φ now becomes

$$\Phi = \sigma^D{:}D - \dot{w} = \sigma^D{:}D - R\dot{p} \qquad R = dw/dp \qquad (12)$$

The yield function depends on the thermodynamic force (σ^D,R) and is postulated as

$$f(\sigma^D,R) = f_o(\sigma^D) - R \leq 0 \qquad (13)$$

The plastic and hardening evolution laws are then given by

$$D = \lambda\ \partial f/\partial\sigma^D = \lambda\ \partial f_o/\partial\sigma^D \qquad \dot{p} = -\lambda\ \partial f/\partial R = \lambda \qquad (14)$$

In particular, it is easily shown that using von Mises' form (11) for f_o results in $p = \bar\varepsilon$ which directly results in (5), (6) with $E\to\infty$ (rigid plasticity).

2.3 Elastic and plastic deformations

It is now well understood that the large strain decomposition of the total deformation in an elastic and a plastic part corresponds to a local relaxation process and therefore to a multiplicative decomposition of the deformation gradient F (Figure 1)

$$F = A \ P \qquad\qquad (15)$$

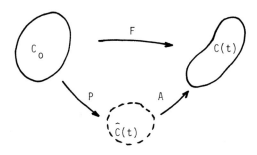

Fig.1 . Elastic and plastic deformation.

The elastic energy w is an isotropic function of the elastic deformation

$$w^e = w^e(E) \qquad\qquad E = \tfrac{1}{2}(A^T A - 1) \qquad\qquad (16)$$

A classical thermodynamical analysis [7], [8] then provides the elastic law and the plastic dissipation as

$$\tau = A \ (\partial w^e/\partial E) \ A^T \qquad\qquad (17)$$
$$\Phi = \tau{:}\bar{D}^p \qquad\qquad \bar{D}^p = R^e \ (\dot{P}P^{-1})^S \ R^{eT} \qquad\qquad (18)$$

where $\bar{D}^p$ is the plastic deformation rate rotated by the elastic rotation tensor R^e obtained from the polar decomposition of A. This is the appropriate eulerian plastic rate of deformation, and finite elastoplasticity is simply obtained by substituting $\bar{D}^p$ for D and τ^D for σ^D in the rigid plastic formulation described above

$$\tau = A \ (\partial w^e/\partial E) \ A^T \qquad\qquad f = f_o(\tau^D) - R(p) \leq 0$$
$$\bar{D}^p = \lambda \ \partial f_o/\partial \tau^D \qquad\qquad \dot{p} = \lambda \qquad\qquad (19)$$

The transformation of these relations in an incremental form similar to (2) requires the decomposition of the rate of deformation tensor D into an elastic and a plastic part, but this is not so easy. Three different solutions for this difficult problem have been proposed in [7], [8] and [9]. For instance, in [7], a general kinematic relation has been obtained as

$$(V^e D V^{e-1})^S = \bar{D}^p + (V^e J V^{e-1})^S \qquad\qquad (20)$$

which is the correct general form of (2) in large elastic-plastic strain, and which reduces to (2) in the case of small elastic strain where

$$A = R^e(1 + E) \qquad\qquad E \ll 1 \qquad\qquad (21)$$

Different derivations of (2) can be found in [8] or [9], but the fundamental results will be the same:

i) The additive decomposition (2) is the limiting case of (19) for small elastic strain which is an appropriate assumption for metal forming processes;

ii) The stress rate which appears necessarily is the Jaumann rate.

The use of the Kirchhoff stress τ instead of the usual Cauchy stress σ, on the other hand, is merely a numerical convenience to ensure a symmetrical formulation [10]: from a practical point of view, the difference between τ and σ only results from the volume change, which is purely elastic, and therefore small.

3. ANISOTROPIC RIGID PLASTICITY
3.1 An example: kinematic hardening

Similarly, the development of anisotropic elastic-plastic models requires first a clear understanding of the rigid plastic underlying behaviour, and we shall, for the moment, focus our attention on this aspect. The concept of anisotropic plasticity, however, is not so clear and may include many different aspects (initial anisotropy, induced anisotropy, anisotropic hardening,...). Even in the small strain case, a reasonably general anisotropic plastic model does not exist: the physical and phenomenological content to be included in the material model must be decided separately for each case. We shall not be concerned here with this - difficult - choice, but rather with the correct large strain formulation of a given content and the associated geometrical problems.

For the presentation of these problems, we shall begin with the simplest case of induced anisotropy: linear kinematic hardening, and Prager's model [11]. Using, for instance, the generalized standard scheme, a tensorial internal variable α is introduced, and the hardening energy is taking as a quadratic function leading to the following expression for the dissipation

$$\sigma:D - X:\dot{\alpha} \geq 0 \qquad\qquad X = \partial w/\partial\alpha = c\,\alpha \qquad\qquad (22)$$

In order to obtain an objective formulation, the derivative $\dot{\alpha}$ can be replaced by the Jaumann rate α^J or more generally by any rotational objective rate $\overset{\circ}{\alpha}$ such that

$$\alpha:\dot{\alpha} = \alpha:\overset{\circ}{\alpha} \qquad\qquad \overset{\circ}{\alpha} = \tilde{Q}(\tilde{Q}^T\alpha\tilde{Q})\dot{}\ \tilde{Q}^T \qquad\qquad (23)$$

where $\tilde{Q}$ is any objective rotation. The thermodynamic force X is then interpreted as the back stress, and the yield condition taken as

$$f(\sigma,X) = f_0(\sigma - X)$$

$$D = \lambda \, \partial f/\partial \sigma \qquad\qquad \overset{\circ}{\alpha} = -\lambda \, \partial f/\partial X = D \qquad\qquad \overset{\circ}{X} = cD \qquad (24)$$

Different derivations of these equations can be found, but the result is always the same: a) the back stress is obtained from the objective integration of the strain rate; b) the corresponding objective rate must be rotational (23) and the usual convected rates are excluded.

This naturally leads to the use of the Jaumann rate (3) α^J which is, among the classical objective rates, the only rotational one, and which also appeared to be the good one in the isotropic case. Unfortunately, this leads to an unreasonable oscillatory behaviour in simple shear [12]. This discovery was the starting point of wide controversies, and quite a great amount of work has been expanded on that subject [13], [14]. The problem however is much more general.

3.2 Rotating frame formulation

Most existing model for anisotropic plasticity can be put in the following small strain form

$$f(\sigma,\alpha,\varepsilon) \leq 0$$

$$\dot{\varepsilon} = \lambda \, h(\sigma,\alpha,\varepsilon) \qquad\qquad \dot{\alpha} = \lambda \, g(\sigma,\alpha,\varepsilon) \qquad (25)$$

where α is an appropriate set of scalar or tensorial hardening variables. This includes for instance the kinematic hardening model discussed above, as well as other anisotropic hardening models. This also includes the standard anisotropic Hill's model with isotropic hardening

$$f(\sigma,p) = \sqrt{\sigma \, H \, \sigma} \; - \; \sigma_0(p) \; \leq 0$$

$$\dot{\varepsilon} = \lambda \, \partial f/\partial \sigma = \lambda \, H\sigma \, / \, \sqrt{\sigma \, H \, \sigma} \qquad\qquad \dot{p} = \lambda = \sqrt{\dot{\varepsilon} \, H^{-1} \, \dot{\varepsilon}} \qquad (26)$$

as well as the model proposed by Boehler to account for both initial and induced anisotropy [15], and which, in small strain, can be written as

$$f(\sigma,\varepsilon) \leq 0 \qquad\qquad \dot{\varepsilon} = \lambda \, h(\sigma,\varepsilon) \qquad (27)$$

This small strain formulation remains correct in "triaxial" large strain for orthotropic materials, i.e. when all tensors remain diagonal in the orthotropy axis

$$F = \begin{vmatrix} e^{\varepsilon_1} & 0 & 0 \\ 0 & e^{\varepsilon_2} & 0 \\ 0 & 0 & e^{\varepsilon_3} \end{vmatrix} \qquad\qquad \sigma = \begin{vmatrix} \sigma_1 & 0 & 0 \\ 0 & \sigma_2 & 0 \\ 0 & 0 & \sigma_3 \end{vmatrix} \qquad (28)$$

Provided the logarithmic strain tensor is used for ε, the formulation (25) can be used without any change. This is an important special case, because it covers most mechanical tests used for model identification.

Extension to the complete large strain case however requires an objective formulation of (25). This is simply achieved by using a rotating frame formalism [16], [17], [18], i.e. by replacing σ, α, ε in (25) by the corresponding rotated tensors $\tilde{\sigma}$, $\tilde{\alpha}$, $\tilde{\varepsilon}$ by an objective rotation Q

$$\sigma = \tilde{Q}\,\tilde{\sigma}\,\tilde{Q}^T \quad , \quad D = \tilde{Q}\,\tilde{D}\,\tilde{Q}^T \quad , \quad \dots$$

$$f(\tilde{\sigma},\tilde{\alpha},\tilde{\varepsilon}) \leq 0 \tag{29}$$

$$\tilde{D} = \lambda\,h(\tilde{\sigma},\tilde{\alpha},\tilde{\varepsilon}) \qquad\qquad \overset{\circ}{\tilde{\alpha}} = \lambda\,g(\tilde{\sigma},\tilde{\alpha},\tilde{\varepsilon})$$

This is the simplest - if not the only - way to reconcile objectivity with the physically significant formulation (25) identified for instance in small strain or triaxial situations. It is also easily shown that this formulation reduces to (23), (24) in case of kinematic hardening.

3.3 Choice of the rotating frame

In this formalism, the choice of the rotating frame is an essential component of the consitutive model. The simplest choice is the corotational or Zaremba's frame which is defined by a vanishing rotation rate $\tilde{W}$ leading to the following differential system

$$d\tilde{Q}/dt = W(t)\,\tilde{Q} \qquad\qquad \tilde{Q}(0) = 1 \tag{30}$$

which determines $\tilde{Q}(t)$. The associated objective rate is the Jaumann rate.

Another simple choice is the "proper rotating frame" defined from the rotation R resulting from the polar decomposition of the deformation gradient F

$$\tilde{Q} = R \qquad\qquad F = V\,R = R\,U \tag{31}$$

As will be discussed later. these two choices can be considered as typical cases, but many other choices have been proposed and may be acceptable like for instance the triaxial frame which is such that D = $\overset{\circ}{\varepsilon}$ or a director frame defined from a material evolution equation postulated for $\tilde{Q}$

$$\overset{\circ}{\tilde{Q}}\,\tilde{Q}^T = \lambda\,l(\tilde{\sigma},\tilde{\alpha},\tilde{\varepsilon}) \tag{32}$$

in agreement with Mandel's approach [20], [13].

Like any phenomenological assumption, this choice has to rely on an experimental basis, but this is not so easy: there are rather few experiments dealing with large strains, and most of them are triaxial and without rotation, so that $\tilde{Q} = 1$ in all cases. As we shall see later, simple shear (or torsion) is the only simple situation resulting in significant differences between all these models, and this is the reason why it has been so extensively analysed. No experimental results are available, but nevertheless this analysis provides a deep insight into the significance. In particular, the oscillatory behaviour associated to the corotational frame and Jaumann's rate (which has been mentioned for the Prager's model and which is in fact much more general [18]) clearly results from a too rapid rotation badly following the material rotations which are much better fitted to the proper rotation frame.

3.4 Practical aspects

To illustrate the importance of these differences, we shall investigate three kinds of situations typical of metal forming and involving rotations. These three situations are described in Figure 2.

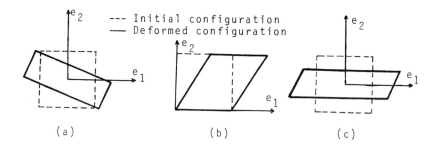

Fig.2 . Typical situations.

Situation a) corresponds to the combination of a rotation with a triaxial deformation

$$F = \begin{vmatrix} \cos \theta & \sin \theta \\ -\sin \theta & \cos \theta \end{vmatrix} \begin{vmatrix} e^{\varepsilon_1} & 0 \\ 0 & e^{\varepsilon_2} \end{vmatrix} \tag{33}$$

There is no material rotation in this case, and it is obvious that all objective rotations $\tilde{Q}$ will coincide with the overall rotation θ.

Situation b) is the well known simple shear, while situation c) describes an off axis tensile test. For definiteness, we shall consider the plane stress response of an anisotropic perfectly plastic material obeying Hill's criterion with

$$r_0 = 1.84 \qquad r_{45} = 1.22 \qquad r_{90} = 1.99$$

In both cases, the rotation $\tilde{Q}$ is defined by the angle θ between the fixed x_1 axis and the rotating orthotropy direction $\tilde{x}_1$. The analysis is quite classical in situation b), more difficult in situation c), but details can be found in [21]. Results are presented in Figures 3 and 4, and they show the evolution of $\tilde{\theta}$ and of the shear stress τ as a function of χ in simple shear, and of the tensile stress σ as a function of the longitudinal strain ε.

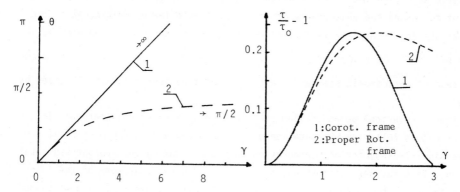

Fig.3 . Simple shear.

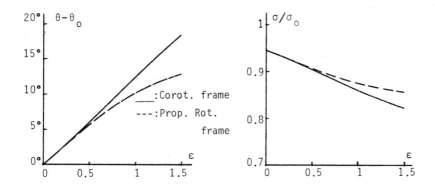

Fig.4 . Off axis tensile test $(\theta_o = 15°)$

These results are only presented for the corotational frame and for the proper
rotation frame, but other rotating frames will lead to similar results. These
two frames can be considered as extreme cases. Many similar results may be presented, and they will lead to the same conclusions: the differences exist, but
are very small for the off axis tensile test. On the contrary, they may be significant in simple shear. It must also be noted that the difference in the
stress response directly follows the difference in the rotation $\overset{\approx}{\theta}$.

4. ANISOTROPIC ELASTO-PLASTICITY
4.1 Rate type formulation

From a practical point of view, elastic deformations may easily be included
in the rotating frame formulation by adding an elastic rate of deformation to
the plastic contribution discussed above.This approach is therefore based on an
additive decomposition of the rate of deformation tensor D as in (2). The plastic rate D^p will then be related to τ as discussed above, while the plastic rate will be given by some hypoelastic relation from the stress rate. Since the

plastic evolution law is to be written in a rotating frame, it is very convenient to ensure the objectivity of this hypoelastic law by writing it in the same rotating frame. The isotropic relation (2) is then simply replaced by

$$\tilde{D} = \tilde{D}^e + \tilde{D}^p = \Lambda \, \dot{\tilde{\tau}} + \lambda \, \partial f / \partial \tilde{\tau} \tag{34}$$

in case of anisotropic standard plasticity, or more generally with $\tilde{D}^p$ given by (29).

From a numerical point of view, this kind of formulation is natural. It simply requires following the rotation $\tilde{Q}$ (i.e. one angle $\tilde{\theta}$ in the most usual two dimensional cases) for all integration points and making the appropriate rotations . This may seem disturbing, but it must be noted that some of the integration algorithms used in the isotropic case, already use this kind of transformation which automatically ensures incremental objectivity [2].

From a fundamental point of view, questions arise about the physical meaning of this decomposition and of the associated elastic strain which can be defined from the integration of $\tilde{D}^e$ [16]. This requires an analysis of the elastic unloading resulting from the hypoelastic law (34) when $\lambda = 0$. This is still an open problem. This elastic unloading response obviously depends on both the chosen rotating frame and the actual deformation, but it seems that this influence is rather weak in case of small elastic deformations [19]. In any case, some work remains to be done on this aspect.

4.2 Released configuration

This kind of difficulty can be avoided by coming back to the kinematical decomposition presented in section 2.3, and which is the appropriate geometrical framework for the formulation of large strain elasto-plasticity. Its general form however [22] is too wide to be of practical use, and some further assumptions or simplifications must be made. A general analysis of the decomposition of D in its plastic and elastic parts can be found in [8], but we shall restrict our attention here to the case of isotropic elasticity which is an usual assumption for metal forming analysis, and which allows a simpler analysis. In this case indeed most of the results described in section 2.3 remains valid: starting from (16), the dissipation is obtained as

$$\Phi = \tau : D - \dot{w} = \tau \, (\dot{A}A^{-1} + A\dot{P}P^{-1}A^{-1}) - (\partial w / \partial E) : A^T \dot{A}$$
$$[\tau - A(\partial w / \partial E)A^T] : \dot{A}A^{-1} + V^e \tau V^{e-1} : R^e \dot{P} P^{-1} R^{eT} \tag{35}$$

Elastic reversibility requires the vanishing of the first term, and gives the elastic law (17). Due to isotropy, V^e commutes with τ so that Φ is still given by (18), and the plastic evolution law will relate τ and $\bar{D}^p$. The rigid plastic formulation discussed in section 3 is then easily extended to the elastic-plastic case by substituting $\bar{D}^p$ for D. The kinematical analysis of [7] and (20) also remain true, so that, in the small elastic strain approximation, the appropriate decomposition is

$$D = \varepsilon^{eJ} + \bar{D}^p \qquad\qquad \varepsilon^{eJ} = \Lambda\, \tau^J \qquad\qquad (36)$$

This shows that in this formulation, the hypoelastic law must be written in terms of the Jaumann rate. This result has been obtained in the case of isotropic elasticity, but a similar result would have been obtained in the general case [8].

5. CONCLUSION

From a theoretical point of view, the formulation of large strain anisotropic elasto-plasticity is not yet fully understood, and many aspects are still controversial and open to discussion. Further the formulation of section 4.2, which at present seems the best founded, is quite complicated, since it requires two different rotations: the Jaumann rotation which is needed for the elastic part cannot be used for the plastic part because it may lead to unreasonable behaviour.

Most fortunately, these problems, though very important from a fundamental point of view, can be forgotten for the practical purpose of numerical simulation of metal forming processes: the differences between the elastic description in sections 4.1 and 4.2 seem to be small, so that the same rotating frame can practically be used for the elastic and plastic parts. It only remains the choice of a rotating frame, but this may also be not so important. Indeed it follows from the results in section 3.4 that the influence of this choice does not exist for the triaxial case, is always very small for the off axis tensile test, and in worse case (simple shear) the differences only become significant for shears of order one which are not so frequent.

It is therefore essential to account for the rotations and to use an objective rotating frame, but the choice of this rotating frame is not crucial, at least for stress analysis (the situation may be different when plastic instabilities are involved). The corotational frame may be the simplest choice.

REFERENCES

1 R.M. McMeeking and J.R. Rice, Finite element formulation for problems of large elastic plastic deformation, Int. J. Solids Struct., 11 (1975) 601-616.
2 H.J. Braudel, M. Abouaf and J.L. Chenot, An implicit and incrementally objective formulation for the solution of elastoplastic problems at finite strain by F.E.M., Rapport GRECO GDE 152/1985.
3 O. Debordes, M. Elmouatassim and G. Touzot, Local numerical integration of large strain elastoplastic constitutive laws, Rapport GRECO GDE 193/1986.
4 J.P. Boehler, Lois de comportement anisotrope des milieux continus, J. Méc., 17 (1978), 153-190.

5 J.J. Moreau, Sur les lois de frottement, plasticité, viscosité, C.R. Acad. Sc., Paris, Série A, t.271 (1970), 608-611.
6 B. Halphen and Q.S. Nguyen, Sur les matériaux standard généralisés, J. Méc. 14 (1975), 39-64.
7 F. Sidoroff, Incremental constitutive equations for large strain elastoplasticity, Int. J. Eng. Sc., 20 (1982), 19-26.
8 J. Mandel, Sur la définition de la vitesse de déformation élastique en grande transformation élastoplastique, Int. J. Solids Struct., 19 (1983), 573-578.
9 V.A. Lubarda and E.H. Lee, A correct definition of elastic and plastic deformation and its computational significance, J. Appl. Mech., 48 (1) (1981), 35-40.
10 F. Sidoroff, Formulations élasto-plastiques en grandes déformations, Rapport GRECO GDE 29/1981.
11 F. Sidoroff, Internal variables and phenomenological models for metals plasticity, M.M.P. 87, Aussois, Avril 1987, to be published.
12 J.C. Nagtegaal and J.E. De Jong, Some aspects of non isotropic work hardening in finite strain plasticity, in: E.H. Lee and R.L. Mallet (Ed), Plasticity of metals at finite strain: Theory, experiment and computation, Div. Appl. Mech., Stanford University and Dept. Mech. Eng., R. P. I., (1982), 65-102.
13 Y. Dafalias, Corotational rates for kinematic hardening at large plastic deformations, J. Appl. Mech., 50 (1983), 561-565.
14 A. Dogui and F. Sidoroff, Kinematic hardening in large elastoplastic strain, Eng. Fract. Mech., 21(4) (1985),685-695. From the International Symposium on Current Trends and Results in Plasticity, Udine, Italy, June 1983.
15 J.P. Boehler and J. Raclin, Ecrouissage anisotrope des matériaux orthotropes prédéformés, Colloque Franco-Polonais, Marseille, Juin 1980, J. Méc. Th. et Appl.,numéro spécial 1982, 23-44.
16 P. Ladevèze, Sur la théorie de la plasticité en grandes déformations, Rapport interne, 9 (1980), Lab. Méc. et Tech., ENSET, Cachan, France.
17 Y.F. Dafalias, A missing link in the macroscopic constitutive formulation of large plastic deformations, Proc. Plasticity Today, Udine, A. Sawczuk and G. Bianchi (Ed), Elsevier Applied Science Publishers, U. K., (1985), 483-502. From the International Symposium on Current Trends and Results in Plasticity, Udine, Italy, June 1983.
18 A. Dogui and F. Sidoroff, Rhéologie anisotrope en grandes déformations, in: Rhéologie des matériaux anisotropes, Proc. Coll. G.F.R., Paris, (1984), 69-78, Ed. CEPADUES, Toulouse, 1986.
19 A. Dogui, Thèse de Doctorat ès Sciences, en préparation.
20 J. Mandel, Plasticité et viscoplasticité, Courses and Lectures, n° 97(1971), I.C.M.S., Udine, Springer, New York.
21 A. Dogui, Cinématique bidimensionnelle en grandes déformations - Application à la traction hors-axes et à la torsion, J. Méc. Th. et Appl., to be published.
22 F. Sidoroff and C. Teodosiu, Microstructure and phenomenological models for metals, Proc. Physical Basis and Modelling of Finite Deformation of Agregates, CNRS, Paris (1985), to be published.

DRAWING OF VISCOPLASTIC TUBES THROUGH ROTATING CONICAL DIES

DAVID DURBAN*

Department of Engineering, University of Cambridge, Cambridge CB2 1PZ, U.K.

SUMMARY

Steady forming processes of viscoplastic tubes drawn through rotating
conical dies are investigated. A detailed analysis is given for power-law
viscous solids. It is shown that the mathematical problem can be reduced to a
system of two non-linear coupled differential equations. An exact solution is
obtained for the Newtonian fluid. For the non-Newtonian model it is possible
to derive an approximate solution valid for relatively small die angles and low
wall friction. The results predict a substantial reduction in the required
drawing tension as compared with the same forming process through stationary
dies.

1.   INTRODUCTION

A common limit on the efficiency of forming processes follows from the
restriction that the external driving stresses should not exceed certain
permissible loading levels. The tension stress required in tube drawing, for
example, should always be kept below the necking stress of the material to
avoid severe damage at the exit from the die. Recent studies on tube forming
processes (ref. 1-6) have shown that for long and tapered conical dies, with
small wall friction, it is permissible to simulate the flow field, within the
working zone, by the simple radial velocity pattern. That kinematical model
predicts a drawing tension stress T expressed by

$$T = \left[1 + \frac{(m/\sqrt{3})}{\alpha_2 - \alpha_1} \right] T_u \qquad (1.1)$$

where α_1, α_2 are the die angles (Fig.1), m is the friction factor at the walls
and T_u denotes the required uniform drawing tension for the same die but with
frictionless walls. It is worth mentioning what within the framework of the

----

* On sabbatical leave (until October 1987). Permanent address: Department
of Aeronautical Engineering, Technion, Haifa 32000, Israel.

radial flow simulation model, restricted to long dies with small angles, the effect of the transition shear zones, at the entry and exit areas, can be neglected.

In drawings of thin walled tubes, where the die angle $(\alpha_2 - \alpha_1)$ is small, we have from (1.1) that the contribution of wall friction to the drawing tension is quite appreciable. Thus, with typical values of $m = 0.06$ and $\alpha_2 - \alpha_1 = 2°$ we have that the friction term in (1.1) accounts for nearly 50% of the required tension T.

This paper explores a possible method of reducing the wall friction contribution to the drawing tension. Assuming that the conical dies (Fig.1) are not stationary but rather rotating, in opposite directions, we show that the friction part in (1.1) can be substantially decreased. That reduction is a direct outcome of the shift in the direction of the shear stresses, along the walls, induced by the rotating dies. The investigation here is for the particular case where no end couples are applied to the deforming specimen, but there should be no difficulty in extending the analysis to similar processes (such as wire drawing through a rotating conical wall) with end torsion.

The problem is formulated in the next section for a power-law viscous solid. That useful model covers a wide range of material response, from the Newtonian fluid to the Mises rigid/perfectly-plastic solid. Purely axisymmetric radial flow of the power-law viscous material has been analysed in the past for the Mises solid (ref.7), the Newtonian fluid (ref.8) and the family of the power law viscous solids (ref.9). The related problem of helical flow of rigid/linear-hardening plastic solids is treated in (ref.10).

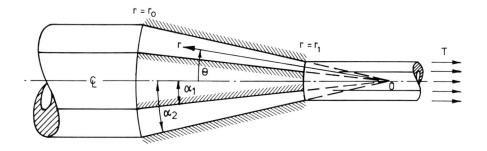

Fig.1. Drawing of tubes through conical dies. The deformation zone is bounded by the conical walls $\alpha_1 \leq \theta \leq \alpha_2$ and by the entry-exit radii $r_1 \leq r \leq r_0$. The walls are rotating in opposite directions.

A judicious choice of the velocity components leads to a system of two coupled non-linear differential equations for the velocity profiles. With the Newtonian fluid it is possible to obtain an exact analytical solution presented in section 3. A consistent first order approximation of that solution, valid for small die angles, is derived as a guiding result for further analysis.

An approximate solution of the governing equations, for the general power law material, is given in section 4 along the lines of the approximate version of the Newtonian fluid solution. The eventual outcome are simple relations for the velocity profiles and the stress components. It is shown that the drawing tension is again given by an expression similar to (1.1) but with a lower wall friction term. The amount of reduction in that term depends on the geometrical parameter (α_2/α_1) as well as on the applied surface torsion at the walls.

2. FORMULATION OF THE PROBLEM

A power-hardening viscous solid is forced to flow through converging conical dies (Fig.1) with the common virtual apex 0. The working zone is bounded by the entry and exit radii $r_1 \leq r \leq r_0$ and by the rough walls $\alpha_1 \leq \theta \leq \alpha_2$. The external die is rotating in the positive direction with respect to the axis $\theta = 0$, while the internal die is rotating in the opposite direction but about the same axis. That rotation induces a circumferential velocity component into the deforming material and we expect the velocity vector to have two components; U in the radial direction and W in the circumferential direction. Both U and W depend only on r and θ and the incompressibility constraint is satisfied with

$$U = -fr^{-2} \qquad f = f(\theta) \qquad (2.1)$$

where f is an unknown function of θ and the negative sign stands for converging flow patterns. It is anticipated that the circumferential velocity would take the similar form

$$W = gr^{-2} \qquad g = g(\theta) \qquad (2.2)$$

where again, g is an unknown function of θ.

The components of the Eulerian strain rate associated with (2.1)-(2.2) are, with a standard notation,

$$\epsilon_r = 2fr^{-3} \qquad \epsilon_\theta = -fr^{-3} \qquad \epsilon_\phi = -fr^{-3}$$

$$\epsilon_{r\theta} = -(1/2)f'r^{-3} \qquad \epsilon_{r\phi} = -(3/2)gr^{-3} \qquad \epsilon_{\theta\phi} = (1/2)(g'-g\cot\theta)r^{-3} \qquad (2.3)$$

where the prime indicates differentiation with respect to θ.

The constitutive relation of a power-law viscous material is

$$\underset{\sim}{S} = \sqrt{2/3} \; \sigma_e \; \underset{\sim}{D}/\sqrt{\underset{\sim}{D} \cdot\cdot \underset{\sim}{D}} \tag{2.4}$$

where $\underset{\sim}{S}$ is the stress deviator tensor, $\underset{\sim}{D}$ is Eulerian strain rate tensor, and σ_e is the invariant

$$\sigma_e = \sigma_o (\sqrt{2/3} \; \sqrt{\underset{\sim}{D} \cdot\cdot \underset{\sim}{D}})^{\frac{1}{n}} \tag{2.5}$$

with σ_o and n denoting material parameters.

Observing the coaxiality of tensors $\underset{\sim}{D}$ and $\underset{\sim}{S}$ we find that the stress components σ_θ and σ_ϕ are identical. Combining (2.4)-(2.5) with the strain rate components (2.3) results in the scalar relations

$$\Sigma_e = \sqrt{3} \; F\Delta\rho^{-3/n} \qquad \Sigma_r - \Sigma_\theta = \sqrt{3} \; F\rho^{-3/n}$$

$$\Sigma_{r\theta} = -\beta F\rho^{-3/n} \qquad \Sigma_{r\phi} = -\gamma_1 F\rho^{-3/n} \qquad \Sigma_{\theta\phi} = \gamma_2 F\rho^{-3/n} \tag{2.6}$$

where $\rho = r/r_1$ is the non-dimensional radial coordinate, all stress components $(\Sigma_r, \Sigma_\theta, \Sigma_{r\theta}, \dots)$ have been normalised with respect to $(2^{1/n}/\sqrt{3})\sigma_o r_1^{-3/n}$, and

$$\beta = (\sqrt{3}/6)f'/f \qquad \gamma_1 = (\sqrt{3}/2)g/f \qquad \gamma_2 = (\sqrt{3}/6)(g'-g \cot\theta)/f \tag{2.7}$$

$$F = f^{1/n}\Delta^{(1-n)/n} \qquad \Delta = (1 + \beta^2 + \gamma_1^2 + \gamma_2^2)^{1/2} \tag{2.8}$$

Finally, we have the three equilibrium equations

$$\rho \frac{\partial\Sigma_r}{\partial\rho} + \frac{\partial\Sigma_{r\theta}}{\partial\theta} + 2(\Sigma_r - \Sigma_\theta) + \Sigma_{r\theta}\cot\theta = 0 \tag{2.9}$$

$$\rho \frac{\partial\Sigma_{r\theta}}{\partial\rho} + \frac{\partial\Sigma_\theta}{\partial\theta} + 3\Sigma_{r\theta} = 0 \tag{2.10}$$

$$\rho \frac{\partial\Sigma_{r\phi}}{\partial\rho} + \frac{\partial\Sigma_{\theta\phi}}{\partial\theta} + 3\Sigma_{r\phi} + 2\Sigma_{\theta\phi}\cot\theta = 0 \tag{2.11}$$

which complete the construction of the mathematical model.

Inserting $(\Sigma_r - \Sigma_\theta)$ and $\Sigma_{r\theta}$ from (2.6) in the radial equilibrium equation (2.9), and intergrating over ρ, gives

$$\Sigma_r = -(n/3)\left[(\beta F)' + (\beta F)\cot\theta - 2\sqrt{3}F\right]\rho^{-3/n} + G(\theta) \tag{2.12}$$

where $G(\theta)$ is an unknown function of θ. The circumferential stress component Σ_θ follows from (2.6) as

$$\Sigma_\theta = -\left\{(n/3)\left[(\beta F)' + (\beta F)\cot\theta\right] - \left[(2n-3)/\sqrt{3}\right] F\right\}\rho^{-3/n} + G(\theta) \quad (2.13)$$

Next, we substitute the shear stress $\Sigma_{r\theta}$ from (2.6) and Σ_θ from (2.13) in the second equation of equilibrium (2.10). It follows that the latter is completely satisfied if $G(\theta)$ is a constant (say $G=A$), and

$$\left\{(n/3)\left[(\beta F)' + (\beta F)\cot\theta\right] - \left[(2n-3)/\sqrt{3}\right]F\right\}' + 3(1-n^{-1})\beta F = 0 \quad (2.14)$$

It remains to consider the third equilibrium equation (2.11). Inserting the shear stresses $\Sigma_{r\phi}$ and $\Sigma_{\theta\phi}$ from (2.6) we find that (2.11) is satisfied provided that

$$(\gamma_2 F)' + 2(\gamma_2 F)\cot\theta - 3(1-n^{-1})\gamma_1 F = 0 \quad (2.15)$$

Thus, the particular velocity field (2.1)-(2.2) is determined by the solution of two non linear differential equations (2.14)-(2.15) with the functions $f(\theta)$ and $g(\theta)$ as unknowns.

3.   SOLUTIONS FOR THE NEWTONIAN FLUID

The deep non linearity of the governing equations (2.14)-(2.15), as well as their highly coupled nature, make any further exact analytical advance a formidable task. However, for a Newtonian fluid, with n=1, it becomes possible to find the exact integrals of these equations. There is little practical significance in that solution but it nevertheless reflects the behaviour which is common to all values of n.

When n=1 we find that F=f and equations (2.14)-(2.15) are reduced to

$$(f'' + f'\cot\theta + 6f)' = 0 \quad (3.1)$$

$$(g' - g\cot\theta)' + 2(g'-g\cot\theta)\cot\theta = 0 \quad (3.2)$$

with the respective solutions

$$f = C_1 + C_2(3\cos^2\theta-1) + C_3\left[(3\cos^2\theta-1)\ln\tan\frac{\theta}{2} + 3\cos\theta\right] \quad (3.3)$$

$$g = C_4\left[(\sin\theta)\ln\tan\frac{\theta}{2} - \cot\theta\right] + C_5\sin\theta \quad (3.4)$$

where $C_1,\ldots,C_5$, are intergration constants. For small angles, where $\theta^2 \ll 1$, we may replace (3.3)-(3.4) by the approximations

$$f = U\left[1-3k_1\left(\frac{\theta^2-\alpha_m^2}{2} - \alpha_m^2 \ln \frac{\theta}{\alpha_m}\right)\right] \tag{3.5}$$

$$g = U\left[\frac{3}{2} k_2\left(\frac{\theta^2-\alpha_o^2}{\theta}\right)\right] \tag{3.6}$$

where U, k_1, k_2, α_m, α_o are new suitably chosen constants.

The shear profiles follow now from (2.7), again for small angles, as

$$\beta = -\frac{\sqrt{3}}{2} k_1\left(\frac{\theta^2-\alpha_m^2}{\theta}\right) \quad \gamma_1 = \frac{3\sqrt{3}}{4} k_2\left(\frac{\theta^2-\alpha_o^2}{\theta}\right) \quad \gamma_2 = \frac{\sqrt{3}}{2} k_2\left(\frac{\alpha_o}{\theta}\right)^2 \tag{3.7}$$

and the stresses can then be explicitly expressed by (2.12)-(2.13) and (2.6). This solution for the stresses contains five constants $(k_1,k_2,\alpha_o,\alpha_m,A)$ which should be determined from the corresponding stress boundary data.

4. APPROXIMATE SOLUTION FOR THE POWER LAW VISCOUS SOLID

Assuming that the radial velocity component is nearly uniform we may integrate the first of (2.7) and write, following (3.5),

$$f = U \exp(2\sqrt{3} \int_{\alpha_m}^{\theta} \beta d\theta) \approx U(1+2\sqrt{3} \int_{\alpha_m}^{\theta} \beta d\theta+\ldots) \tag{4.1}$$

where U and α_m are integration constants. For long dies with small angles and low wall friction it is permissible to neglect the contribution of the shear stresses to the effective stress. Thus $|\Sigma_{r\theta}|$, $|\Sigma_{r\phi}|$, $|\Sigma_{\theta\phi}|$ are expected to be much smaller than Σ_e. It follows from (2.6)-(2.7) that $|\beta|$, $|\gamma_1|$, $|\gamma_2|$ are much smaller than unity so that $f\approx U$, by (4.1), and similarly from (2.8)

$$\Delta \approx 1 \qquad F \approx U^{1/n} \tag{4.2}$$

Equation (2.14) can therefore be replaced by

$$(\beta'+\beta\cot\theta)' - C_n\beta = 0 \qquad C_n = (4n-3)(n-3)/n^2 \tag{4.3}$$

This equation has an exact solution in terms of the Legendre functions. For small angles, however, we can use with sufficient accuracy the first order approximation

$$\beta = -\frac{\sqrt{3}}{2} k_1\left(\frac{\theta^2-\alpha_m^2}{\theta}\right) \tag{4.4}$$

where k_1 and α_m are integration constants. Likewise, the differential equation (2.15) is now simplified to

$$(\gamma_2{}' + 2\gamma_2\cot\theta)' - 3(1-n^{-1})\gamma_1 = 0 \qquad (4.5)$$

A second equation for γ_1 and γ_2 is obtained from the compatibility requirement of relations (2.7), namely

$$\gamma_1{}'-2\gamma_1(\cot\theta-2\sqrt{3}\beta)-3\gamma_2 = 0 \qquad (4.6)$$

Note however that within the framework of our assumptions the β-term in (4.6) is negligible in comparison with $\cot\theta$. The compatibility equation is therefore reduced further to the form

$$\gamma_1{}'-2\gamma_1\cot\theta-3\gamma_2 = 0 \qquad (4.7)$$

Equations (4.5) and (4.7) admit an exact solution expressible by the Legendre functions. For our purpose however it would be sufficient to take the first order approximation, obtained for small θ,

$$\gamma_1 = \frac{3\sqrt{3}}{4}\, k_2\!\left(\frac{\theta^2-\alpha_o^2}{\theta}\right) \qquad \gamma_2 = \frac{\sqrt{3}}{2}\, k_2 \left[\frac{9(n-1)}{8n}\,\theta^2+\left(\frac{\alpha_o}{\theta}\right)^2\right] \qquad (4.8)$$

where k_2 and α_o are integration constants. The validity of these approximate solutions, (4.4) and (4.8), is supported upon comparison with the consistent first order approximations (3.7) for the Newtonian fluid (n=1).

Turning to the boundary data we impose first the requirement that there are no end couples acting on the material. Thus, over any spherical cross section within the working zone

$$\int_o^{2\pi} \int_{\alpha_1}^{\alpha_2} \sigma_{r\phi} r^3 \sin^2\theta d\theta d\phi = 0 \qquad (4.9)$$

For small angles that condition can be rewritten as

$$\int_{\alpha_1}^{\alpha_2} \gamma_1\theta^2\, d\theta = 0 \qquad (4.10)$$

resulting in, via the first of (4.8),

$$\alpha_o^2 = \frac{1}{2}(\alpha_1^2+\alpha_2^2) \qquad (4.11)$$

An immediate consequence of (4.11) is that γ_2 from (4.8) can be approximated by the simple expression

$$\gamma_2 = \frac{\sqrt{3}}{2} k_2 \left(\frac{\alpha_o}{\theta}\right)^2 \tag{4.12}$$

Over the curved surfaces (at $\theta=\alpha_1,\alpha_2$) it is assumed that the friction factors (m_1,m_2) are specified. Each of these factors determines the ratio of the resultant surface shear stress $(\sigma_{r\theta}^2+\sigma_{\theta\phi}^2)^{1/2}$ to the effective shear stress $\sigma_e/\sqrt{3}$ at the surface. Using (2.6) and (4.2) we get the two algebraic equations

$$\theta=\alpha_1 : \beta^2+\gamma_2^2=m_1^2 \quad , \quad \theta=\alpha_2 : \beta^2+\gamma_2^2=m_2^2 \tag{4.13}$$

or, with the aid of (4.4) and (4.12),

$$k_1^2\left(\frac{\alpha_1^2-\alpha_m^2}{\alpha_1}\right)^2 + k_2^2\left(\frac{\alpha_o}{\alpha_1}\right)^4 = \frac{4m_1^2}{3} \qquad k_1^2\left(\frac{\alpha_2^2-\alpha_m^2}{\alpha_2}\right)^2 + k_2^2\left(\frac{\alpha_o}{\alpha_2}\right)^4 = \frac{4m_2^2}{3} \tag{4.14}$$

The shear factors are expected to be much smaller than unity, (a representative value would be 0.05) and are not necessarily equal. Here however we shall proceed with the assumption that $m_1=m_2=m$. Conditions (4.14) are supplemented by the relation

$$\frac{\sqrt{3}}{2} k_2 = \xi m \tag{4.15}$$

where ξ is a known parameter which reflects the relative rotation of the material. Inserting (4.15) back in (4.14) we find that

$$k_1 = \left(\frac{2m}{\sqrt{3}}\right) \frac{X_2\alpha_2+X_1\alpha_1}{\alpha_2^2-\alpha_1^2} \qquad \alpha_m^2 = \alpha_1\alpha_2 \frac{X_2\alpha_1+X_1\alpha_2}{X_2\alpha_2+X_1\alpha_1} \tag{4.16}$$

with $\qquad X_1 = \left[1 - \left(\frac{\alpha_o}{\alpha_1}\right)^4 \xi^2\right]^{1/2} \qquad X_2 = \left[1 - \left(\frac{\alpha_o}{\alpha_2}\right)^4 \xi^2\right]^{1/2}$ (4.17)

All four integration constants $(\alpha_o,\alpha_m,k_1,k_2)$ are now determined and the corresponding expressions for the stresses follow with no difficulty. From (2.12) we get the radial stress

$$\Sigma_r = \frac{2n}{\sqrt{3}} U^{1/n}(1 + \frac{k_1}{2})\rho^{-3/n} + A \tag{4.18}$$

In pure drawing where $\Sigma_r(\rho=\rho_o)=0$ we find that the drawing tension $T\equiv\Sigma_r(\rho=1)$ is simply $\qquad T = (1+\frac{k_1}{2})T_u \quad \text{with} \quad T_u = \frac{2n}{\sqrt{3}} U^{1/n}(1-\rho_o^{-3/n})$ (4.19) where k_1 is given by (4.16). Relation (4.19) is similar to the radial flow solution (1.1) but has a different wall-friction term. It is a matter of ease to verify that when $\xi=0$ the present solution is reduced exactly to (1.1). A reasonable measure of the reduction in the wall friction contribution to the drawing tension would be the ratio $\eta = k_1/k_1(\xi=0)$. That ratio can be put in the form, using (4.11) and (4.16)-(4.17),

$$\eta = \frac{X_2\delta+X_1}{\delta+1} \qquad \delta = \frac{\alpha_2}{\alpha_1} \qquad (4.20)$$

where $\quad X_1 = \left\{ 1 - \left[\frac{\xi(1+\delta^2)}{2}\right]^2 \right\}^{1/2} \qquad X_2 = \left\{ 1 - \left[\frac{\xi(1+\delta^2)}{2\delta^2}\right]^2 \right\}^{1/2} \qquad (4.21)$

Typical values of η are displayed in Fig. 2, for different δ, over the entire range of parameter ξ. The latter is bounded by

$$0 \leq \xi \leq \frac{2}{1+\delta^2} \qquad (4.22)$$

and the lowest value of η, for a given δ, is obtained at the upper bound of (4.22), namely $\qquad \eta_{min} = \frac{(\delta^4-1)^{1/2}}{\delta(\delta+1)} \qquad (4.23)$

These minimum values of η decrease as δ approaches unity (Fig. 2).

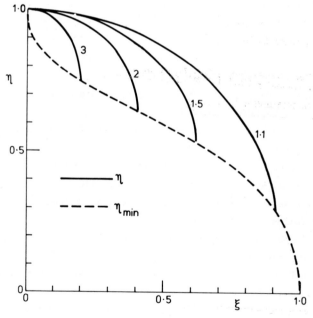

Fig 2. Reduction in wall friction contribution to the drawing tension as measured by the ratio $\eta = k_1/k_1(\xi=0)$. Numbers at the full lines indicate the corresponding values of $\delta = \alpha_2/\alpha_1$. The broken line shows the theoretical minimum values of η.

Note that, for small angles and low friction factors, the shear profiles $(\beta, \gamma_1, \gamma_2)$ are indeed very small in agreement with our basic assumption. The resulting spiral motion of the deforming material is fairly slow; The velocity ratio g/f obtained from the second of (2.7), (4.8) and (4.15), is

$$\frac{g}{f} = \sqrt{3}\xi m\left(\frac{\theta^2 - \alpha_o^2}{\theta}\right) \tag{4.24}$$

with purely radial flow at $\theta = \alpha_o$.

Similar investigations can be carried out, as in (ref. 4-6), for other material models characterized by the coaxiality relation (2.4). For the particular case of rigid/perfectly-plastic solids it is possible to extend the analysis to composite multilayered tubes along the lines of (ref .3).

ACKNOWLEDGEMENT

I am indebted to the kind assistance and hospitality of the Engineering Department at Cambridge University.

REFERENCES

1   D. Durban, Drawing of Tubes, J.Appl.Mech, 46 (1980) 736-740.

2   D. Durban, Radial Flow Simulation of Drawing and Extrusion of Rigid/Hardening Materials, Int.J.Mech.Sci., 25 (1983) 27-39.

3   D. Durban, Drawing and Extrusion of Composite Sheets, Wires and Tubes, Int.J.Solids Structures, 20 (1984) 649-666.

4   D. Durban, Rate Effects in Steady Forming Processes of Plastic Materials, Int.J.Mech.Sci., 26 (1984) 293-304.

5   D. Durban, On Some Simple Steady Forming Process of Viscoplastic Solids, Acta Mech., 57 (1985) 123-141, 62 (1986) 184.

6   D. Durban, On Generalized Radial Flow Patterns of Viscoplastic Solids with Some Applications, Int.J.Mech.Sci, 28 (1986) 97-110.

7   R.T. Shield, Plastic Flow in a Converging Conical Channel, J.Mech.Phys.Solids, 3 (1955) 246-258.

8   W.N. Bond, Viscous Flow Through Wide-Angled Cones, Philosophical Magazine and Journal of Science, Sixth Series, 50 (1925) 1058-1066.

9   G. Devries, D.B. Craig and J.B. Haddow, Pseudo-Plastic Converging Flow, Int. J.Mech.Sci., 13 (1971) 762-772.

10   D. Durban and M.R. Sitzer, Helical Flow of Plastic Materials, Acta. Mech., 50 (1983) 135-140.

Computational Methods for Predicting Material Processing Defects, edited by M. Predeleanu   103
Elsevier Science Publishers B.V., Amsterdam, 1987 — Printed in The Netherlands

INFLUENCE OF THE BLANK CONTOUR ON THE THICKNESS VARIATION IN SHEET METAL FORMING

P. DUROUX[1], G. de SMET[1], J.L. BATOZ[2]

[1]C.E.D., USINOR ACIERS, BP 2, 60160 Montataire (France)

[2]Division Modèles Numériques en Mécanique, Université de Technologie de Compiègne, BP 233, 60206 Compiègne (France)

SUMMARY
    A finite element formulation is presented for the analysis of thin elasto-
plastic isotropic HSS metal sheets subjected to hydrostatic pressure and contact
without friction in deep drawing. The influence of the shape of the blank contour on the thickness variation and on the flow of the material is particularly
studied.

INTRODUCTION

    Deep drawing is one of the oldest and most important processes for sheet me-
tal forming (refs. 1-2). It is widely used in the automobile industry for the
mass production of car panels at low cost. The structural components thus fabricated can lead to light-weight and safer automobile bodies. The components
that play a leading role in terms of safety are subjected to severe mechanical
constraints and materials with high elastic limits (High Strength Steel) must be
considered. Moreover the manufacturers need to produce geometrically complex
parts with a minimum number of operations obtain precise geometrical characteristics and without defects.

    The deep drawing method where the blank is fixed along the periphery to
obtain the workpiece by stretching the sheet is not adequate and efficient for
HSS materials. The maximum strain at fracture in HEL materials is less than
20 %. In this case it is not possible to control the thickness variation, the
spring back or the residual stresses which mainly determine the final stiffness
of the structural component. The deep drawing process must be combined with
bending and stretching processes to permit the flow of the material in the case
of irregular automobile body parts. The development of new methodology of
drawing is essential for HSS materials. For example the "isoforming" process
(ref. 3) is such that sheet parts can be obtained with a minimum thickness variation and with a residual stress distribution having a stiffening effect.

    It is necessary to analyse the flow behaviour of the material in order to
help the preparation of the drawing tool box.

This paper deals with a non linear finite element analysis of thin sheets and its application to the modelling of "isodrawing" type problems. The blank is considered as a membrane, the elasto-plastic law is based on Von Mises yield criterion with nonlinear isotropic hardening. The influence of the shape of the blank (circular and square) is studied and the results are correlated with experimental observations.

FINITE ELEMENT FORMULATION

Continuum mechanics aspects

If we consider a plane state of stress and a membrane behaviour the expression of the principle of virtual work is given by :

$$W = h \int_S <D*> \{\sigma\} \, dS - \int_S w* \, p \, dS = 0 \tag{1}$$

for any $\{u*\} = \{0\}$ on C_u

h, S   actual thickness and surface area of the deformed sheet at time t

p   normal pressure

$<u*> = <u* \ v* \ w*>$ virtual displacement (or velocity) field with $u*$, $v*$ components in a tangent plane and $w*$ along the normal direction

$<D*> = <D*_{11} \ D*_{12} \ 2D*_{12}>$ virtual strain rates compatible with $<u*>$

$<\sigma> = <\sigma_{11} \ \sigma_{22} \ \sigma_{12}>$ Cauchy stresses in the membrane

C_u is the part of the boundary of S with prescribed displacements

The sheet undergoes large displacements, large rotations and large strains. The kinematic is described by the deformation gradient tensor [F], the orthogonal rotation tensor [R] and the symmetric strain tensor [U] such that :

$$[F] = [R] \, [U] \tag{2}$$

The constitutive equation must involved quantities independant of rigid body rotations (objectivity). In this study we consider the following relations :

$$\{\sigma_R\}_{t+\Delta t} = \{\sigma_R\}_t + \int_t^{t+\Delta t} [H_{ep}(\sigma_R)] \{D_R\} \, dt \tag{3}$$

where $\{\sigma_R\}$ and $\{D_R\}$ are the components of the following tensors : (ref. 4)

$$[\sigma_R] = [R]^T \, [\sigma] \, [R] \tag{4}$$

$$[D_R] = [R]^T \, [D] \, [R] = 1/2 \, ([\dot{U}] \, [U]^{-1} + [U]^{-1} \, [\dot{U}]) \tag{5}$$

$[H_{ep}(\sigma_R)]$ is the elasto-plastic matrix. Von Mises yield criterion and non linear isotropic hardening are considered.

The following expression verify the objectivity condition on a finite step :

$$\{\sigma_R(t + \Delta t)\} = \{\sigma_R(t)\} + [H_{ep}(\sigma_R)] \{D_R^{1/2}\} \tag{6}$$

If $\langle\Delta u \ \Delta v \ \Delta w\rangle$ are the increments of displacements between t and $t + \Delta t$ and $\langle x \ y \ z\rangle$ the updated coordinates at mid-step $(t + \Delta t/2)$ in a coordinate system associated with the rotation at mid step then :

$$\langle D_R^{1/2}\rangle = \langle \frac{\partial\Delta u}{\partial x} \quad \frac{\partial\Delta v}{\partial y} \quad \frac{\partial\Delta u}{\partial y} + \frac{\partial\Delta v}{\partial x}\rangle \tag{7}$$

This quantity is decomposed in elastic (small strains) and plastic parts :

$$\langle D_R^{1/2}\rangle = \langle D_R^e\rangle + \langle D_R^p\rangle \tag{8}$$

The integration of plastic strains are obtained by a semi-implicit scheme. The thickness variation is calculated assuming total incompressibility :

$$\det [F] = \det [U] = 1 \tag{9}$$

Finite element model

The blank is discretized with flat 3 nodes (CST) triangular elements with 3 dof per node.

Eq. 1 is now expressed as :

$$W = \sum_e W^e = 0 \tag{10}$$

with

$$W^e = - \langle u_n^*\rangle \{r\} \tag{11}$$

$$\langle u_n^*\rangle = \langle u_1^* \ v_1^* \ w_1^* \ u_2^* \ v_2^* \ w_2^* \ u_3^* \ v_3^* \ w_3^*\rangle \tag{12}$$

$\{r\}$ is the residual vector :

$$\{r\} = [T_R]^T (- [B_R]^T \{\sigma_R\} h + \{N\} p) S \tag{13}$$

$[T_R]$ orthogonal transformation matrix between global coordinate and local co-rotated (with $[R]$) coordinate systems

$[B_R]$ is the strain matrix computed according to Eq. 7

$\{N\}$ contains the linear interpolation functions in area coordinates

h, S   are the thickness and area of the element ($h.S$ = constant)

After assembling Eqs 10 and 11 gives :

$$\{R(U_n)\} = \{0\} \tag{14}$$

where $\{R(U_n)\}$ is the global residual vector and $\{U_n\}$ is the global displacement vector.

Eq. 14 is solved by Newton-Raphson iteration strategies, i.e. :

$$[K_T^i] \{\Delta U_n\} = \{R^i\} \qquad (15)$$

$$\{U_n^{i+1}\} = \{U_n^i\} + \{\Delta U_n\}$$

$[K_T^i]$ is the global tangent matrix.

In this study the following tangent stiffness matrix is used (based on Truesdell derivative of Cauchy stress rather than Green-Naghdi derivative.

$$[k_t] = [T_R]^T \, ([B_R]^T \, [H_{ep}] \, [B_R] + [B_\phi]^T \, [\sigma_R] \, [B_\phi]) \, [T_R] \, S \, h \qquad (16)$$

where

$$<u,_x \quad u,_y \quad v,_x \quad v,_y \quad w,_x \quad w,_y> \; = \; <u_n> \, [B_\phi]^T \qquad (17)$$

and

$$[\sigma_R] = \begin{bmatrix} [\underset{\sim}{\sigma}] & & \\ & [\underset{\sim}{\sigma}] & \\ & & [\underset{\sim}{\sigma}] \end{bmatrix} \; ; \; [\underset{\sim}{\sigma}] = \begin{bmatrix} \sigma_{R\,11} & \sigma_{R\,12} \\ \sigma_{R\,12} & \sigma_{R\,22} \end{bmatrix} \qquad (18)$$

Eqs 10 to 18 corresponds to an updated Lagrangian formulation where the reference configuration is modified at each iteration.

The following operations are performed for each element at each iteration for a given estimation of $\{\Delta u_g\}$ (global) of the element between t and $t + \Delta t$

- compute the total stresses $\{\sigma_R\}_{t+\Delta t}$
  - define $[T_R(t + \Delta t/2)]$ using global coordinates $\{X_{t+\Delta t/2}\}$ in $t + \Delta t/2$
  - define local coordinates

  $$\{x_{t+\Delta t/2}\} = [T_R(t + \Delta t/2)] \, \{X_{t+\Delta t/2}\}$$

  - define displacements increments

  $$\{\Delta u_\ell\} = T_R(t + \Delta t/2) \, \{\Delta u_g\}$$

  - define strain increment between t and $t + \Delta t$

  $$\{D_R^{1/2}\} = [B(x_{t+\Delta t/2})] \, \{\Delta u_\ell\}$$

  - integration of constitutive equations (semi implicit scheme)

  $$\{\Delta \sigma_R\} = [H_{ep}] \, \{D_R^{1/2}\}$$

- addition of stresses

$$\{\sigma_R\}_{t+\Delta t} = \{\sigma_R\}_t + \{\Delta\sigma_R\}$$

- compute the transformation matrix $[T_R(t + \Delta t)]$ using global coordinates in $t + \Delta t$
- define the local coordinates

$$\{x_{t+\Delta t}\} = [T_R(t+\Delta t)] \{X_{t+\Delta t}\}$$

- define matrices $[B_R]$, $[B_\phi]$, $[k_t]$ and $\{r\}$

In the problems considered in the next section axisymmetric contact conditions exist between the sheet wall and the rigid blank-holder. In our computations we assume contact without friction. Gap or contact elements are formulated and the contact conditions are satisfied using a penalty approach. The elements are able to simulate perfect sliding with large relative displacements.

In the examples considered the central normal displacement is controlled and the corresponding equilibrium pressure is computed (ref. 5). The procedure used to obtain the complete pressure-displacement curves is a particular case of the control arc-length method (ref. 6).

At each iteration the following convergence criterion is considered :

$$\frac{\langle\Delta U_n\rangle \{\Delta U_n\}}{\langle U_n\rangle \{U_n\}} < 10^{-5}$$

where $\langle\Delta U_n\rangle$ is the global displacement vector at iteration i (solution of Eq. 15) and $\langle U_n\rangle$ is the global displacement vector between step t and step $t + \Delta t$.

NUMERICAL APPLICATIONS

Test problems

They deal with the "isoforming" process : an hydroforming process where the fluid is replaced by an incompressible soft elastomeric material.

The drawing of a bulge is simulated with flow of the material under the blank holder (Figure 1). The bulge is mainly obtained by stretching. This type of deformation is often easily obtained when the material has high planar anisotry, however this property is not present in HSS materials. It is thought that the shape of the blank can have a similar effect by perturbing (geometrically) the flow.

A circular and a square blank contour are considered (Figure 2). The circular case corresponds to zero planar anisotropy. The finite element meshes considered on a quarter of the sheets are presented in Figure 3. The involve 224 (circular) and 720 (square) elements. In the square test high deformation gradients are expected and a fine mesh is justified. The elastomeric material is replaced by an hydrostatic pressure and friction is neglected.

Results and discussion

In Figure 4 we present the workpieces at different level of deformation showing the flow of the material. In figures 5 and 6 the variation of the thickness and of the principal strains along some typical directions are presented. In accordance with our expectations the minimum thickness is greater in the square specimen, therefore it is possible to obtain a deeper workpiece in this case. However the increase of the thickness at mid side could lead in practice to wrinkling under the blank holder.

The numerical results are in qualitative agreement with experimental observarions done at our Laboratory. It is quite difficult to verify the symmetry and to obtain very precise results by experimental testing. In the numerical model the thicknesses are overestimated. By neglecting the friction the flange is able to slide more easily, the thickness at the central zone decreases less than in reality and it increases more at the middle of the side. This has been observed when we have tried to test a specimen with a Teflon sheet : a fold (wrinkling) appeared very repidly.

These numerical difficulties and the need to simulate other situations involving high compressive stresses and bending have motivated the development of a shell element taking bending and large strains into account. The increase of nodal variables and of numerical integration points makes the use of these elements very expensive (on a standard bulge-test problems the shell element is ten times more expensive than the membrane element for the same mesh of 25 elements on a quarter).

Comments on the numerical aspects

The computations are performed on a DEC/VAX 8600 in double precision. Our subroutines are compatible with the general subroutines of MEF (ref. 7). The computer time is of the order of 2 hours 30 for the circular blank and of 8 hours for the square case (this is quite important with regard to the geometrical simplicity of the problems).

The use of membrane finite elements is rather difficult in terms of convergence : the tangent stiffness is ill conditionned, singularities and zero energy modes are present especially with fine meshes and when compression exists.

CONCLUSION

A finite element analysis has been performed to study the deep drawing of two HSS metal sheets in the isoforming process. The model is based on triangular membrane elements and elasto-plasticity (with the Green-Naghdi derivative of Cauchy stresses). The influence of the shape of the contour of the blank on the thickness variation and on the strain distribution has been studied in details. It is observed that the geometry of the blank can partly replace the very low planar anisotropy existing in HSS materials.

REFERENCES

1 Handbook of metal forming, K. Lange Ed., Mac Graw Hill, 1985, Chapter 20.
2 Computer Modeling of Sheet Metal Forming Process, Theory, verification and application, N.M. Wang and S.C. Tang Eds., The Metallurgical Society, 1985.
3 G. de Smet, Emboutissage des aciers H.L.E., Metal Tribune, N°3, USINOR publication, 1986, pp. 14 à 15.
4 P. Pegon, J.P. Halleux and J. Donea, A discussion of stress rates and their application in finite strain plasticity, Proc. NUMETA'85 Conf., Swansea, 7-11 jan. 1985, J. Middleton and G.N. Pande, eds., A.A. Balkema, Rotterdam, 1985, 315-325.
5 J.L. Batoz, G. Dhatt, Incremental displacement algorithms for non linear problems, Int. J. Num. Meth. Eng., 1979, 14, 1262-1267.
6 J.L. Batoz, G. Dhatt, M. Fafard, Algorithmes de calcul automatique des configurations prê et post-flambement, Colloque Tendances Actuelles en Calcul des Structures, Bastia, 6-8 novembre 1985.
7 G. Dhatt, G. Touzot, Une présentation de la méthode des éléments finis, Maloine Ed., 1981.

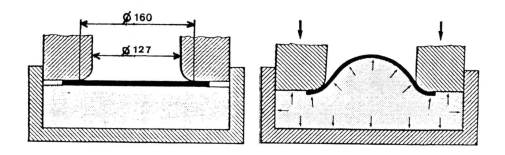

Figure 1. Isoforming process

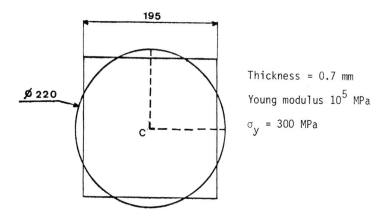

Thickness = 0.7 mm

Young modulus 10^5 MPa

σ_y = 300 MPa

Figure 2. Circular and square specimen

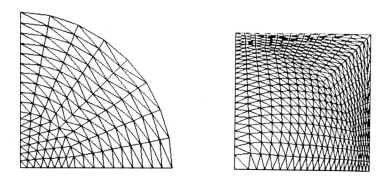

Figure 3. Finite element meshes

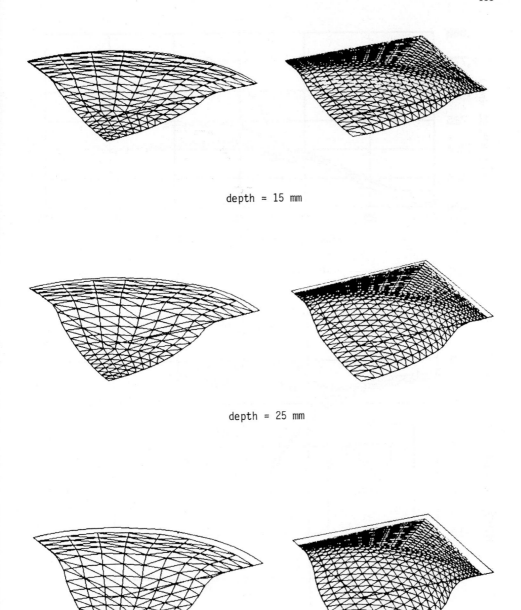

depth = 15 mm

depth = 25 mm

depth = 33 mm

Figure 4. Three deformation levels

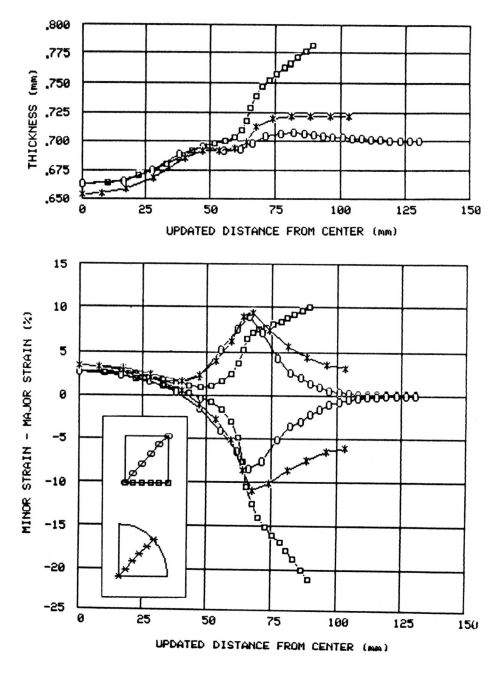

Figure 5. Thickness and principal strains

Computational Methods for Predicting Material Processing Defects, edited by M. Predeleanu   113
Elsevier Science Publishers B.V., Amsterdam, 1987 — Printed in The Netherlands

CHARACTERIZATION AND MODELING OF SPREAD IN ROLLING ANISOTROPIC STRIP

A.S. EL-GIZAWY[1], R.M. GRIFFIN[2], and J.A. SCHEY[3]

[1]Department of Manufacturing, Bradley University, Peoria, Illinois 61625 USA

[2]New Brunswick Power, Fredericton, New Brunswick, Canada

[3]Department of Mechanical Engineering, University of Waterloo, Waterloo, Ontario, Canada N2L 3G1

SUMMARY

The phenomenon of lateral flow (spread) in rolling of anistropic strips was investigated using four different materials. The process variables studied were the starting width-to-thickness ratio, roll surface finish, lubrication, and the degree of the homogeneity of deformation. The study revealed that high r-value hexagonal metals exhibit similar spread characteristics to those of isotropic metals, but that in cubic metals with r less or equal to unity, spread generally decreases with decrease in r.

A new model for predicting spread was developed. The model includes the powerful geometric factor w/L and accounts for the effects of friction and the angle of inclination at the plane of entry through a non-dimensional factor L_f/L related to the length of the forward-slip zone.

NOMENCLATURE

h - specimen thickness
L - projected length of the arc of contact corrected for rolling flattening
L_b- length of the zone of backward slip
L_f- length of the zone of forward slip
r - plastic strain ratio
R - radius of work rolls corrected for roll flattening
S - spread coefficient; $S = \varepsilon_w/\varepsilon_t$
S_f- forward slip in rolling
w - width of specimen
ε_w- true strain in width direction
ε_t- true strain in thickness direction
μ - coefficient of friction
$\emptyset$ - angle of no slip or neutral angle

Subscripts

0 - original state
1 - final state

INTRODUCTION

Rolling of slabs with high thickness-to-width ratios often results in significant lateral flow (spread). This changes the deformation mode from the usually assumed plane-strain flow to a three-dimensional one. Free spread could seriously affect product quality not only from the final dimensional of point of view but also in terms of integrity of the product (1). Spread causes bulging and development of secondary tensile stresses along the longitudinal edges of the strip which often lead to edge cracking, particularly in the hot rolling of metals of limited ductility or in cases when the ductility of material is exhausted during cold working. There have been a number of attempts to quantify the effect of various process variables on spread. Several empirical models have been developed (2-4) but the very wide range of rolling conditions occurring in industrial hot and cold rolling practice limits their validity. Analytical modeling of spread has proven to be very difficult without unrealistic assumptions. Some analytical techniques (5-7) have been devised but they still cannot cover all variables. Friction remains one of the most unpredictable influences; also, most of the reported work has dealt with isotropic material and, until now, little work has been published on the effect of plastic anisotropy on spread. Interactions between the effects of anisotropy and those of roll pass geometry and friction complicate the picture. The present work aimed at investigating the process experimentally with a view to deducing a semi-empirical formula based on some of the earlier analytical work and with full consideration of the actual friction condition and the effects of plastic anisotropy. In this way, a better understanding of the process can be obtained and perhaps the basis for a fundamental, analytical model can be laid.

EXPERIMENTAL

Materials

Four different materials were investigated; 1020 mild steel, 6061-T6 Al alloy, 5052-H32 Al alloy and commercial purity titanium. The crystal structure of these materials varies from body-centred cubic for mild steel and face-centred cubic for aluminum alloys to hexagonal close-packed for titanium. The 5052 Al alloy was brought into the program to compare its behavior with the 6061 Al alloy in which the second phase precipitated upon age hardening was suspected to give anomalous effects. The 5052 Al alloy is a solid-solution alloy and develops its strength through work hardening.

Plane-Strain Compression Testing

These experiments were planned to give information on through-thickness mechanical properties and on differences in spread development for different materials. The instantaneous L/h ratio was held to less than four to guard against the development of a large friction hill and to greater than two to prevent greatly inhomogeneous deformation.

After experimentation with several lubricants, a lubrication scheme using SAE 5 oil supplied on both faces of a thin (3μm) polyethylene sheet proved to be the most successful. Compression was carried out in increments to allow relubrication of the specimen and to measure the spread profile. The average width was calculated based on the assumption of a parabolic spread profile. A ram velocity of 1 mm/s was maintained for all compression tests.

Uniaxial Tension Testing

Tensile tests were performed on specimens having 50 mm gage length, machined according to ASTM standards from the original plate at three different angles (0, 45, and 90 degrees) to the rolling direction in order to determine the degree and nature of the prevailing plastic anisotropy. Each specimen was stretched by approximately 10%. The plastic strain ratio (r-value) was obtained from the width strain to thickness strain ratio.

Rolling Experiments

Two sets of steel rolls with 150 mm diameter and 200 mm length were used. One set was ground smooth (0.17μm AA) and the other set was roughend (3.1μm AA) by shot blasting. Three basic geometries of rolling specimens were used in order to vary the width-to-thickness ratio (w_0/h_0 = 1, 3, and 6 or 10). A taper was machined on the leading edge for ease of entrance into the roll gap at high reductions. All rolling specimens were finish machined on their edges and then ground on the top and bottom faces to a surface finish of 0.5μm RMS to insure the same surface roughness effects on friction for all experiments.

The process variables investigated were the starting width-to-thickness ratio, roll surface finish, lubrication, and the degree of the homogeneity of deformation as indicated by the h/L ratio.

A lubricant of one percent oleic acid in mineral spirits was used with the smooth rolls. This lubricant was chosen to give boundary lubrication without hydrodynamic contribution which could have been very sensitive to speed. No lubricant was used with rough rolls.

To maintain a constant value of h/L throughout a rolling sequence, roll flattening and mill spring were calculated using the load from the previous pass. After each pass the specimens' average thickness, average maximum width, and average minimum width were measured using a micrometer.

In all rolling experiments, the coefficient of friction was calculated from measurements of forward slip (8).

RESULTS AND DISCUSSION

Anisotropy of Deformation

Since spread is a function of strain, the strain ratio r is an obvious parameter to evaluate material effects. If the tensile strain ratios were transferable to rolling, then a material with r = 0 would be expected to show no spread, whereas a material with r>1 should spread more than an isotropic material. There is, however, no evidence or theoretical basis for a direct correlation between tensile and compressive r-values. Likewise, there is no information on how the rigid material adjacent to the deformation zone affects spread of an anisotropic material in a typical rolling situation.

Comparison Between Strain Ratios from In-Plane Tension and In-Plane Compression

The measured r-values in uniaxial tension ranged from less than 0.5 to greater than 3.5 depending on the material and the direction of testing (Table 1). The cubic metals (fcc and bcc) showed less plastic anisotropy than titanium (hcp), titanium showed also substantial planar anisotropy. Only mild steel (bcc) had $r_0 \simeq r_{45} \simeq r_{90}$.

TABLE 1

Comparison between strain ratios obtained from in-plane tension and in-plane compression

MATERIAL	TENSION			COMPRESSION*	
	r_0	r_{45}	r_{90}	r_{c0}	r_{c90}
6061-T6 Aluminum	0.40	0.62	0.42	0.45	0.59
5052-H32 Aluminum	0.33	1.06	0.84	0.85	0.78
Mild Steel	0.81	0.8	0.77	0.80	0.80
Titanium	1.19	-	3.55	1.15	1.31

*Data from experimental work of Chitkara (9).

The strain ratio in compression was defined by $r_c = \varepsilon_{90}/\varepsilon_0$. Cubes of the investigated materials were compressed by Chitkara (9) along different axes to determine the strain ratio parallel to the rolling direction r_{c0} and perpendicular to the rolling direction r_{c90} (Table 1). Comparison of the r-values in the plane of the sheet obtained in tension and compression (Table 1) indicates that the values are generally consistent except for titanium which has $r_{c0} \simeq r_{c90}$ perhaps because of twinning during compression. Thus, strain ratios obtained from in-plane compression tests, under conditions of low friction, are similar in magnitude to conventional tensile r-values for cubic materials but can be very different for hexagonal materials.

Spread in Plane-Strain Compression

Width strain during frictionless plane-strain compression was found to be a function of workpeice material and of starting width. A spread parameter suggested by Hill (10) $S = \varepsilon_w/\varepsilon_t$ was calculated for each experiment and plotted against w_0/L for different materials (Fig. 1). Spread showed a strong dependence on the geometrical parameter w_0/L and could be approximated by an exponential relationship of the form:

$$S = A \exp \left(B\frac{w_0}{L} \right) \tag{1}$$

where the constant A is a function of the material type and takes values ranging from 0.5 to 1.0 while the constant B is material insensitive and takes a value approaching (-0.5).

As expected from the r-values (Table 1), the spread of 6061-T6 Al was the low-est (Fig. 1). The spread of titanium was close to that of steel even though r_c-values in both directions were higher than for steel. This indicates that crystal structure is also important, and r-values for hexagonal metals cannot be used to predict the strain behavior in compression. If titanium is repre-sentative of hcp metals with low c/a ratios then it is suggested that the spread of these metals will be similar to that of an isotropic material during compression.

Spread in Rolling

Spread during rolling is affected by geometric, friction, and material parameters.

(i) Effects of Geometry on Spread. Spread increases with Δh, R/h and decreases with increasing strip width or, rather, w/L ratio which defines the relative size of the deformation zone. As in simple forging with overhanging platens, the ratio w/L governs to a large degree the direction of material flow (11). In rolling, in the presence of the nondeforming material adjacent to the deform-ation zone, most of the widening of the strip takes place close to the neutral plane where there is less restraint from the rigid-nondeformed material (Fig. 2). When w/L > 1, spread decreases with increasing w/L (Fig. 3) in a manner similar to that observed in plane-strain compression (Fig. 1). In rolling, however, there are some differences which must be explained: the spread curve for roll-ing is lower, is not monotonic, and shows greater curvature for w/L > 1.

The most likely explanation involves the angle of inclination caused by the curvature of the rolls, as seen more clearly by examining Fig. 4a and 4b in which each curve represents spread for a constant value of inclination, expressed as $\Delta h/L$.   The curves rise continuously as w/L decreases, unlike the curve for constant h/L in Fig 3. It will be noted that inclination has a very strong effect on spread when w/L is small but diminishes in importance as w/L increases. The curves also show that higher inclination decreases spread. By comparing the differences in spread between the first and second passes, it is clearly evident that changes in inclination could cause the peak in the spread curve with constant h/L. With h/L = 0.3, the value of $\Delta h/L = 0.20$, for the first pass, while for the second pass $\Delta h/L = 0.11$. With h/L = 0.5, the value of $\Delta h/L = 0.15$ for the first pass and $\Delta h/L = 0.10$ for the second pass. The difference in inclinations for the first and second passes for h/L = 0.3 is approximately twice that for   h/L = 0.5. This explains why the effect is much more dramatic for   h/L = 0.3.

(ii) Effects of Friction. Values of the coefficient of friction calculated from measurements of forward slip during rolling different materials are summarized in Table 2.

TABLE 2

Coefficients of friction μ calculated from measured forward slip, S_f, under conditions of homogeneous deformation (h/L = 0.3) and close to plane strain deformation (w/h = 6 or 10)

MATERIAL	SMOOTH ROLLS-LUBRICATED REDUCTION %				ROUGH - DRY REDUCTION %			
		25	20			25	20	
	40	40	25	20	40	40	25	20
6061-T6 Al	0.15	0.13	0.12	0.10	0.37	0.31	0.31	-
5052-H32 Al	0.16	0.15	0.13	0.11				
Mild Steel	0.10	0.10	0.08	0.08	0.39	0.39		
Titanium	0.27	0.27	0.23	0.15				

All results indicate that spread increases dramatically with increasing friction when w/L is small but, for w/L > 3, the effect becomes less pronounced (Fig. 5). For rolling with smooth rolls the differences in spread for h/L = 0.5 and h/L = 0.3 are significant, but rolling with rough rolls completely masks this effect; the curves for h/L = 0.3 and h/L = 0.5 are the same for high friction.

Development of spread along the arc of contact has been studied experimentally by Klimenko and Grashcenko (12). They concluded that spread occurs primarily in the backward slip zone. An increase in friction causes the neutral plane to shift towards the entry plane which shortens the zone of backward slip and lengthens the zone of forward slip. The frictional forces along the longer forward slip zone retard the forward flow and force the material to extrude sideways. The length of forward slip zone is thus considered a valuable quantity in predicting spread, especially under conditions of sliding friction at the roll-workpiece interface.

(iii) Effect of Material Properties. In general, the aluminum alloys exhibited less spread than the mild steel, as was found also in plane-strain compression tests (Fig. 6). The low r-value in the 0-direction causes the 6061-T6 and 5052-H32 Al alloys to have less spread in this direction.

Greater spead was measured for titanium compared to mild steel (Fig. 6). At first glance one would attribute this to the higher r-value of titanium, but this interpretation is contradicted by the plane-strain compression results (Fig. 1). The effect is more likely due to differences in friction. Friction in rolling titanium was more than double that of friction in rolling steel, even though the same condition were maintained for each. Titanium gives higher adhesion to the steel rolls, hence friction was also higher. Rolling of mild steel on rough rolls with $\mu = 0.39$ gave similar spread as rolling of titanium on smooth rolls with $\mu = 0.27$.

DEVELOPMENT OF SPREAD MODEL

The discussion in the previous section suggests that inclination ($\Delta h/L$) and w/L are influential geometric parameters and must be included in any spread model. Frictional effects must be quantified and included in the model as well. The approach of Bakhtinov (13) was to introduce the length of the forward slip zone L_f. This quantity contains within it both μ and Δh and, if divided by L, a nondimensional parameter L_f/L takes the form:

$$L_f/L = 0.5 - \Delta h/4\mu L \tag{2}$$

which contains μ and the angle of inclination at the plane of entry ($\Delta h/L$). Note the limiting values of L_f/L: When friction is high, L_f/L approaches 0.5, i.e., the neutral plane in the middle of the roll gap; when friction is low or inclination is high, L_f/L approaches zero, i.e., the neutral plane is at the exit, which represents the commencement of skidding. The minimum friction condition ($L_f/L = 0$) does not correspond to zero spread. In this case, the workpiece is drawn into the roll gap, with spread occurring up to the point where skidding begins. One way of satisfying this condition in a spread model is to include the factor L_f/L in an exponential term multiplying the exponential term containing w/L (the exponential relationship between spread and the parameter w/L was confirmed before from the results of plane-strain compression tests, Eqn. (1), where there was no angle of inclination and interface friction was negligible).

Accordingly, the spread model is simply:

$$S = \frac{\varepsilon_w}{\varepsilon_t} = A \exp \left[B\, \frac{w}{L} + C\, \frac{L_f}{L} \right] \tag{3}$$

where A, B, and C are constants, with values depending on the properties and structure of the workpiece material.

Table 3 shows the values of coefficients A, B, and C obtained from regression analysis of the experimental results according to Eqn. (3).

As shown in Fig. 7, values of spread calculated according to Eqn (3) are in close agreement with experimental values for all tested materials.

TABLE 3

Summary of coefficients determined from first-pass data. Smooth rolls;
h/L = 0.5 and 0.3

MATERIAL		COEFFICIENTS		
		A	B	C
6061-T6	0°	0.023	-0.630	10.349
	90°	0.017	-0.646	12.439
5052-H32	0°	0.005	-0.913	17.605
	90°	0.013	-0.922	15.063
Mild Steel	0°	0.286	-0.563	3.347
Titanium	0°	1.187	-1.147	-1.197
	90°	0.020	-0.923	9.921

CONCLUSIONS

1. High r-value hexagonal metals have similar spread characteristics to
    isotropic metals, but for cubic metals with r less than or equal to unity,
    spread generally decreases with decrease in r.

2. The rigid material adjacent to the deformation zone affects spread in the
    same way, no matter what material is being tested, as long as the parameter
    w/L is kept constant.

3. Inclination is another important geometric parameter. It affects the
    position of the neutral plane and thus it should appear in a spread model,
    in conjunction with friction, as a non-dimensional factor L_f/L.

4. There appears to be an upper limit to the effect of friction which occurs
    when the natural plane is situated at the center of the arc of contact.

5. The proposed semi-empirical model agrees well with experimental results
    obtained on different materials.

ACKNOWLEDGEMENTS

The authors are indebted to the National Science and Engineering Research
Council of Canada for financial support of this work and to George Roberts for
assistance with the experiments. The authors are also grateful to
Ms. Jennie Hale for typing the manuscript.

REFERENCES

1. J. A. Schey, Workability in Rolling, in: G. Dieter (Ed.), Workability
    Testing Techniques, ASM, Ohio, 1984, p. 269.
2. L. G. Sparling, Formula for Spread in Hot Flat Rolling, Proc. Inst. Mech.
    Engr, 175, 1961, p. 604.
3. A. Helmi and J. M. Alexander, Geometric Factors Affecting Spread in Hot
    Rolling of Steel, J. Iron Steel Inst. (London), 206, 1968, p. 1110.
4. N.R. Chitkara and W. Johnson, Some Experimental Results Concerning Spread
    in the Rolling of Lead, Trans, ASME, Basic Eng., 88, 1966, p. 489.
5. R. Kummerling and H. Lippmann, On Spread in Rolling, Mechanics Research
    Communications, 2, 1975, p. 113.

6. L. X. My, G. D. Lahoti, J. T. Black and T. Altan, Development and Evaluation of an Interactive Graphics Program for Plate Rolling, Mech. Working Tech., 4, 1980, p. 105.
7. K. F. Kennedy and W. R. D. Wilson, A Refined Analytical Model for the Prediction of Side Spread in the Rolling of Flat Products, Proc. 10th NAMRC, SME, 1982, p. 149.
8. J. A. Schey, Tribology in Metal Working: Friction, Lubrication and Wear, ASM, Ohio, 1983, p. 249.
9. N. R. Chitkara, Unpublished Work: University of Waterloo, 1981.
10. R. J. Hill, A General Method of Analysis of Metal Working Processes, J. Mech.Phys. Solids, 11, 1963, p. 305.
11. J. A. Schey, Introduction to Manufacturing Processes, 2nd Ed., Mc-Graw-Hill, 1987, p. 221.
12. P. L. Klimenko and V. I. Garashcenko, Generalized Functions for Describing the Change of Width and Stretching in the Rolling Process, Metallurgiya I. Kikoskhimiya, 60, 1979, p. 19
13. Z. Wusatowski, Fundamentals of Rolling, Pergamon Press, 1969.

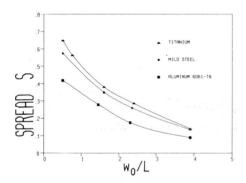

Fig. 1 Spread in plane-strain compression of different materials

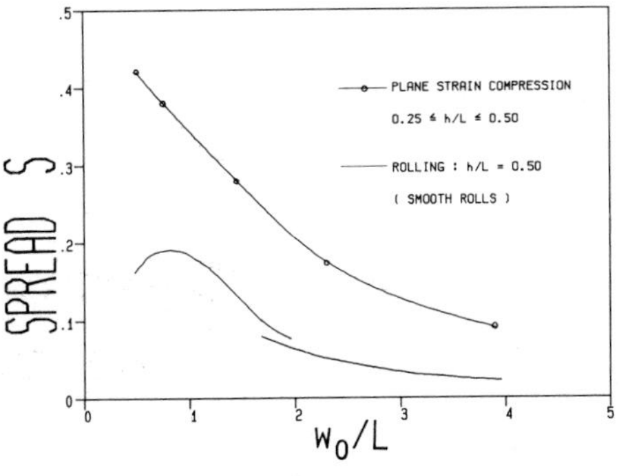

Fig. 3 Comparison of spread in plane-strain compression with that in rolling for 6061-T6 Al alloy.

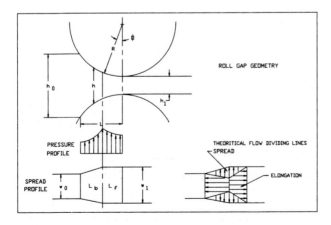

Fig. 2 Features of rolling process.

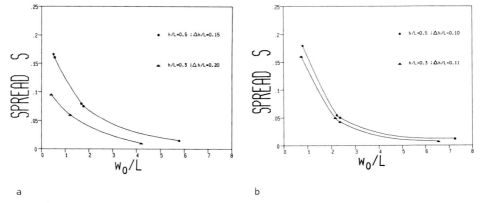

a   b

Fig. 4   Spread curves of 6061-T6 Al alloy, smooth lubricated rolls.

a - first pass rolling data   b - second pass rolling data

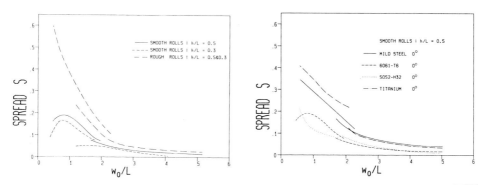

Fig. 5   Comparison of spread in rolling 6061-T6 Al alloy on smooth
and rough rolls.

Fig. 6   Comparison of spread in rolling of different materials at
0-degree with the original
rolling direction.

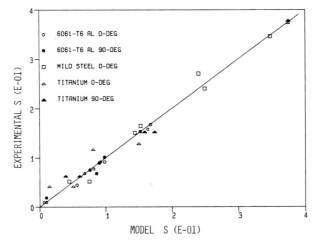

Fig. 7   Comparison between
experimental and calculated
values of spread according
to Eqn. 3.

Computational Methods for Predicting Material Processing Defects, edited by M. Predeleanu
Elsevier Science Publishers B.V., Amsterdam, 1987 — Printed in The Netherlands

NUMERICAL ANALYSES OF STRAIN RATE AND TEMPERATURE EFFECTS ON LOCALIZATION OF
PLASTIC FLOW AND DUCTILE FRACTURE - APPLICATIONS TO METAL FORMING PROCESSES

J.C. GELIN - Laboratoire de Génie Mécanique - Université de Valenciennes
F 59326 VALENCIENNES
et GRECO CNRS Grandes Déformations et Endommagement

ABSTRACT
        The analysis of damage in rate sensitive and temperature sensitive materials is of a primary importance to be able to predict the loss of strength capacity, due to localization of plastic flow or due to ductile fracture.
This paper gives an extension of the Gurson model to deal with rate sensitive and temperature effects. The model, which is mainly based on growth and coalescence of voids is applicable in warm conditions where the steels begin to exhibit a viscoplastic behaviour.
        A large strain implicit algorithm is presented for the integration of the constitutive equations and the elastic-viscoplastic resulting finite element analysis is shown to be inconditionally stable.
        Numerical applications are given for industrial problems. First application concerns the analysis of fracture in upsetting tests using a plastic flow localization criterion. The second example concerns an axisymmetric wire drawing problem, where the influence of strain rate and temperature increasing are analysed.

INTRODUCTION

        The analysis of metal deformation at warm temperature involves strain rate effects and temperature effects. It is well known that the temperature has a significant importance on plastic flow and can lead to a softening of the material. This is principally due to the dislocations motions, which are controlled by a large variety of microstructural effects, some of these effects are thermally dependent and lead to an increasing dependence of plastic flow with the temperature.

        The most important data known up to day concerning temperature effects, come from uniaxial tensile tests, torsion or compression tests, see Lemaître and Chaboche [1]. These results are not sufficient to give a very good knowledge of hardening and softening effects, but in a first approximation the generalization of results to multiaxial stress states and to large strains is straightforward. An experimental procedure based on upsetting tests and tensile tests have been developped by the author [2] to obtain a good representation of the flow stress versus effective strain, effective strain rate and temperature.

        From the numerical point of vue, strain rate effects have been described in the context of sheet metal forming by Hutchinson and Neale [3] in an analysis of the growth of long wavelength nonuniformities. Another approach using a modified form of the Gurson potential have been used to describe localized necking in biaxially streched high-strength metal sheets by Needleman and Tvergaard [4].

Constitutive equations involving strain rate effects and temperature effects, in conjunction with a softening effect due to the growth of microscopic voids have long been known to be numerically very stiff. But recently, various authors, see for example Pierce and al. [5], Gelin [6], Lush and Anand [7], Gelin [8], have proposed implicit time integration procedures which allow large time increments without encountering stability problems. These implicit procedures have been applied to the case of voids containing materials, see Needleman and Tvergaard [9], Gelin [6], to analyse the influence of strain rate on ductile fracture.

In this paper, we present a large strain implicit algorithm for the integration of the constitutive equations above described, with natural respect of material objectivity when large rotations occur. A fully implicit procedure is used which necessitates an iterative technique to compute the end state from the initial state. We also present a criterion for plastic flow localization in a void containing material.

Then numerical applications are given in the field of industrial problems. First application concerns the analysis of fracture in upsetting tests, using a plastic flow localization criterion when the strain paths are known at the surface of the cylinder. The other example concerns an axisymmetric wire drawing problem, where the influence of strain rate and temperature increasing are analysed.

CONSTITUTIVE EQUATIONS

Attention in this paper is focussed on modeling isotropic metal with isotropic hardening for which the state variables are the Cauchy stress tensor σ, an isotropic hardening parameter α, the absolute temperature θ, and the void volume fraction f_v. Starting from the definition of free energy potential, we can write, from the assumption that elastic and hardening-damage parameters are uncoupled

$$\psi = \psi^e(e,\theta) + \psi^p(\alpha, f_v, \theta) \tag{1}$$

where ψ^e is the portion of ψ associated with reversible transformations, while ψ^p is associated with irreversible transformations, in equation (1) e is an elastic strain measure to be defined. From equation (1), it results immediatly

$$\sigma = \rho \frac{\partial \psi}{\partial e} = \rho \frac{\partial \psi^e}{\partial e} (e, \theta) \tag{2}$$

In the case of small elastic strains, the elasticity tensor C^e is generally written in the form

$$C^e = 2G(\theta) (II - \frac{1}{3} 1 \otimes 1) + K(\theta) 1 \otimes 1 = \rho \frac{\partial^2 \psi^e}{\partial e \partial e} \tag{3}$$

where $K(\theta)$ and $G(\theta)$ are the classical bulk modulus and shear modulus (dependent on temperature).

The viscoplastic strain rate, noted as D^p is given by the following relation

$$D^p = \dot{\gamma} \frac{\partial \phi}{\partial \sigma} \tag{4}$$

where ϕ is a viscoplastic potential depending on Cauchy stress tensor invariants, hardening and void volume fraction. A simple way in the case of pressure dilatant material is to use a relation of the following type :

$$\phi = \phi (J_1, J_2', g(\alpha), f_v) \tag{5}$$

where

$$J_1(\sigma) = \mathrm{tr} \ \sigma$$

$$J_2'(\sigma) = \frac{1}{2} \|s\|^2 = \frac{1}{2} \|\sigma - \frac{1}{3} \mathrm{tr} \ \sigma \ 1\|^2$$

$g(\alpha)$ is the stress associated to α

We obtain immediatly from the macroscopic normality rule on the aggregate

$$D^p = \dot{\gamma} (\frac{\partial \phi}{\partial J_2'} \frac{\partial J_2'}{\partial s} + \eta 1) \tag{6}$$

where $\frac{\partial \phi}{\partial J_2'} \frac{\partial J_2'}{\partial s}$ is a deviatoric term, and η is a term depending on $J_1 = \mathrm{tr} \ \sigma$.

Generally $\dot{\gamma}$ is a function of the internal variables defining the system at time t, such that

$$\dot{\gamma} = \dot{\gamma}(g(\alpha), \theta, f_v) \tag{7}$$

this parameter is the equivalent viscoplastic strain rate on the aggregate in opposite to $\dot{\alpha}$ on the matrix.

From the equivalence between plastic work dissipated on the matrix and on the aggregate we can write

$$\sigma : D^P = \sigma : \dot{\gamma} \frac{\partial \phi}{\partial \sigma} = (1 - f_v) \, g(\alpha,\theta)\dot{\alpha}$$

thus it results immediatly :

$$\dot{\gamma} = (1 - f_v) \, g(\alpha,\theta)\dot{\alpha} \, [\sigma : \frac{\partial \phi}{\partial \sigma}]^{-1} \tag{8}$$

$\dot{\gamma}$ is known as a function of void volume fraction and current state of stress.

The rate of void volume fraction is given by the conservation of mass law and we have

$$\dot{f}_v = (1 - f_v) \, \mathrm{tr} \, D^P = 3 \, \dot{\gamma} \, (1 - f_v) \, \frac{\partial \phi}{\partial J_1} \tag{9}$$

Furthermore it is assumed that most of the plastic work accumulated by a material particle in the matrix is dissipated in the form of heat. Assuming that the process is quasi-adiabatic, this dissipation can be expressed as :

$$\dot{\theta} = \frac{\xi}{\rho C_v} \, g(\alpha,\theta)\dot{\alpha} \tag{10}$$

where ρ and C_v are respectively, the mass density and specific heat at constant volume, ξ represents the fraction of plastic work which is converted directly into heat (generally this coefficient is in the range 0.9 up to 1.0).

Equations (8), (9) and (10) represent the evolution equations for the above described constitutive model.

To describe completly the model it is now necessary to choose a visco-plastic potential, the modified Gurson [10] potential can be choose in this way, and

$$\phi = \frac{3J'_2(\sigma)}{[g(\alpha,\theta)]^2} + 2 \, f_v q_1 \, \cosh \, [\frac{J_1(\sigma)}{2g(\alpha,\theta)}] - [1+q_3 f_v^2] = 0 \tag{11}$$

where $J_1(\sigma)$ and $J'_2(\sigma)$ have been previously described. The parameters q_1 and q_3 were introduced by Tvergaard [11] who remarked that the critical strains obtained when localization of plastic flow take place are over-estimated by the original Gurson yield function, these coefficients have been used by Needleman and Tvergaard [4] in an analysis of limits to formability in rate sensitive metal sheets.

Having described the viscoplastic potential, it is now necessary to know the evolution equation for $\dot{\alpha}$ which is the hardening rate of the matrix.

A very useful constitutive function is the power law function

$$\dot{\alpha} = \dot{\alpha}_o \, [\frac{g(\alpha)}{g(\alpha_o)}]^{1/m} \tag{12}$$

where $g(\alpha)$ is the flow strength of the matrix, m is the strain rate sensivity exponent and $g(\alpha_0)$ is the flow strength corresponding to the reference strain rate $\dot{\alpha}_0$.

The influence of temperature on flow strength of the material can be expressed with an exponential function of the following type :

$$g(\alpha,\theta) = g_0(\alpha) \ (\theta/\theta_0)^p \tag{13}$$

where p is the temperature sensivity exponent (p<o), θ_0 is a reference temperature.

Then, we can write the flow strength of the matrix on the following form

$$g(\alpha,\theta) = g_0(\alpha) \ \left(\frac{\dot{\alpha}}{\alpha_0}\right)^m \ \left(\frac{\theta}{\theta_0}\right)^p \tag{14}$$

Now, the evolution equations are entirely known as functions of current state of stress and materials parameters.

ANALYSIS OF PLASTIC FLOW LOCALIZATION

Employing the above constitutive equations, an analysis of plastic flow localization in shear bands can be led, based on the three dimensional theory of shear band localization, see for example Rice [12], Needleman and Rice [13].

Only the basic concepts are recalled here. Consider a volume of material subjected to a deformation gradient $F^0 = 1 + \nabla u^0$.

Consider a narrow band of material where the deformation gradient takes the form

$$F^b = F^0 + g \otimes n \tag{15}$$

where n is the unit normal to the narrow band, $(\ )^b$ denotes quantities within the band, and $(\ )^0$ denotes quantities outside the band.

The rate version of equation (15) is immediatly given by

$$\dot{F}^b = \dot{F}^0 + \dot{g} \otimes n \tag{16}$$

Compatibility conditions require that equilibrium must be satisfied accross the band, such that

$$n.T^0 = n.T^b \Rightarrow n.\Delta T = 0 \tag{17}$$

where T is the first Piola Kirchhoff stress tensor.

To proceed further it is interesting to write equation (17) in an incremental form as :

$$n \dot{\Delta T} = 0 \tag{18}$$

Writing the constitutive equations in an incremental form, the equation (18) becomes

$$n.H^b : \dot{F}^b - n.H^o : \dot{F}^o = 0$$

or

$$(n.H^b.n).\dot{g} = n.(H^o - H^b) : \dot{F}^o \qquad (19)$$

where H^b and H^o are the fourth order tensors that relate $\dot{T}^b$ to $\dot{F}^b$ and $\dot{T}^o$ to $\dot{F}^o$.

The condition $\det(n.H^b.n)=0$ is the localization condition when the material is homogeneous, e. g. when localization takes place from homogeneous state.

When an imperfection is assumed in the band, it is more convenient to use an elastic unloading condition outside the band as localization condition.

NUMERICAL IMPLEMENTATION OF VISCOPLASTIC CONSTITUTIVE EQUATIONS

The equations governing the viscoplastic behaviour have been described previously, see equations (8), (9), (10), (11) and (14). From numerical point of vue, we have to compute the Cauchy stress tensor and internal parameters that satisfy the constitutive equations.

We proceed from an updated Lagrangian manner. If F_n is the total gradient of deformation at the start of a loading increment, the displacement finite element method consists to calculate u between time t_n and time $t_n + \Delta t = t_{n+1}$. We have the relation :

$$F_{n+1} = F_u F_n = (1 + \frac{\partial \Delta u}{\partial x_n})F_n \qquad (20)$$

First, an elastic stress state prediction is made, in using the basic multiplicative decomposition relation $F=F^eF^p$,

$$F^{p(o)}_{n+1} = F^p_n \text{ and } F^{e(o)}_{n+1} = F_u F_n F^{p^{-1}}_{n+1} \qquad (21)$$

Then, a trial elastic strain is deduced from (21)

$$e^{(o)}_{n+1} = \frac{1}{2} \ln B^{e(o)}_{n+1} \qquad (22)$$

and we have immediatly

$$\tau^{(o)}_{n+1} = \rho^o \frac{\partial \psi^e}{\partial e} \Rightarrow \sigma^{(o)}_{n+1} = J^{-1}_{n+1} \tau^{(o)}_{n+1} \qquad (23)$$

From the first step of the algorithm, and due to the fact that the current configuration remains fixed, we have

$$D = D^e + D^p = 0 \qquad (24)$$

and it results immediatly, if C^e is the elasticity tensor :

$$\frac{D\sigma}{Dt} = - C^e : D^p = - C^e : \dot{\gamma} \frac{\partial \phi}{\partial \sigma} \qquad (25)$$

where $\frac{D\sigma}{Dt}$ is the convective contravariant derivative of the Cauchy stress tensor.

The rate of change of the viscoplastic potential ϕ is given by :

$$\frac{\partial \phi}{\partial t} = \frac{\partial \phi}{\partial \sigma} : \frac{\partial \sigma}{\partial t} + \frac{\partial \phi}{\partial g} \dot{g} + \frac{\partial \phi}{\partial f_v} \dot{f}_v + \frac{\partial \phi}{\partial \theta} \dot{\theta} \qquad (26)$$

it is exactly the counterpart of the consistency condition in rate independent plasticity.

From equations (8), (9), (10) it results that

$$\frac{\partial \phi}{\partial t} = \dot{\gamma}[- \frac{\partial \phi}{\partial \sigma} : C^e : \frac{\partial \phi}{\partial \sigma} + \frac{\partial \phi}{\partial g} h_1 + \frac{\partial \phi}{\partial f_v} h_2 + \frac{\partial \phi}{\partial \theta} h_3] = \dot{\gamma} A \qquad (27)$$

where h_1, h_2 and h_3 are the following quantities :

$$h_1 = g'(\alpha) [\sigma : \frac{\partial \phi}{\partial \sigma}] \{(1-f)g(\alpha)\}$$

$$h_2 = 3(1-f_v) \frac{\partial \phi}{\partial J_1} \qquad (28)$$

$$h_3 = \frac{\xi}{\rho C_v} g(\alpha) [\sigma : \frac{\partial \phi}{\partial \sigma}]/\{(1-f)g(\alpha)\}$$

Following the work of Simo and Taylor [14], we can define a relaxation time parameter by the expression

$$\bar{t} = (\phi/\dot{\gamma})/A \qquad (29)$$

and we obtain immediatly from equation (26) :

$$\frac{\partial \phi}{\partial t} + \frac{\phi}{\bar{t}} = 0$$

The above equation is integrated along an iteration and we obtain

$$\phi_{n+1}^{(i+1)} = \phi_{n+1}^{(i)} \exp (-\Delta t/\bar{t}) \qquad (30)$$

so that

$$\Delta t = \bar{t} \ln \frac{\phi_{n+1}^{(i)}}{\phi_{n+1}^{(i-1)}}$$

The end of integration scheme is when $\Sigma \Delta t^{(i)} = \bar{\tau}$ where $\bar{\tau}$ is the time step

length previously fixed. This algorithm have been implemented in our finite
element program and the results obtained on elementary tests are inconditio-
naly stable with respect to the time step.

NUMERICAL RESULTS

Ductile fracture at the equatorial surface of a compressed cylinder

Figure 1a shows the typical geometry of such a specimen where a crack
is generally observed on the equatorial surface of the deformed cylinder. The
equatorial strain paths have been computed using an elasto-plastic finite element approach and figure 2 shows the typical equatorial strain paths obtained,
corresponding at two friction conditions between the dies and the specimen
(dry friction case and sticking friction case).

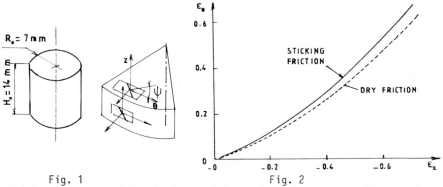

Fig. 1
Initial geometry and initial
mesh for the finite element
simulation

Fig. 2
Equatorial strain paths corresponding to the
upsetting test.

Using the plastic flow localization approach we have studied the influence of material parameters on the circumferential strain at fracture obtained by plastic flow localization. The initial void volume fraction f_0 is e-
qual to $1/5 \ 10^{-5}$ and the flow strength of the material is given by $\sigma_0 = 872 \bar{\epsilon}^{-0.09}$
$\bar{\epsilon}^{-0.012} (\theta/\theta_0)^{-4.2}$, with $\theta_0 = 450°C$, $\dot{\bar{\epsilon}}_0 = 10^{-2} s^{-1}$. An imperfection $\Delta f_0 = 10^{-3}$ is assumed in a narrow band inclined from an angle ψ relatively to x_1.

In figure 3, it is clear that the increasing of strain rate sensitivity
exponent increases the equatorial strain at fracture. The influence of temperature exponent is also clear, a low temperature exponent leads to an increasing
of circumferential strain at fracture and an high temperature exponent leads
to a decreasing of circumferential strain at fracture.

Development of damage in wire drawing process

A typical wire drawing process have been studied with the model described in this paper. The geometry and the finite element mesh are shown in figure 4. The semi-cone angle is 11,30° and reduction in area is 44 %.

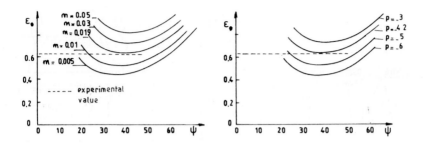

Fig. 3. Influence of strain rate sensivity exponent and temperature exponent on circumferential strains at fracture.

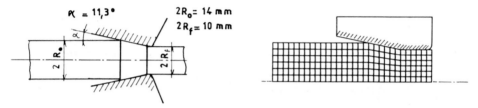

Fig. 4. Geometry and finite element mesh used for the numerical simulation of wire drawing.

The computation have been carried out in using two different drawing speeds, 0.05ms^{-1} and 1m s^{-1}. The flow strength of the material is given by the relation $g(\alpha) = 872 \, \alpha^{0.09} \, \alpha^{0.012}$. The deformed mesh corresponding to a stroke equals $0.4 \, r_f$ and $1 \, r_f$ are shown in figure 5(a,b).

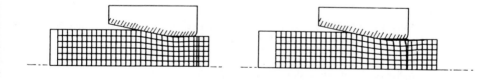

Fig. 5. Deformed mesh corresponding to various drawing lengths.

The axial stress distribution and the development of void volume fraction are shown in figure 6. Fig. 6a corresponds to the axial stress distribution, Fig. 6b corresponds to the distribution of void volume fraction

for the drawing speed equal $0.05ms^{-1}$ and Fig. 6c corresponds to the distribution
of voids volume fraction corresponding to the drawing speed equal $1ms^{-1}$.

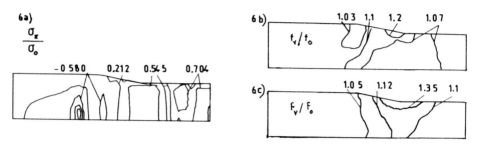

Fig. 6. Distribution of axial stress (6a) and distribution of void volume
fraction (6b, $v=0.05ms^{-1}$, 6c-$v=1.0ms^{-1}$)

REFERENCES

[1] J. Lemaitre, J.L. Chaboche - Mécanique des Matériaux Solides. Dunod,
Paris, 1985.

[2] J.C. Gelin, J. Oudin, Y. Ravalard - New upsetting and tensile testing procedures for metals in warm or hot forging conditions, Annals of the C.I.
R.P., 33(1) (1984) 155-159.

[3] J. W. Hutchinson, K. W. Neale - Sheet Necking III. Strain rate effects.
In : D.P. Koistinen and N.M. Wang (Eds.), Mechanics of Sheet Metal Forming,
1978, 269-283.

[4] A. Needleman, V. Tvergaard - Limits to formability in rate sensitive metal
sheets, In : J. Carlsson and N.G. Ohlson, Mechanical Behaviour of Materials,
Pergamon Press, Oxford, 1984, 51-65.

[5] D. Pierce, C.F. Shih and A. Needleman - A tangent modulus method for rate
dependent Solids, Comput. and Structures, 18(5), (1984) 875-887.

[6] J.C. Gelin - Application of a thermo-viscoplastic model to the analysis
of defects in warm forming conditions, Annals of the C.I.R.P., 35(1) (1986).

[7] A. Lush, L. Anand - Implicit time-integration procedures for a set of internal variable constitutive equations for hot working. In : K. Mattiasson
and al., Numerical Methods in Industrial Forming Processes, A.A. Balkema-
Rotterdam, 1986, 131-137.

[8] J.C. Gelin - Application of implicit methods for the analysis of damage
with temperature effects in large strain plasticity problems. In : C.
Taylor and al., Numerical Methods for Non-Linear Problems, Pineridge Press,
1986, 494-507.

[9] A. Needleman, V. Tvergaard - Material strain rate sensivity in round tensile bar, In : J. Salençon (Ed.), 1985, 251-261.

[10] A.L. Gurson - Continum theory of ductile rupture by void nucleation and
growth : Part I - Yield criteria and flow rules for porous ductile media.
Trans. A.S.M.E., J. of Eng. Mat. Tech., 99 (1977), 2-15.

[11] V. Tvergaard - On localization in ductile materials containing spherical
voids. Int. J. Fracture, 18 (1982), 237-252.

[12] J.R. Rice - The localization of plastic deformation. In : W.T. Koiter
(Ed.), Theoretical and Applied Mechanics, North Holland, 1 (1976) 207-
220.

[13] A. Needleman, J.R. Rice - Limits to ductility set by plastic flow localization. In : D.P. Koistinen and N.M. Wang, Mechanics of sheet Metal Forming, Plenum Press, (1978) 237-264 .

[14] J.C. Simo, M. Ortiz - A unified approach to finite deformation elasto-
plastic analysis based on the use of hyperelastic constitutive equations.
Comput. Meth. Appl. Mech. Engng., 49, (1985) 221-245.

Computational Methods for Predicting Material Processing Defects, edited by M. Predeleanu
Elsevier Science Publishers B.V., Amsterdam, 1987 — Printed in The Netherlands

SOME PHYSICAL DEFECTS ARISING IN LASER, PLASMA AND WATER JET CUTTING

S K GHOSH[1], J E BEITIALARRANGOITIA[2] and G E GARCIA DE VICUNA[3]

[1,2,3]Department of Mechanical and Computer-Aided Engineering, North
Staffordshire Polytechnic, Beaconside, Stafford, ST18 0AD (England)

SUMMARY
    Nowadays several modern technologies are employed in wide areas of
manufacturing due to the new and extended possibilities and advantages offered
over conventional techniques. Laser, plasma and water jet can be considered
some of these advanced manufacturing techniques which offer the possibility
of dealing with almost any kind of material.

    In the cutting and drilling processing of material with these technologies
there are many variables and factors which are difficult to be controlled and
have to be optimised to obtain the required component quality.

    In this paper a review is presented of the most common effects/defects
that appear in the manufacturing processes with each of the aforementioned
techniques. Discussion is also given on numerical actions necessary either
to avoid or partially eliminate the presence of undesired defects in the
finished products, as far as possible.

INTRODUCTION

    A brief introduction to different technologies is given first, before
discussing the effects/defects that arise in material processing using these
techniques.

    A laser is a source of light which is nearly monochromatic, nearly
unidirectional, has a spatial distribution imposed by the optical resonator
and can be very energetic. The main characteristics are the beam intensity
and the exposure time, which determine the specific application. The laser
beam is a very precise working tool of high flexibility free from wear-out;
see Figure 1, (ref 1-3). Laser cutting produces narrow heat-affected zones
and is appropriate for fast accurate cutting of complex shapes.

    The process of plasma arc is based on a high powered electric arc between
the plasma torch and the workpiece which melts or vapourises the metal to be
cut. The molten material is blown away with a high pressure gas jet. Its
choice (O_2, N_2, H/Air mix or air) depends on the equipment used and the material
to be cut. It is capable of cutting any material electrically conducting
and there is no minimum thickness, only the cut edge quality is deteriorated
with reduce thickness; see Figure 2 (ref 4). The kerf produced in plasma arc
cutting is always cone-shaped, narrowing at the bottom of the cut and the
top edge is slightly radiussed.

    In water jet cutting a high-velocity water jet strikes the surface of its target in such a way that its velocity is virtually reduced to zero, converting most of its kinetic energy in pressure energy. Erosion will occur if the local fluid pressures exceed the strength of the target material (ref 30).

LASER PROCESSING

The basic factors to be determined in an analysis of the power balance in laser-material interactions are the fraction of laser beam power absorbed, the material ejected from the specimen as a consequence of the absorbed power and the fraction of material removed as molten material (ref 5). Material removal begins with vapourisation; as the vapourisation proceeds, the absorption of the laser beam increases rapidly as the depth of the crater increases. The temperature increases rapidly, and at the same time, the thermal diffusion process begins as the local temperature rises. The behaviour of material removal in the liquid phase depends on characteristics of the pulse, and material constants such as the thermal conductivity, the difference between vapourisation and melting temperatures, and the ratio of the heats of vapourisation and liquifaction (ref 6).

Drilling

Drilling takes place generally by repetitive pulsing using from one to a number of pulses per hole. The material inside the hole melts and is removed by vapourisation, which makes the process violent and, therefore difficult to control (ref 7).

For expulsion of metal in liquid form, a mechanism involving radial liquid movement caused by the evaporation presure is considered. The expulsed liquid must come from the bottom of the holes and not from the walls. The expulsed liquid jet forms the envelope of a cone (ref 8).

To increase the depth of laser drilling the laser pulse duration must be increased, but an increase in the pulse duration results in a larger volume of the molten material leading to poorer accuracy. In this case one should determine the optimum ranges of pulse duration and power density at which the volume of the molten material is minimised. To increase the depth of drilling, the multipulse technique (ref 9) should be used.

Heat Loss (ref 6)

There are several causes of heat loss. Whilst the beam cuts a hole, part of the beam energy is lost. As workpiece velocity is increased the fraction of beam energy lost is reduced until the point is reached when the plate is no longer pierced. However, heat is lost by conduction through the workpiece and because the power per unit area of metal surface reduces with increasing speed, the temperature reached is also reduced.

Hole Requirements

In the hole requirements (diameter, depth, taper, recast, microcracking and angle) the laser variables which have an effect are pulse energy, pulse duration, number of pulses, focal length of lens, beam diameter and beam quality (ref 10). Figure 3 shows the faults most likely to occur in laser drilling. Figure 4 shows how a typical laser drilled hole might appear in cross-section in a metal sample (ref 11). The diameter of the hole is dependent on the incident beam diameter. Holes smaller than 0.0005" (0.013 mm) diameter are difficult due to the inability to maintain depth of focus at this spot size (ref 1).

Taper is a result of erosion caused by the expulsion of molten and vapourised material from the hole. The degree of it can be controlled by the number of laser pulses, pulse energy level, and optical system design. Figure 5 shows the relationship found for different materials between thickness and hole taper (refs 10, 12).

Any molten residue of vapourised material that is not completely expelled, but resolidifies inside and around the hole, is known as recast. Thermally induced stresses during the solidification process produce microcracking, which is propagated into the parent material at grain boundaries, weakens the final product. Recast and microcracking are usually reduced by selecting power densities that effectively expel molten and vapourised material along with pulse widths short enough and repititive rates slow enough to minimise heating of the surrounding parent material (ref 10);(see Figure 6 in this context).

Drilling perpendicular to the surface the maximum amount of expelled debris is directed back to the focussing lens, so it is necessary to protect the lens. As the drilling angle with respect to the surface becomes smaller, the volume of debris directed toward the lens decreases, but the focusing geometry becomes more complicated.

At high energy levels there is a tendency to cause more deformation on the top surface. Generally short pulse duration (0.4 mS) will produce higher quality holes, but will require more pulses to drill through. Longer pulse durations (to 3 mS) will remove more material per pulse, but will produce holes of inferior quality.

Nozzle Design Effects

At a given supply pressure there is an optimum nozzle diameter. The maximum cutting speed for a given nozzle will be proportional to the distance between the pressure axis and the point of intersection of the horizontal line through the minimum pressure required to remove molten material from the kerf in the time available and the nozzle pressure distribution curve.

In the pressure range of up to 4 bar, the maximum cutting rates for high quality cuts increases with increasing pressure. Over 4 bar a further increase in pressure will allow no or very little increase in cutting speed. At a pressure higher than 5 bar a new phenomenon is observed. This is self-burning marks in the kerf; these are very small and regular marks compared to those seen at low speed.

Pulses Effects (ref 13)

For a very short illumination duration the temperature rise approaches a linear dependence on time. For longer light pulses, a larger portion of the laser energy is lost to the substrate by heat conduction during the pulse. Varying the laser pulse intensity, the machining spot area varies nearly linearly with the pulse intensity over a certain range above some threshold value, see Figure 7.

Various Improvements (refs 1,7,14)

(a) use of faster jets: with disappointing results due to excessive cooling of the molten product and shock waves interfering with the fine focus on the laser beam.

(b) use of multiple jets: concentric jets that can increase the jet potential core simultaneous use of jets above and below, cross blowing beneath to remove the underbead and cross blowing over a venturi to introduce suction from below.

(c) changes in gas composition: an alternative method of introducing more energy is by adding an electric arc to the laser interaction zone.

(d) combination of laser types: the penetration of thin plates can be speeded up significantly by the application of a trailing pulse to a cw beam. Here the advantage remains in the elimination of the need to liquify all the metal and a single final pulse is adequate.

(e) trepanning: this operation involves the laser beam being focused more tightly so that its diameter is much smaller than the hole to be produced. The beam is then used to cut around the profile of the hole. This involved more pulses than the direct drilling technique, which meant that drilling times could be reduced substantially. The energy transmitted to the workpiece is reduced and with it the likelihood of microcracking because it used a small spot size. A coaxial assist gas is required to remove debris from the drilling zone. By this way any shape hole could be produced.

Hard Brittle Materials (ref 15)

The drilling of hard brittle-like materials by laser is attractive due to the extremely difficult task for the current tool technology. The maximum hole depth achievable is limited by the amount of energy lost due to reflections from the hole wall and by the decrease in the aperture of the hole as a result of vapour from the bottom of the hole being cooled by, and depositing on, the wall of the hole.

Common Metals

With common metals, generally, when the laser pulse energy is kept
constant, the largest hole depths are produced for low-melting point materials.
Drilling with a superposition of pulses usually yields holes whose sides have
less taper than those of holes drilled with one pulse of higher energy (ref 3).

Laser Cutting (ref 16)

There are three different versions of laser cutting:

(a) laser sublimation cutting, where the focused laser beam heats the
material to its evaporation temperature. A jet of inert gas carries the vapour
out of the cutting front.

(b) Laser fusion cutting, where a stronger inert gas jet is used to blow
the molten material out of the kerf. The material has to be heated only above
its melting point. The required cutting energy per unit length is less, and
higher cutting rates can be reached.

(c) Laser gas cutting, where oxygen gas is used reacting exothermically
with the material when the ignition temperature is reached.

Quality of Cut (refs 17-19)

The best quality is judged by a clean, smooth cut without any dross or
burr, small heat-affected zone (HAZ), sharp edges and fine striation. The
quality is shown as a function of cutting speed and oxygen pressure. The
parameters which affect the quality of the cutting are the power of laser,
the feed rate, metal thickness, nozzle design and the gas used in the jet.

The geometrical properties of a cutting surface are characterised by the
surface roughness, inclination of the cutting surface, roundings at the upper
and lower cutting edges, presence of squeezings, blowpipe bead and clinging
dross.

With laser gas cutting at low cutting speeds self-burning of the material
occurs. By increasing the speed, the self-burning is increasingly suppressed,
and the quality of the cut improves until the optimal speed is reached.
Exceeding this speed, the quality deteriorates rapidly, producing blowpipe
bead clinging at the rear surface of the plate.

The kerf width depends on the cutting speed, oxygen pressure, polarisation
and the focal spot. The main problem in producing fine cuts lies in the proper
extraction of molten material through the narrow kerf width. As soon as a
considerable momentum transfer of slag is directed towards the kerf wall the
material flow will be unbalanced and an enhanced dross formation or even an
overgrowing of the kerf is the result.

The striations which occur on the edge of laser cut material are a
function of the viscosity of the slag associated with each material. The
formation of striation is caused not only by blowing out the droplets from the

front of the cut but also by difference in the velocity of the moving laser beam and the velocity of the oxidation front. Two types of observations are generally made: if the process is unstable, the characteristics of the striations suddenly change from fine striations to rough striations, and in some cases, different striations are observed on the two sides. Figure 8 shows some examples at different pressures.

Effects due to Gas Flow (refs 1, 9)

The oxygen jet has two effects: it increases the absorptivity of the material and removes the oxide film formed on the surface and the melt from the cutting zone until the material is completely cut. The quality of cutting is better at high gas flow rates and smaller distances between the nozzle and the material surface.

At high gas pressures, however, a density gradient field (DGF) is formed above the workpiece and lies within the region covered by the gas jet, see Figure 9. The shape and size of the DGF depend on gas pressure, distance between nozzle and workpiece and nozzle diameter.

The type of gas used in the jet affects how much heat is added to the cutting action. There is a large difference between using oxygen and using argon in cutting any metal. However, with some materials oxygen is too reactive and causes a ragged edge.

Improvements (refs 1,6)

An increase in metal removal rate could be achieved by the use of vibration during laser cutting. This would either shake away molten material or cause a better penetration of unreacted oxygen to the hot surface.

By adding an arc to a laser there is an increase in speed or penetration without significant loss of quality. The arc is not of much help when it is used on the same side as the laser because it flickers from one side of the kerf to the other and so damages the cut edges. The arc is much more stable at the underside of the cut and initially keeps the dross more fluid and thus extends the velocity (figure 10).

To lower the dross melting point and reduce its viscosity to increase fluidity some promising results have been obtained by blowing slagging agents such as lime into the kerf. However, this process also increases the thermal load. An exothermically reacting slagging agent would be more effective but probably too costly for the overall process.

Different Materials (ref 20)

4340 Steel Alloy : For oxygen assisted cutting the material can be severed at substantially higher speeds but a speed reduction and increased gas pressure is required to prevent tenacious dross at the base of the cut. In inert-gas-assisted cutting, optimum speeds are of the order of 50-60% of oxygen-assisted rates, and a generally smoother edge is obtained.

Titanium alloy: inert-gas-assisted cutting gives a generally smooth cut surface with less apparent heat damage and easy removal of adherent slag at the lower lip of that cut. With helium assist, some of the irregularities in the cut surface are removed to give a relatively smooth edge.

Waspaloy : Oxygen-assisted cutting results in a smaller lower edge burr than for inert-gas assist. The slag generated during inert-gas-assisted cutting could readily be chipped from the edge, leaving a relatively smooth cut surface.

Wood : Cuts are of good quality with some tendency for charring to occur on the side of the sheet not subjected to the gas assist. The sides of the cut are parallel and the kerf width can be adjusted by a change of laser power or gas flow. There appears to be little gain in using inert gases such as nitrogen or helium as the assist gas.

Plastics : The penetration is found to be greater with a gas assist only at high incident laser powers. However, the quality of the cuts produced is significantly better with the gas assist even at low laser powers.

PLASMA PROCESSING

The main relationships amongst many process variables leading to optimum cut quality are gas type, gas flow rate, cutting speed and stand-off distance.

The different regions defined below are sufficient to describe adequately the quality of a cut surface:

Top Edge : (a) edge condition - square or rounded

          (b) top dross

Top Side : (a) surface condition - smooth or rough

          (b) bevel

          (c) straightness

Bottom Side : (a) surface condition - smooth or rough

          (b) bevel

          (c) straigtness

Bottom Edge : (a) edge condition - square or rounded

          (b) bottom dross

An example of a 'perfect cut' is shown in Figure 11. The major areas of imperfection include surface roughness, dross adherence, bevel and edge rounding. Figure 12 illustrates the various possible cut imperfections (ref 26).

In the variations of cut geometry (width of cut at top b_t, and at bottom b_b) in relation to the field intensity, the width of cut at the bottom reacts most markedly to alter the field intensity H, and the difference between b_t and b_b decreases with increase of H. The external magnetic field has virtually no influence on the variation of b_t, but it affects b_b considerably.

Thus, when a magnetic field acts on a plasma arc, the electroconductive part of the arc discharge moves across the face of the cut edge, and there is a corresponding movement of the stream of gas issuing from the column of the arc discharge. It is this difference between the amplitudes of the arc oscillations across the thickness of the metal being worked which ensures that the cut edges are parallel.

The higher the cutting speed, the larger the difference between b_t and b_b. This explains the need to increase the value of H_{opt} with increase of the cutting speed. Thus, by varying the power of the deflecting system it is possible to control the geometry of the cut at a constant linear heat input of the process (ref 27).

The highly efficient method of air-plasma cutting is associated with the generation of a large amount of dust and gases and intensive radiation and noise. To eliminate these shortcomings, plasma cutting is carried out above a water pool or the plasma torch and sheet to be cut are immersed in water.

The width of cutting above water is usually 10-20% smaller than in normal conditions. The degree of non-perpendicularity of the cut surface slightly decreases, and the minimum values are obtained when cutting is carried out with the workpiece touching the surface of the water.

The process of cutting above water is characterised by higher susceptibility to formation of the flash on the edges in comparison with conventional cutting. To suppress the flash formation process, the cutting speed above water should be reduced by 5-15%.

The hardness of the heat affected zone (HAZ) metal in cutting above water is increased. The duration of the cathodes of the plasma torches is almost identical in both cases: conventional cutting and cutting with the workpiece touching the surface of the water (ref 28).

Programming Difficulties

Most of the programming problems of plasma arc are caused directly or indirectly by the speed with which the plasma cuts or by the bevel generated on the workpiece by the plasma arc.

One major problem is the rounding of outside corners and the burning-off of sharp-end points on plasma cut parts caused by an error associated with the NC machines. This error occurs because the flame cannot keep up with the programmed path of the torch and when a direction change takes place, the flame tends to round the corner. The faster the machine moves, the more pronounced the rounding of corners becomes.

The cutting of inside peripherical corners and of inside cutouts is another programming problem related to the following error. This error can be corrected by using programmable pause or dwell codes.

The location of a pierce-point and burn-in and burn-out is often the most difficult step of programming for any plasma cut part.

NC flamecut machines with kerf compensation capabilities greatly simplify part-programming but they also create other problems. It is an additional factor which helps multiply the effort of following error on inside corners and cutouts. Kerf compensation is also the major reason for having a burn-out (ref 29).

Some Common Materials

Mild Steel

In general, additional hydrogen leads to less bevel, but with the disadvantage of possible heavy dross and some undercutting. Also, the dross adheres more tightly than when nitrogen is used alone. Another advantage can be obtained with proper control of the stand-off distance. The closer the torch tip is positioned to the workpiece, the less is the bevel. The speed and amount of hydrogen can often be determined by observing the dross ejection pattern. The optimum pattern is vertical with larger droplets actually being projected in the forward direction.

Stainless Steel

The total included kerf angle can be reduced to as small as 1° by utilising slightly higher gas flow and a closer stand-off distance. In general, added hydrogen reduces the top side roughness problem, but leads to roughness of the bottom side and to increased dross adherence. Best cuts are associated with the dross ejection pattern. In some cases the dross issues directly downwards but displaced slightly behind the centreline of the torch nozzle.

Aluminium

Aluminium is more sensitive to a hydrogen content with a torch positioned vertically at a fixed stand-off distance. There is one best concentration of hydrogen for producing a vertical cut. Where a single surface is of interest, the higher hydrogen concentrations can be used for straight-line cutting, the torch can be slanted from the vertical to offset the bevel.

Stand off distance does not greatly affect bevel. Aluminium is not too sensitive to speed, the only adverse effect of very low speed is an undercutting of the top side of the inside surface. High speed produces excess dross adherence on the outside.

WATER JET CUTTING

In this process, the initiation and the propagation of the cracks are
the most important steps, taking into account the characteristic loading
as a result of the jet formation the jets can be divided into 3 types:
a) continuous jets, b) mixed jets, and c) impact jets (Figure 13).

The continuous pressure acting at te material surface results from the
continuous running of the jet. The impact loading can be characterised by
the formation of a peak pressure which lasts only for an extremely short time.
The impact pressure peak normally reaches a level which is a multiple of the
steady state pressure. Between these two extreme types of jets there is a
whole spectrum of others characterised by a mixture of impact pressure and
steady state pressure (ref 31).

There are three different types of loading: impact pressure, stagnation
pressure and shear stress. In the first moment after the impact of the liquid
particle on the solid body the liquid produces a very high pressure. Its
duration is of a few microseconds, depending mainly on the impact velocity.
The rapid drop in impact pressure is characterised by the initialisation of the
radial flow of the liquid with a very high speed. By that the pressure decreases
to the much lower stagnation pressure in the impact centre, while the radial
flow of the liquid loads in shear the surface of the solid body. This process
will be preserved as long as the liquid impinges continuously (ref 32). However
a compact liquid jet is generated in most of the commonly used jet cutting
devices due to its easier generation.

Processing Effects

Angle of impact: the most eminent ability of cutting is produced when the
jet is impinged perpendicularly to the workpiece surface. As the angle of
incidence changes from the normal, the damage is concentrated on the downstream sides of the ring fracture, and the ring crack is elongated in the
same direction. There is a definite maximum around 17° where damage falls off
very steeply on either side of this angle (ref 33, 34).

Nozzle Standoff : As the distance between the nozzle and the workpiece
is increased, the opening of the groove cut or of the hole pierced becomes
to be enlonged and irregular through the rate of material removal by jet
is increased within a certain range of the distance, See Figure 14. The jet
pressure drops more rapidly with distance and the slot depth varies inversely
with the standoff ratio. Most effective cutting is achieved at small standoff
ratios (refs 33, 35).

Nozzle Shape : The cutting effectiveness of a jet, particularly at large
standoff distances, is very dependant on the shape of the nozzle. It is
generally believed that the nozzle should have an entrance cone angle of

10° to 14°, followed by a straight section of about 2 to 4 diameters long. The surface should be polished. The nozzle should be supplied by a cylindrical feeder section of diameter at least 10 nozzle diameters and with a length preferably at least 40 to 50 feeder diameter long (ref 35).

Jet Size : The width of groove is proved to be on the average proportionalto the jet diameter and consequently to the diameter of the nozzle opening. The cutting efficiency in terms of the generated surface per unit of input power is higher when using smaller nozzles than in the case of larger nozzles. Concerning the cutting efficiency estimated by the shock removal per unit of input power, however, longer jets would represent higher efficiency (ref 33)

Exit Pressure and Workpiece Materials : For all material, there is a threshold pressure required to initiate cutting, which is typically 20 to 50% of the compressive strength. As pressure is raised towards the compressive strength, the cutting depth increases with a power of the ratio of jet pressure to compressive strength greater than 1 (1.5 to 2). As pressure is raised to exceed the compressive strength, the depth of cut increases in most cases approximately linearly with pressure (ref 35).

Material Permeability : Intrinsic speed is proportional to the ratio of permeability to porosity for a given grain size. The supermeable materials, such as metal and plastics have smaller slot depths which correlate approximately directly on the square root of the ratio of jet density to target density. The permeable materials have deeper slots (ref 35).

Temperature Rise during Machining : The temperature rises near the working point. Increasing rate of the temperature rise with the increase of the exit pressureis seen to be lower than that of the input power. This phenomenon is possibly explained by the increase of an ability of the higer powered jets to cool the working point. Though the degree of the temperature rise should depend on the thermal and mechanical properties of work materials, it must have a bearing on the machining mechanism, see Figure 15 (ref 33).

Material temperature can increase significantly during jet cutting. The temperature increases depend on the jet kinetic energy, the mass and kind of material being cut and the proximity to the kerf produced by the jet (ref 36).

Feed rate : Material removal increases with the traverse   speed within the range wherein the quantity of the ratio is large enough, Figure 16. For the materials, which are more machinable under the conditions examined, the relation between the depth of groove cut and the traverse   speed are given in Figure 17 (ref 33).

Abrasive Waterjets (refs 37, 38)

High pressure waterjets are used in industrial applications to cut materials such as plastics, leathers, corrugated board and asbestos.

Limitations exist when the available waterjet cutting devices are not capable of cutting metals. The introduction of abrasives into the jet stream is an excellent way to increase the cutting capabilities of high-pressure waterjets (Figure 18).

There are two ways to form an abrasive jet, depending on whether the abrasives are mixed upstream or downstream of the high-pressure nozzle. The general advantages of downstream mixing are the restriction of wear to the mixing area where can be minimised, and the adaptation of the system with the state of the arc of high-pressure pumps. Upstream mixing is classified as either direct or indirect pumping. The advantages of this approach lie in its more effective cutting capabilities and the compactness of the nozzle head for applications such as small-hole drilling.

The parameters that affect the abrasive jet phenomenon are listed below:

(a)   hydraulic parameters : waterjet pressure and waterjet nozzle diameter

(b)   abrasive parameters : abrasive material, particle size, mass flow rate, and particle shape;

(c)   mixing parameters: mixing method, abrasive condition and mixing chamber dimensions;

(d)   cutting parameters: traverse rate, standoff distance, number of passes, angle of attack and target material; and

(e)   cutting results: cut depth, kerf width and kerf quality.

Effect of Abrasive Material and Size

Some conclusions can be drawn by comparing abrasive flow rates with depths of cut for different abrasive materials (Figure 12). Low dust concentrations in the carrier gas result in more erosion than higher concentrations for the same total abrasive mass flow rate. The effect of granite particle size on depth of cut is shown in Figure 20.

Effect of Abrasive Flow Rate

The relationship between flow rate and depth of cut is initially linear. However, this linearity terminates at higher abrasive flow rates as particle velocity will decrease more rapidly than the number of impacts will increase. Also, higher abrasive flow rates result in increased interference between particles, which reduces the effective number of impacts, alters favourable angles of attack, and reduces local impact velocities. The effect of abrasive flow rate on depth of cut is shown in Figure 21.

Effect of Pressure

An optimum operating pressure should be determined to compromise between the rate of cutting and power requirements. The maximum efficiency value occurs at a pressure equal to $3 P_c$ where P_c is a critical threshold pressure at which no cutting will occur. This threshold pressure appears to be independent of the other parameters. Figure 22 shows the depth of cut as a function of water pressure.

Effect of Traverse Rate

The requirements to achieve a certain depth of cut at a certain power level requires optimisation of the traverse rate and the number of passes, which also includes the effect of the standoff distance. The traverse rate afects the cutting mechanism; cutting wear may occur more often at low traverse rates and deformation wear at high traverse rates. Figure 23 shows the relationship between the depth of cut h and the traverse rate u.

Effect of Standoff Distance

The rapid ineffectiveness of the jet with increased standoff distance can be explained by arguing that the effectiveness of the abrasive jet cutting results from the superimposed hydrodynamic pressure on the impacting particle. At large standoff distances, the liquid phase of the abrasive jet breaks up into droplets and the solid particles become freer to rebound upon impact resulting in shallower penetration. Figure 24 shows the effect of the standoff distance on the depth of cut.

Effect of Number of Passes

In general, the depth of cut increases with the number of passes, as shown in Figure 25. However, it is the rate of increase that is the important issue.

Effect of Waterjet Nozzle Diameter

The depth of cut varies linearly with the waterjet nozzle diameter, and the jet power E is directly proportional to the square of the waterjet nozzle diameter for a given pressure. This suggests that increasing the nozzle diameter is not an effective procedure. Figure 26 shows the effect of the water jet diameter on the depth of cut.

Material Effects

For brittle materials, a cutting nozzle mode with large angles of attack will contribute more to penetration than one with small angles of attack. For ductile materials, particle impact at small angles of attack will be most significant in determining depth of cut. However, all cutting zones will be of equal significance if a normally brittle material behaves in ductile manner under certain conditions of erosion. The mechanism by which a high-velocity solid particle erodes ductile metal depends on its angle of impingement. At shallow angles erosion proceeds as cutting wear; at large angles, the erosion is due to deformation wear.

Kerf Quality

The wedge shape of the kerfs, usually seen in water-jet cutting of soft materials, it also observed for steel cutting with abrasive jets. However,

146

the wedge angle is less in abrasive jet cutting.

CONCLUSIONS

New cutting technologies have prominent advantages that permit to deal with any type of material to achieve very accurate components. However, the large number of parameters that have influence in the processing soon becomes a difficulty. For each specific application it is necessary to obtain the optimal characteristics largely by trial and error procedure. Careful examination of each process discussed above is a must especially if the component is specialised and costly. It is hoped that the present survey will furnish readers with a substantial background.

REFERENCES

1   M Bass, Laser Materials Processing , North-Holland Publising Company
    1983.
2   John E Harry, Industrial lasers and their applications McGraw-Hill,
    1974.
3   W W Duley, CO_2 Lasers Effects and Applications, Academic Press, 1976.
4   J E Barger, Plasma arc cutting, SME Technical paper, MR76-712, 1976,
    15p.
5   M K Chun and K Rose, Interaction of High Intensity laser beams with
    metals, Journal of Applied Physics, Vo. 41, No. 2, Feb. 1970 pp 614-
    620.
6   S J Ebeid and C N Larsson, Ultra-sonic assisted laser machining,
    Proceedings of the 18th International Machine Tool Design and Research
    Conference, held in London 14th-16th September 1977 pp 507-514.
7   Anon, Why a laser is better than EDM for drilling, Production Engineer,
    December 1983 pp 13-14.
8   M Von Allmen, Laser drilling velocity in metals, Journal of Applied
    Physics, Vol 47, No. 12, Dec. 1976 pp 5460-5463.
9   N Rykalin, A Uglow and A Kokora, Laser machining and welding, MIR
    Publisners, Moscow 1978.
10 W B Tiffany, Drilling, marking and other applications for industrial
    Nd:YAG Lasers, SPIE Vol 527, Applications of High Power Lasers, 1985
    pp 28-37.
11 Steve Bolin, Laser Drilling and Cutting, SME Technical Paper MR81-365
    1981, 10p.
12 BrianF Scott, Laser Machining and Fabrication - A review, Proceedings
    of the 17th International Machine Tool Design and Research Conference.
    Held in Birmingham, 20th-24th September 1976, pp 335-339D.
13 D Maydan, Micromachining and Image Recording on Thin Films by Laser
    Beams, The Bell System Technical Journal Vol 50, No 6, July-August
    1971, pp 1761-1789.
14 L C Towle, J A McKay and J T Schriempf, The penetration of thin metal
    plates by combined cw and pulsed laser radiation, J. Applied Physics,
    Vol 50 N6 June 1979 pp 4391-4393.
15 Francis P Gagliano, Robert M Lumley and Laurence S Watkins, Lasers in
    Industry, Proceedings IEEE Vol 57 Feb 1969 pp (114-117).
16 I Decker, J Ruge and W Atzert Physical models and technological aspects
    of laser gas cutting, Industrial Applications of High Power Laser, SPIE
    Vol 455 1983 pp 81-87.
17 Flemming BachThomassen, Flemming O Olsen, Experimental Studies in Nozzle
    Design of Laser Cutting, Proceedings SPIE. Industrial Applications
    of High Power Lasers 1984 pp 169-181.

18   N Forbes, The role of the gas nozzle in metal-cutting with CO_2 lasers, Laser 75, Optoelectronics Conference Proceedings, Munich 24/27 June 1975 pp 93-95.

19   Erich H Berloffa, J Witzmann, Laser materials cutting and related phenomena, Industrial applications of high power lasers, Proceedings SPIE Vol 455 1983 pp 96-101.

20   J Huber, M Warren, Production Laser Cutting, Applications of laser in materials processing, 1979, pp 273-290.

21   Anon, Designing in Profile - What are the alternatives? Engineering Materials and Design, March 1984, pp 44-49.

22   D R Martyr, Laser cutting - a new tool for shipbuilding. Proceedings of the 1st International Conference on Lasers in Manufacturing, Brighton IFS, Nov 1983 pp 28-32.

23   S R Bolin, The effects of F# and Beam divergence on quality of holes drilled with pulsed Nd:YAG Lasers, LIA Vol 31 ICALEO 1982 pp 135-140.

24   F O Olsen, Investigations in optimising the laser cutting process Laser in materials processing, Conference Proceedings, American Society for Metals, 1983, pp 64-80.

25   J Clarke and W M Steen, Arc augmented laser cutting, Laser 79 Opto-Electronics, Conference Proceedings 1979 pp 247-253.

26   James A Browning and Charles Hazard, Quality Determinants in the Plasma Cutting Process, British Welding Journal, Sept 1963, pp 462-469.

27   E P Kharitonov et al, Magnetic control of a plasma arc during the cutting of metals, Welding Production Vol. 19 No 12, 1972, pp 73-77.

28   G I Lashchenko, M T Lysenko and V M Bogdano, Plasma arc cutting of metal above a water pool surface, Welding Production, April 1985, pp 13-17.

29   John E Barger, Plasma arc cutting, SME Technical paper MR76-712 1976 15p.

30   D H Saunders, Water as a cutting tool, Engineering, April 1977 pp 297-301.

31   F Erdmann Jesnitzer, H Louis and J Wiedemeir, Material behaviour, material stressing, principle aspects in the Application of High Speed Water Jets, 4th International Symposium on Jet Cutting Technology, April 1978, Paper E3, pp 29-43.

32   E F Beutin, F Erdmann-Jesnitzer, H Louis, Material behaviour in the case of High-Speed liquid jet attacks, 2nd International Symposium on Jet Cutting Technology, April 1974, Paper C1, pp 1-18.

33   Imanaka et al. Experimental study of machining characteristics by liquid jets of high power density up to 10^8 Wcm^{-2}, 1st International Symposium on Jet Cutting Technology, April 1972, Paper G3, pp 25-35.

34   M C Rochester and J H Brunton, High Speed impacts of liquid jets on solids, 1st International Symposium on Jet Cutting Technology, April 1972, Paper A1, pp 1-23.

35   W C Cooley, Correlation of data on jet cutting by water jets using dimensionless parameter. 2nd International Symposium on Jet Cutting Technology, April 1974, Paper H4, pp 39-48.

36   Neusen, K F and S W Schramm, Jet induced target material temperature increases during jet cutting, 4th International Symposium on Jet Cutting Technology, April 1978, Paper E4, pp 45-52.

37   M Hashish, Steel cutting with abrasive waterjets, 6th International Symposium on Jet Cutting Technology, April 1982, Paper K3, pp 465-486.

38   M Hashish, A Modelling Study of Metal Cutting with Abrasive Waterjets, Journal of Engineering Materials and Technology Transactions of the ASME, Vol 106, January 1984, pp 88-100.

148

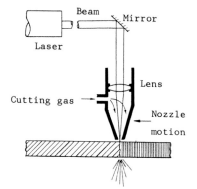

Figure 1. Coaxial gas jet laser
cutting. (Ref. 22)

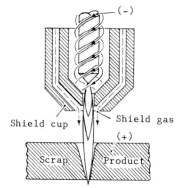

Figure 2. Plasma torch using swirled
cutting gas. (Ref. 21)

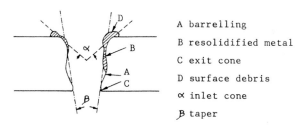

A barrelling

B resolidified metal

C exit cone

D surface debris

∝ inlet cone

β taper

Figure 3. Faults in a laser-drilled hole
(Ref. 12)

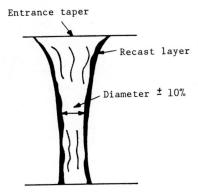

Figure 4. Typical cross section of
laser drilled hole. (Ref. 11)

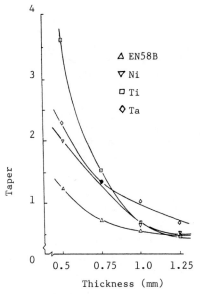

Figure 5. Taper versus thickness. (Ref. 12)

150

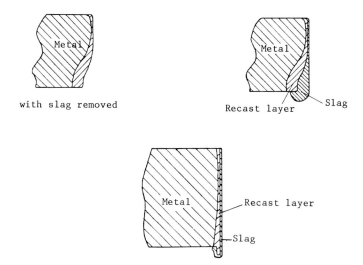

Figure 6. Recast produced (Ref. 18)

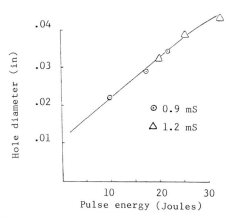

Figure 7.a Variation in hole diameter (Ref. 23)

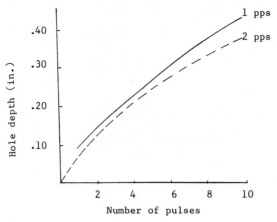

Figure 7.b Hole depth vs number of pulses (Ref.23)

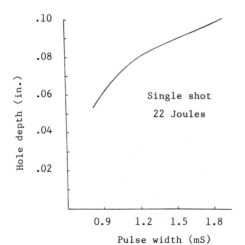

Figure 7.c Variation of hole width (Ref. 23)

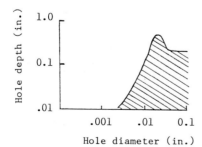

Figure 7.d Depth/diameter envelope in ferrous
         alloys (Ref. 23)

152

High pressure, high speed

Medium pressure, high speed

Low pressure, medium speed

Figure 8. Striations (Ref. 24)

Figure 9. Presence of a density
gradient field. (Ref. 1)

Laser power = 1600 W
Arc power = 1750 W
Oxygen pressure = 68.9 kN/m

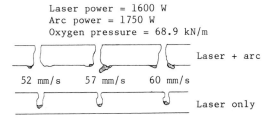

Laser + arc

52 mm/s   57 mm/s   60 mm/s

Laser only

Figure 10. Variation in cut geometry with cutting
speed for 4mm mild steel. (Ref. 25)

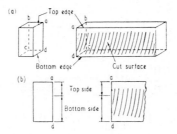

Figure 11. The hypothetical 'perfect'
cut. (Ref. 21)

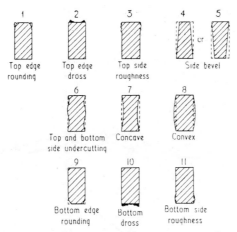

Figure 12. Certain common types of cut
imperfections. (Ref. 21)

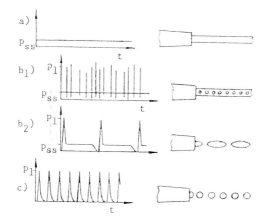

Figure 13. Schematically loading of a jet :

    a) Continuous jet

    b) Mixed jet with examples

      b1 submerjed jet

      b2 pulsed jet

    c) Impact jet

      (Ref. 31)

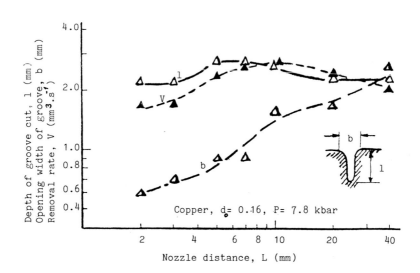

Figure 14. Effect of nozzle distance. (Ref.33)

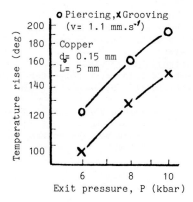

Figure 15. Temperature rise during machining.(Ref.33)

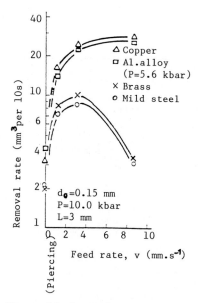

Figure 16. Removal rate at different feed rates. (Ref. 33)

156

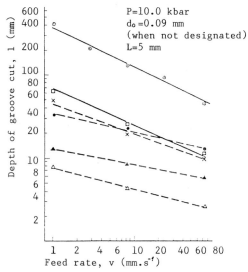

⊙ Unsulphurized rubber (d_o=0.13 mm)
□ Asbestos cement (d_o=0.11 mm)
✕ Polyester
● Cercidiphyllum japonicum
▲ Glass-fibre reinforced polyester resin
△ Glass-fibre reinforced epoxy resin

P=10.0 kbar
d_o=0.09 mm
(when not designated)
L=5 mm

Depth of groove cut, l (mm)

Feed rate, v (mm.s^{-1})

Figure 17. Groove depth at different feed rates.
(Ref. 33)

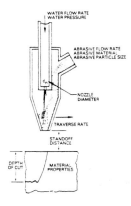

Figure 18. Abrasive jet nozzle.
(Ref. 38)

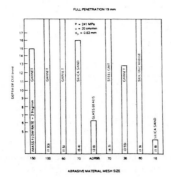

Figure 19. Effects of various abrasive materials and particle sizes on depths of cut in mild steel. (Ref. 38)

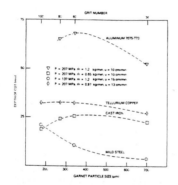

Figure 20. Effect of garnet particle size in depth of cut. (Ref. 38)

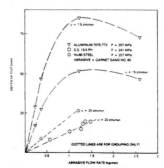

Figure 21. Effect of abrasive flow rate on depth of cut. (ref. 38)

158

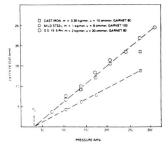

Figure 22. Effect of water pressure
on depth of cut. (Ref. 38)

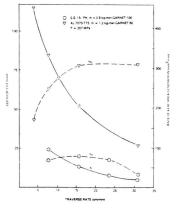

Figure 23. Effect of traverse rate
on cutting results. (Ref. 38)

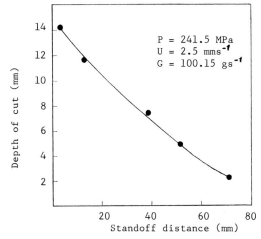

Figure 24. Effect of standoff distance on
depth of cut. (Ref. 37)

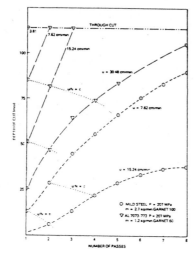

Figure 25. Effect of passes on depth
of cut. (Ref. 38)

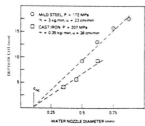

Figure 26. Effect of waterjet
nozzle diameter on depth of
cut. (Ref. 38)

Computational Methods for Predicting Material Processing Defects, edited by M. Predeleanu   161
Elsevier Science Publishers B.V., Amsterdam, 1987 — Printed in The Netherlands

THEORETICAL PREDICTION OF THE LIMIT CURVES FOR SIMULATION OF PLASTIC INSTABILITY

J.J. GRACIO[1], J.V. FERNANDES[1] and A. BARATA DA ROCHA[2]

[1]Dep. Eng. Mecânica F.C.T.U.C., 3000 Coimbra - PORTUGAL
[2]Dep. Eng. Mecânica F.E.U.P., Rua dos Bragas, 4099 Porto - PORTUGAL

SUMMARY

In sheet metal forming operations, the forming limit is usually governed by strain localisation due to plastic instability and fracture. A great number of parameters may influence the forming limitations such as strain hardening, strain rate sensitivity, damage, anisotropy and strain path. The recent advances in finite plasticity, damage modelling and instability theory offer new tools for the prediction of the Forming Limit Diagrams (FLDs).

In the present work, a series of tests to assess strain path effects in metal deformation were performed with different materials. A numerical simulation of necking phenomena under complex strain paths using Hill's theory of orthotropic anisotropy and time dependent constitutive relations was carried out. The validity of the model was tested through the comparison of theoretical results with experimentally determined FLDs.

INTRODUCTION

The analysis of failure by plastic instability in sheet metal forming is often evaluated from strain analysis using the concept of Forming Limit Diagrams (FLDs). FLDs, which represent the relationship between limiting major and minor principal strains in the plane of the strained sheet, were introduced by Keeler (ref. 1) and Goodwin (ref. 2).

The usual way to analyse plastic instability phenomena is to use the mathematical theory of plasticity. Many authors have used computational methods for predicting material processing failure. Swift (ref. 3) first describes diffuse necking in thin sheets under plane stress states assuming that plastic instability occurs at a load maximum for proportional loading. The limit strains for localized necking condition have been derived by Hill (refs. 4-5). However, localized necking cannot be predicted for biaxial stretched sheets using Hill's analysis.

Marciniak and Kuczinsky (refs. 6-8) have developed a theory (the M-K theory) based on the assumption that necking develops from local regions of initial heterogeneity. More recently, several authors (refs. 9-17) have given different approaches to sheet necking description. However, all of the forementioned works considered proportional straining, i.e., the ratio of minor to major strain is constant at any stage of deformation.

Unfortunately, this hypothesis is not valid in the press shop since the strain paths are usually non-linear. Industrial stampings of complex shape often involve multistage forming operations and linear strain paths can no longer be observed (refs. 18-21). Several authors (refs. 22-29) have shown that strain path is probably the most sensitive parameter controlling plastic instability. It is therefore of major importance to study the influence of strain history on plastic instability phenomena.

Barata da Rocha, Barlat and Jalinier (refs. 30-34) have recently proposed a theoretical approach to plastic instability phenomena in order to predict FLDs under complex strain paths.

In this work, we study the FLDs of anisotropic materials under complex strain paths. Several experiments on copper and steel have been performed. The mechanical and rheological parameters experimentally determined were introduced in the numerical analysis, in order to test the model both qualitatively and quantitatively. Finally, a comparison is presented between experimental and theoretical results for a wide range of complex strain paths.

THEORETICAL ANALYSIS OF THE FORMING LIMIT DIAGRAMS

The detailed description of the theoretical model is beyond the scope of this paper. The analytical developments can be found elsewhere (refs. 30-34). The analysis is based on the growth of an initial defect in the form of a narrow band inclined at an angle Ψ_o with respect to the principal axis (Fig. 1).

A plane state of stress is applied to the homogeneous region (zone a). The directions 1,2 represent the principal loading axis and are superposed to the directions x,y of the principal anisotropy reference axis.

The yield surface of anisotropic materials is described using Hill's criterion for plane stress states:

$$f = \sigma_e^2/2 = [(F\sigma_{yy}^2 + G\sigma_{xx}^2 + H(\sigma_{xx} - \sigma_{yy})^2 + 2P\sigma_{xy}^2)]/2 \tag{1}$$

where F,G,H and P are anisotropy coefficients.

Assuming isotropic work-hardening and strain rate hardening we can represent the behaviour of the material in the form:

$$\sigma_e = K(\bar{\varepsilon} + E_o)^n \dot{\bar{\varepsilon}}^m \qquad (2)$$

where n is the strain hardening exponent, m is the strain rate sensitivity, E_o is the prestrain, K is a constant, $\bar{\varepsilon}$ is the effective strain, $\dot{\bar{\varepsilon}}$ is the effective strain rate and σ_e is the effective stress.

Constitutive equations using the above yield function and associated flow rule, are expressed in the principal axis of orthotropic symmetry:

$$d\varepsilon_{ij} = d\lambda' \, \partial f / \partial \sigma_{ij} \qquad (3)$$

where $d\lambda'$ is a proportionality factor. Similarly, equilibrium and compatibility equations are expressed in these principal axis which are coincident with the principal axis of stress in the homogeneous region.

The initial value of the geometrical defect (ref. 32):

$$F_o = e_o^b / e_o^a \qquad (4)$$

can be calculated by statistical analysis in terms of an equivalent defect based on the existence of damage due to the concentration of cavities nucleated around particles. The numerical solution of the differential equations is obtained by a fourth order Range - Kutta method.

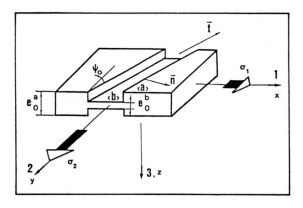

Fig. 1.
Geometric configuration
of the plastic
instability model

EXPERIMENTAL METHODS AND MATERIALS

Three different materials (sheets 1 mm thick) were used to study the effect of non-proportional loading on the shape and position of the FLDs: copper DHP-99.92% pure and two kinds of steel (rimmed steel-steel A and aluminium killed steel-steel B). The chemical composition of the steel sheets is given in table 1.

TABLE 1
Chemical analysis of the steel sheets (wt. %)

	C	Mn	P	S	N	Al	Si	Fe
Steel A:	0.10	0.35	0.035	0.035	0.208	–	0.03	Bal.
Steel B:	0.059	0.297	0.011	0.013	0.005	0.072	0.025	Bal.

The n-value was determined in uniaxial tension for the rolling direction using the experimental effective stress – effective strain curve. This value was obtained by fitting eqn. (2) to the experimental curve.

Several tension tests were carried out at different directions relative to the rolling direction (α). For each direction, the r- value was determined through the following expression:

$$r(\alpha) = \frac{2r_o\, r_{90}\, \cos^2 (2\alpha) + r_{45}\, (r_o + r_{90})\, \sin^2 (2\alpha)}{r_o\, (1 - \cos(2\alpha)) + r_{90}\, (1 + \cos(2\alpha))} \qquad (5)$$

Fig. 2 shows the evolution of r-value as a function of the angle to rolling direction. The useful material parameters are summarized in table 2.

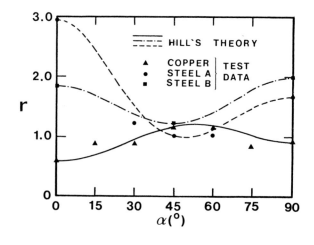

Fig. 2.
Evolution of r-value
as a function of α

TABLE 2
Material parameters of the three different sheets

Mat.	σ_y(MPa)	n	m	r_0	r_{45}	r_{90}
Copper:	77	0.366	–	0.535	1.176	0.917
Steel A:	198	0.151	0.012	2.94	1.01	1.67
Steel B:	170	0.23	0.012	1.84	1.22	1.99

Uniaxial tensile tests were performed on ISO 50 samples (75mm x 12.5mm). For copper and steel A, plane strain conditions were obtained with TPE 1 type specimens (ref. 35); for steel B, the strain paths between uniaxial tension and plane strain were obtained by tension of large samples. Biaxial stretching was performed by using a hydraulic bulge test (the Jovignot test) with rectangular, elliptical and circular dies.

Mechanical tests along the complex strain paths were obtained with the following procedure:
- balanced biaxial stretching followed by uniaxial tension was achieved through a prestrain with the bulge test followed by tensile test of specimens cut from the prestrained sheet;
- uniaxial tension followed by plane strain was achieved through a prestrain of tensile specimens 40 mm wide (the ratio length to width was kept similar to ISO 50 specimens in order to obtain equivalent strain paths) followed by plane straining of TPE 1 specimens cut from the prestrained sheet;
- uniaxial tension followed by balanced biaxial stretching was achieved through a prestrain of wide samples (500mm x 220mm) to allow the preparation of the bulge test specimens; it has been observed that the strain path followed in this case was very close to that followed in a ISO 50 sample.
In all cases, the main principal stress direction was always parallel to the rolling direction.

A photo-sensitive-resist method was used to print the surface of the test-pieces. The pattern used was a grid of circles of 2 mm diameter and the strains were calculated from measurements of distorded grids using a travelling microscope having an accuracy of $\pm$ 1 μm.

The onset of localized necking was determined from strain distribution profiles near the necking region by using Bragard method and following the procedure recommended by the International Deep Drawing Research Group (IDDRG) at Zurich in 1973 (ref. 36).

RESULTS AND DISCUSSION

The experimental FLDs of Copper DHP in linear and complex strain paths is presented in Fig. 3A. As it can be observed the limit strains strongly depend on the strain path. The linear strain paths in the vicinity of plane strain lead to quite different levels of limit strains (points P_1 and P_2 in Fig. 3A). It must be noted that the limit strain corresponding to point P_1 was obtained with a notched specimem (TPE 1) while the limit strain corresponding to point P_2 was obtained with the bulge test (rectangular die). It is well known that notched specimens lead to lower values of the FLD due to strong strain gradients. Moreover, the size of the grid used to measure the strains was relatively large (ϕ = 2mm) which contributes to a low level of the limit strains (ref. 35). The relatively large size of the grid, the strong gradients obtained with TPE 1 specimens and the adoption of Bragard method (Zurich - IDDRG) may explain the different values of the limit strains obtained in the vicinity of plane strain.

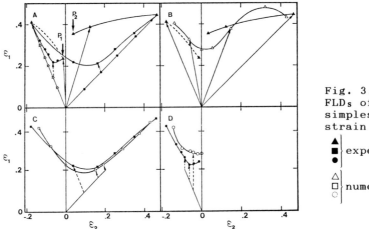

Fig. 3.
FLDs of copper in simples and complex strain paths:

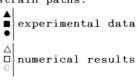 experimental data

numerical results

Fig. 3B shows the experimental FLD of copper in linear strain paths and the corresponding theoretical prediction. The value of the initial defect (F_0=0.979) was determined by fitting the general level of the theoretical FLD under linear strain path to the experimental one. This initial defect was kept constant for all the other strain paths. The analysis of Fig. 3B shows that the simulation does not predict the form of the FLD for positive minor strains. The form of the theoretical FLD near balanced biaxial tension comes from the high value of r_{45}. This subject has already been discussed in a previous paper (ref. 31) in which it has been shown that the limit strains in balanced biaxial tension strongly depend on r_{45} value. The discrepancy between experimental and theoretical results in this region cannot be explained by an incorrect description of r-value evolution. In fact, the analysis of Fig.2 shows that the evolution of r-value with the angle to rolling direction is in good agreement with Hill's theory. However, the model assumes isotropic work-hardening and Hill's yield surface. More experimental work must be carried out in order to validate these hypotesis.

Figs. 3C and 3D show the FLDs of copper in balanced biaxial tension followed by uniaxial tension and uniaxial tension followed by plane strain respectively. For these cases an excelent agreement is obtained between experimental and theoretical results. In fact, premature plastic instabilities are obtained for sequences consisting of prior balanced biaxial stretching followed by uniaxial tension and uniaxial tension followed by plane strain, both in theoretical and experimental curves. The maximum discrepancy observed in complex strain paths is observed for strain paths consisting of uniaxial tension followed by plane strain. However, as it has been refered previously, the second strain path was obtained with TPE 1 specimens, which lead to low levels of limit strains.

The experimental FLDs of steel A are shown in Fig. 4A. In the vicinity of plane strain, different levels of limit strains were obtained for notched and bulge (rectangular) specimens. This discrepancy has already been explained for the case of copper.

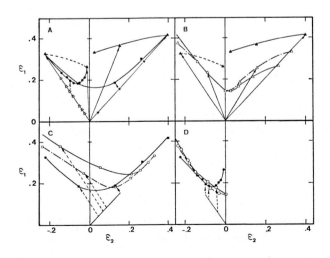

Fig. 4.
FLDs of steel A in simples and complex strain paths:

▲
■ experimental data
●

△
□ numerical results
○

The experimental FLDs of steel A and corresponding theoretical FLDs are presented in fig. 4B for the case of linear strain paths. The value of the initial defect for this case was chosen to be $F_0=0.990$ in order to fit the general level of the theoretical FLD under linear strain path to the experimental one. In this case, the theoretical curve presents strong discrepancy relatively to the experimental one. In fact, the high level of r_0 value ($r_0=2.94$) leads to a high value of the theoretical limit strain in uniaxial tension ($\epsilon_1=0.62$), while the maximum major strain obtained experimentally is of the order of $\epsilon_1=0.33$. The mixed line in Fig. 4B represents the theoretical prediction taking into account the $\bar{r}$ mean value ($\bar{r}=(r_0+2r_{45}+r_{90})/4$). It is obvious that in this case the limit strain in uniaxial tension is greatly reduced. Of course, the value of the initial defect can be chosen in order to fit the theoretical curve to the experimental one for the minor negative strain region. However, if the simulation is carried out, the model will predict excessively low values of the limit strain in balanced biaxial stretching. As it has been mentioned previously, this discrepancy may come from anisotropic work-hardening. In fact, it has been shown in previous works (ref. 30,37) that accurate prediction of FLDs require an exact characterization of the material behaviour. In particular, an increase or decrease of the strain hardening exponent with increasing biaxiality will greatly influence the final limit strain value. It must be noted that the n-value was measured in uniaxial tension and was kept constant in all the strain path simulations. Several experimental works (ref. 38) show that the n-value may strongly depend on the strain path. The fact that the experimental FLD of steel A presents lower limit strains in tension than in balanced biaxial stretching suggests that more experimental work should be carried out in order to study the strain hardening behaviour of this material in more detail.

Figs. 4C and 4D show a comparison between the experimental and theoretical FLDs under complex strain paths. A better agreement is obtained in this case than in the case of linear strain paths.

Fig. 5A shows the experimental FLDs of steel B in linear and complex strain paths. Figs. 5B, 5C and 5D show the corresponding theoretical prediction for a value of the initial defect F_0 of 0.995. For this steel, an excelent agreement is obtained both in linear and complex strain paths. Moreover, in this case it has been verified (ref. 39) that the n-value is identical in uniaxial tension (n=0.23) and balanced biaxial stretching (n=0.25). For the case of complex strain paths it has been previously observed for this steel (ref. 39) that the n-value slightly decreases during deformation for sequences consisting of uniaxial tension followed by balanced biaxial tension (n=0.20 in balanced biaxial tension after a prestrain of $\bar{\varepsilon}=0.22$ in uniaxial tension).

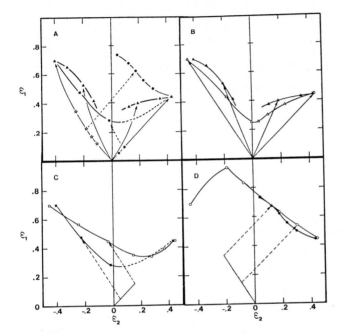

Fig. 5.
FLDs of steel B in simples and complex strain paths:

▲ ■ ● experimental data

△ ▷ □ ○ numerical results

On the contrary, the n-value strongly decreases for the sequence balanced biaxial tension followed by uniaxial tension (n=0.12 in uniaxial tension after a prestrain of $\bar{\varepsilon}=0.10$). This fact was not considered in the simulation. Therefore, the slight discrepancy observed in Fig. 5C can be explained by this phenomena.

CONCLUSIONS

Forming Limit Diagrams in linear and complex strain paths were determined experimentally for three different materials. A theoretical analysis of necking phenomena was used in order to predict the FLDs of these materials. It has been suggested that accurate prediction of FLDs require an exact characterization of the material behaviour through adequate constitutive equations. A successful correlation is observed between the experimental FLDs in complex strain paths and the computed limit strains.

170

ACKNOWLEDGEMENTS
    Part of this work was carried out at the Laboratoire de Genie Physique et Mecanique des Materiaux - GPM2 (Grenoble, France). The authors would like to express their gratitude to Prof. B. Baudelet, Director of GPM2 and to Dr. J.M. Jalinier of RENAULT - DAST.

REFERENCES
1. S.P. Keeler, Sheet Met. Ind., 42 (1965) 683.
2. G.M. Goodwin, Metall. Ital., 8 (1968) 767.
3. H.W. Swift, J. Mech. Phys. Solids, 1 (1952) 1.
4. R. Hill, J. Mech. Phys. Solids, 1 (1952) 19.
5. R. Hill, The Math. Theory of Plast., Oxford Univ. Press, London, 1950.
6. Z. Marciniak and K. Kuczinski, Int. J. Mech. Sci., 9 (1967) 609.
7. Z. Marciniak, K. Kuczinski, T. Pokora, Int. J. Mech. Sci., 15 (1973) 789.
8. Z. Marciniak, Mechanics Sheet Metal Forming, Plenum Press, New-York / London, 1978, p. 215.
9. J.W. Hutchinson and K.W. Neale, Mechanics Sheet Metal Forming, Plenum Press, New-York/London, 1978, p. 127.
10. J.W. Hutchinson and K.W. Neale, Proc. IUTAM Symposium, Lehigh University, Pensylvania, 1980, p. 238.
11. S.Storen and J.R. Rice, J. Mech. Phys. Solids, 23 (1975) 421.
12. R. Hill and J.W. Hutchinson, J. Mech. Phys. Solids, 23 (1975) 239.
13. K.W. Neale and E. Chater, Int. J. Mech. Sci., 22 (1980) 563.
14. K.W. Neale and E. Chater, Proc. ICM 4, Stockolm, 1983, p. 681.
15. J.L. Bassani, J.W. Hutchinson and K.W. Neale, Proc. IUTAM Symp., Tutzing, 1978, p. 1.
16. K.S. Chan, D.A. Koss and A.K. Gosh, Metall. Trans., 15A (1984) 323.
17. K.W. Neale, E. Chater, Coll. Int. CNRS,Villard de Lans, 1981.
18. S.S. Hecker, Proc. 7th IDDRG Congress, Amsterdam, 1972, 15.1.
19. M.M. Fourdain, X. Josselin and R. EL Haik, Proc. 7th IDDRG Congress, Amsterdam, 1972, 15.1.
20. C. Lefevre, R. EL Haik, E. Levrez and M. Bourgeon, Proc. 12th IDDRG Congress, 1982, p. 129.
21. I. Aoki and T. Horita, Proc. 11th IDDRG Congress, Metz, 1980, p. 553.
22. J. Gronostajski, A. Dolny and T. Sobis, Proc. 12th IDDRG Congress, Italy, 1982, p. 39.
23. T. Kobayashi, H. Ishigaki and T. Abe, Proc. 7th IDDRG Congress, Amsterdam, 1972, 8.1.
24. A. Ranta-Eskola, Proc. 11th IDDRG Congress, Metz, 1980, p. 453.
25. T. Kikuma and K. Nakazima, Trans. Iron Steel Inst. Japan, 11 (1971) 827.
26. D.J. Loyd and H. Sang, Metall. Trans., 10A (1979) 1767.
27. W.B. Hutchinson and R. Arthey, Scr. Metall., 10 (1976) 673.
28. J.V. Laukonis and A.K. Gosh, Metall. Trans., 9A (1978) 1849.
29. R. Arrieux, C. Bedrin and M. Boiven, Proc. 12th IDDRG Congress, Italy, 1982, p. 61.
30. A. Barata da Rocha and J.M. Jalinier, TISIJ, 24 (1984) 132.
31. A. Barata da Rocha, F. Barlat and J.M. Jalinier, Proc. 13th IDDRG Congress, Melbourne, 1984.
32. A. Barata da Rocha, F.Barlat, J.M. Jalinier, Mat. Sci. Eng., 68 (1984) 151.
33. A. Barata da Rocha, F. Barlat, J.M. Jalinier, Int. Conf. on Computer Modelling of Sheet Metal Forming Process, Detroit, 1985.
34. A. Barata da Rocha, J. M. Jalinier, Int. Symposium on Plastic Instability (Considere Memorial), Paris, 1985.
35. G. Pomey and G. Sanz, Aptitude a l'emboutissage des toles minces, Vol. 3, IRSID - OTUA, Paris, 1976
36. P. Parniere , G. Sanz, in B. Baudelet (Ed.), Mise en forme des metaux et alliages, Editions CNRS, Paris, 1976, p. 305.
37. J.H. Schmitt, F. Barlat, A. Barata da Rocha and J.M. Jalinier, Mem. et Etudes Sci., Rev. Metall. (Sept. 1984) 462.
38. D. Rault and M. Entringer, Mem. Sci. Rev. Metall., 77 (1980) 535
39. J.V. Fernandes, Doctoral Thesis, Univ. Coimbra, Portugal, (1984).

HOMOGENIZATION OF A PLASTIC POROUS MEDIA. INFLUENCE OF SECONDARY CAVITIES

T. GUENNOUNI and D. FRANCOIS
Laboratoire des Matériaux and GRECO Grandes Déformations et Endommagement, Ecole Centrale Paris, F 92295 Chatenay-Malabry cedex, France

SUMMARY
    Using an homogenization technique the plastic behaviour of a porous material containing two populations of cavities is constructed. The corresponding load function is given. It is shown that the interaction between the main and the secondary cavities increases greatly the relative growth rate of the porosity. This effect is the more important the higher the stress triaxiality ratio or the larger the ratio between the main and the secondary porosities.

INTRODUCTION
    Generaly speaking in metal forming operations it is important to know the evolution of damage. For instance, in sheet metal forming, the limit curves might be related to the initiation and growth of cavities at inclusion sites (ref. 1). Indeed it can be shown that there is a strong interaction between strain hardening and damage. A better describtion of the evolution of the cavities would thus provide a tool for the analysis of forming limit curves.
    This evolution was considered by various authors (refs. 2-4) who studied the growth of a single cavity in an infinite medium. A more elaborate treatment was given by Gurson (ref. 5) who proposed a yield criterion and an associated growth law rule for rigid prefectly plastic material taking into account the interaction between cavities.
    However it seemed to us that this criterion which violates the homogeneity condition (ref. 6), could be improved. We then used an homogenization technique and finite element calculations in order to find the yield criterion of a medium containing a set of periodical cylindrical cavities (ref. 7). We showed that the solution could be reasonably well extended to the case of spherical cavities. It was found that the Gurson's yield criterion predicts too stiff a porous medium at intermediate strees triaxiality ratios and thus underestimates the cavities growth rate.
    Experimentaly it was shown for many materials that the cavities growth rate was rather high (refs. 8-9) and it was often proposed that this was due to an interaction with secondary cavities which are seen on fractographs as

small dimples surroundings the large ones triggered by the main inclusions. Such are the observations of ductile fracture of Hancock and Mackenzie (ref. 10), showing that it takes place by plastic instability between large cavities triggered itself by microinstability between secondary holes. The nucleation of the large cavities occurs for comparatively small plastic deformations whereas the nucleation of the smaller secondary ones occurs for larger plastic strains.

On the other hand, in the case of materials containing a low volume fraction of inclusions, Thomson and Hancock (ref. 11) noticed that their statistical distribution led to a rather high number of small inclusions in some areas. A better modeling of there materials consists in considering a few main cavities surrounded by much smaller secondary ones.

In the present work the interaction between main and secondary cavities is studied and more precisely the macroscopic plastic potential of a porous material taking into account the formation of secondary cavities is constructed.

STATEMENT OF THE PROBLEM

In order to study the interaction between main and secondary cavities, a model square cell made of a porous medium of porosity f_0 representing the secondary cavities containing a central hole with radius r representing the primary cavities is considered (fig. 1). This cell is loaded on its external boundary Γ_e with a macroscopic stress Σ. The velocity field v and the microscopic σ stress field obey the usual equilibrium equations with boundary conditions :

$\sigma . n = \Sigma . n$ along Γ_e

$\sigma . n = O$ along Γ_i

in the case of a hollow cell containing a fluid at zero pressure. In these equations n stands for the external normal to the given boundary. The behaviour of the matrix is supposed to be rigid perfectly plastic with a yield function proposed by the authors (ref. 7) for a medium containing cylindrical pores.

$$\sqrt{\frac{\sigma'^2_{eq}}{B^2(f_0)} + \frac{\sigma'^2_m}{A^2(f_0)}} \leq 1 \tag{1}$$

where σ'_{eq} is the reduced equivalent Mises stress and σ'_m the reduced mean hydrostatic stress, both relative to the shear yield stress k of the dense material. Notice that the compressibility of the matrix comes from the hydrostatic component of the stress tensor in the yield function (equation1). $A(f_0)$ and $B(f_0)$ are adjusted by

$$A\ (f_0) = -\ (1- 1{,}92f_0 + 5{,}57f_0{}^2 - 6{,}03f_0{}^3\ )\ Ln\ f_0$$
$$B\ (f_0) = \sqrt{3}\ (1- 1{,}33\ f_0{}^{0{,}6}\ ) \qquad when \quad f_0 \le 0{,}1256$$
$$B\ (f_0) = 2{,}06\ exp\ (-4{,}9\ f_0)\ -0{,}044 \qquad otherwise$$

$$\tag{2}$$

The macroscopic strain rate is defined as

$$\dot{E}_{ij}\ (v) = \frac{1}{2\,Y} \int_{\Gamma_e} (\,v_i\,n_j + v_j\,n_i\,)\ d\Gamma \tag{3}$$

One wishes to build the constitutive equations of the homogeneous equivalent material which relates the macroscopic variables Σ and $\dot{E}$ taking into account the interaction between main and secondary cavities.

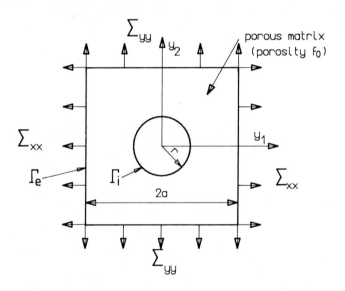

Fig 1. Elementary cell with compressible matrix

Most numerical methods fail when one tries to approximate $\dot{E}(v)$ through (3) because the unicity and regularity of the microscopic velocity field are lost. Attempt is then made to compute directly the macroscopic yield function Φ by building the domain K of the admissible macroscopic stress defined such that there exists an admissible microscopic stress which equilibrates the macroscopic stress Σ:

K = { Σ, there exists a microscopic stress tensor σ verifying (1) such that
$\quad\ \sigma \in S(\ \Sigma)$ }

where S (Σ) is the statically admissible microscopic stress domain

$S(\Sigma) = \{ \text{div } \boldsymbol{\sigma} = O, \quad \boldsymbol{\sigma}.n = \Sigma.n \quad \text{along } \Gamma_e, \quad \boldsymbol{\sigma}.n = O \text{ along } \Gamma_i \}$
The constitutive equations of the homogeneous equivalent material is obtained by noting that the Hill maximum work principle is kept when using the macroscopic variables Σ and E.

Using the same approach as in ref. 7, the macroscopic yield domain K of the basic cell is sought in plane strain with uniform macroscopic stresses applied on the external sides.

NUMERICAL RESULTS

On figure 2 are plotted the numerical results obtained for a 1% porosity of the matrix. The main porosity is varied according to the size ratios r/a= 0.2; 0.3; 0.4 and 0.5 (r/a=0.2 and 0.5 are only represented). Various macroscopic stress ratios Σ_m/Σ_{eq} are considered. On figure 2 is also shown the yield criterion given by relation (1) either neglecting the porosity of the matrix (i.e incompressible medium) or assuming that the effect of its porosity is simply additive, in which case the total porosity f_t of the cell is given by

$$f_t = f + f_0 - f f_0$$

(4)

where f represents the main porosity resulting from the central hole. It can be seen that the interaction between the main and secondary cavities is the more noticeable the smaller their size or the higher the stress triaxiality ratio. The numerical results are well approximated by a quadratic criterion:

$$\Phi(\Sigma_m, \Sigma_{eq}) = \sqrt{\frac{\Sigma'^2_{eq}}{\overline{B}^2(f, f_0)} + \frac{\Sigma'^2_m}{\overline{A}^2(f, f_0)}} \leq 1$$

(5)

In order to study the interaction between main and secondary cavities, the main porosity dependence and the matrix porosity dependence of the two parameters $\overline{A}$ and $\overline{B}$ of the proposed criterion were investigated. Numerical calculations were carried out for various values of the size ratio r/a and of f_0 and for two macroscopic stress states: a purely deviatoric one ($\Sigma_m=0$) and a purely hydrostatic one ($\Sigma_{eq}=0$). On figure 3 are plotted the evolutions of $\overline{A}$ and $\overline{B}$ as a function of f_0 for various r/a ratios. $\overline{A}$ is seen to be very porosity dependent for small r/a ratios whereas $\overline{B}$ is not. As far as the yield criterion is concerned the interaction between main and secondary cavities is larger for high stress triaxiality ratios or low main porosities.
The evolution of $\overline{A}$ is well approximated by

$$\overline{A}(f, f_0) = -(1-1.92 f_t + 5.57 f_t^2 - 6.03 f_t^3) \, \mathrm{Ln} \, (f + f_0 - 1.25 \, (f f_0)^{0.7} + 0.227 \, (f f_0)^{0.33})$$

(6)

As for $\overline{B}$, the interaction effect can be neglected

$$\overline{B}(f, f_0) = B(f_t)$$

(7)

where f_t and B are given by formulae (4) and (2) respectively. Relations (6) and (7) reduce to relation (2) when the matrix is incompressible ($f_0=0$).

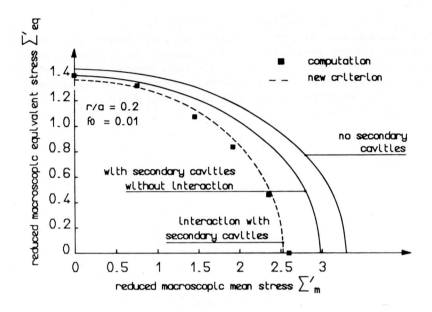

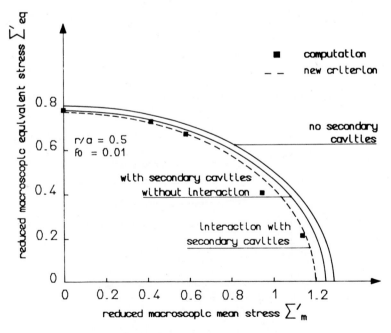

Fig. 2. Influence of the interaction between the main and secondary cavities on the yield criterion for r/a =0.2 and 0.5.

In the case of low porosities formulae (6) and (7) reduces to

$$\overline{A} = -Ln\ f_{eq}$$
$$\overline{B} = \sqrt{3}$$

with $\qquad f_{eq} = f + f_0 - 1.25(ff_0)^{0.7} + 0.227(ff_0)^{0.33}$

which corresponds to the equivalent porosity taking into account the interaction between main and secondary cavities. It is much larger than the total porosity when f is low. Thus a 1% main porosity and a 0.1% secondary porosity corresponds to a 4.5% equivalent porosity! . It emphasizes the large effect of this interaction in the case of ductile fracture of materials with a low inclusion content.

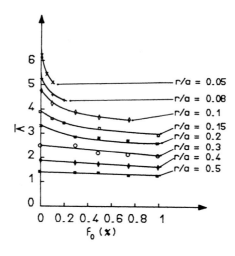

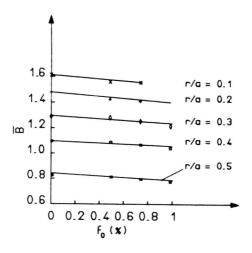

Fig. 3. Yield criterion parameters $\overline{A}$ and $\overline{B}$ versus void volume fraction of the matrix f_0 for different r/a ratios

CONSEQUENCES CONCERNING THE CAVITIES GROWTH RATE

The mass conservation is written:

$$\dot{f}_t = (1 - f_t) \, \dot{E}m \tag{8}$$

$\dot{E}_m$ being the mean inelastic macroscopic strain rate. This deformation takes into account the main cavities (represented by the central hole) and the secondary cavities (represented by the porous matrix).

The macroscopic behaviour of the equivalent homogeneous medium is written

$$\left. \begin{aligned} \dot{E}_m &= \lambda \, \frac{\partial \Phi}{\partial \Sigma_m} \\[2mm] \dot{E}_{eq} &= \lambda \, \frac{\partial \Phi}{\partial \Sigma_{eq}} \end{aligned} \right\} \tag{9}$$

λ being a positive real. Taking into account the criterion (5), by elimination of λ, the relation (9) yields

$$\dot{f}_t = (1 - f_t) \, (\frac{B}{A})^2 \, T \dot{E}_{eq}$$

where $T = \Sigma_m / \Sigma_{eq}$ is the stress triaxiality ratio.

Figure 4 shows the ratio $\dot{f}_t / (f_t \, T \, \dot{E}_{eq})$ as a function of the porosity f of the main cavities assuming that the matrix is incompressible $(f_0 = 0)$ or weakly compressible $(f_0 = 2.10^{-3})$. As expected, this ratio is much dependent on the porosity f_0 for low main porosity f.

On figure 5 is shown the evolution of the relative cavity growth rate $\dot{f}_t / (f_t \, \dot{E}_{eq})$ as a function of the triaxiality ratio T for a weakly compressible matrix $(f_0 = 10^{-3})$ and for a 10^{-2} and a 5.10^{-2} main porosity. It is again found, as expected, that the interaction effect between main and secondary cavities is the higher the lower the main porosity.

DISCUSSION

Yamamoto (ref 12) used a modeling rather similar to ours using Gurson's criterion . In that case the main cavities were represented by a thin layer whose porosity was much larger than in the surrounding matrix. This hypothesis yields a better agreement with experimental fracture strains than Gurson's with a uniform porosity. Tvergaard (ref 13) using also Gurson's criterion and a similar approach to ours, studied the influence of the interaction between main and secondary cavities on the strain localization conditions and the coalescence of cavities. Thomson and Hancock (ref 14) analysed the stress and strain distribution in a cell containing an inclusion of larger porosity than the surrounding matrix. They showed that strain softening in the inclusion is insufficient to produce a macroscopic cracking corresponding to the fracture of the cell. Li and Howard (ref 15) used the same elementary cell axisymmetrical geometry in order to study the effect of

178

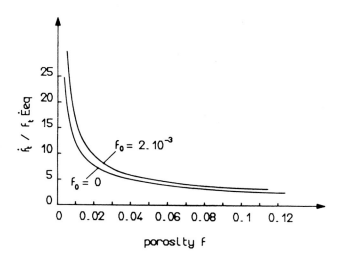

Fig. 4. Variation of the ratio between the relative cavities growth rate $\dot{f}_t/f_t$ and the product of the macroscopic equivalent strain rate $\dot{E}_{eq}$ by the stress triaxiality ratio T versus the main volume fraction for an incompressible matrix ($f_0=0$) and for a compressible matrix ($f_0=0.002$).

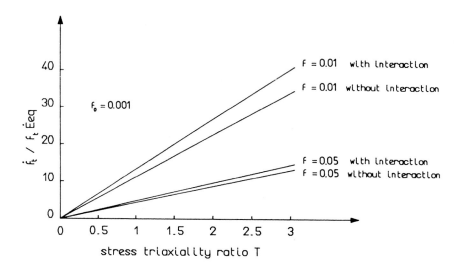

Fig. 5. Influence of the interaction between main and secondary cavities on the variation of the ratio between the relative cavity growth rate $\dot{f}_t/f_t$ and the macroscopic equivalent strain rate $\dot{E}_{eq}$ versus the stress triaxiality ratio T for two initial main void volume fractions (f=0.01 and 0.05) with a compressible matrix ($f_0=0.001$).

the secondary cavities on the fracture strains. The effect was taken into account by assuming a certain softening of the matrix past a critical stress. These authors showed that a critical cavity growth rate was valid only for high stress triaxiality ratios.

The results obtained here suggest that a better modeling of ductile damage using a phenomenological approach would consist in adopting several damage threshholds each corresponding to the nucleation of a certain type of cavties. Overcoming each threshhold would accelerate the damage rate of the material.

CONCLUSIONS

We have shown that the yield criterion of a material containing main cavities surrounded with a population of smaller secondary cavities is of the form:

$$\Phi(\Sigma_m, \Sigma_{eq}) = \sqrt{\frac{\Sigma'^2_{eq}}{B^2(f, f_0)} + \frac{\Sigma'^2_m}{A^2(f, f_0)}} \leq 1$$

Σ'_{eq} and Σ'_m being the macroscopic equivalent and mean stresses relative to the shear Von Mises stress of the solid matrix, A and B being function of the porosities f and f_0.

There is a strong interaction between main and secondary cavities for low porosities. The cavities growth rate is also much enhanced for low values of the porosity. It is given by

$$f_t = (1 - f_t)(\frac{B}{A})^2 T E_{eq}$$

T being the stress triaxiality ratio.

REFERENCES

1   F. Moussy , Les différentes échelles du développement de l'endommagement dans les aciers. Influence sur la localisation de la déformation à l'échelle macroscopique, Proc. int. seminar on Plastic Instability, Considère Memorial, Ecole Nationale des Ponts et Chaussées, Paris, September 9-13,1985, Presses de l'Ecole Nationale des Ponts et Chaussées,pp   263-272.
2   F.M. McClintock , A Criterion for Ductile Fracture by the Growth of Holes, J. App. Mech., 35 (1968) 363-371.
3   J.R. Rice and D.M. Tracey, On the Ductile Enlargement of Voids in Triaxial Stress Fields, J. Mech. Phys. Solids , 17 (1969) 2O1-217.
4   B. Budiansky, J.W. Hutchinson and S. Slutsky, Void Growth and Collapse in Viscous Solids, in H.G. Hopkins and M.J. Swell (Ed.), Mechanics of Solids, the Rodney Hill 6Oth Anniversary Volume, Pergamon Press, Oxford, 1982.

180

5   A.L. Gurson ., Continuum Theory of Ductile Rupture by Void Nucleation and Growth:Part I Yield Criteria and Flow Rules for Porous Ductile Media, Trans ASME, J. Eng. Mat., January 1977,.pp 2-15

6   T. Guennouni, Yield Criteria of Heterogeneous Materials With Rigid-Plastic Constituents. Case of Porous or Cracked Materials, J. de Mécanique théorique et appliquée, 6 (4) (1987)

7   T. Guennouni and D. François, Constitutive Equations for Rigid Plastic or Viscoplastic Materials Containing Voids, J. Fatigue Fracture Engineering Material and Structures (in Press).

8   Y.Q. Sun, J.M. Detraux, G. Touzot and D. François, Mécanisme de la rupture ductile dans la fonte à graphite sphéroïdal ferritique, Mémoires et études scientifiques Revue de Métallurgie, Avril, 1983

9   F. MUDRY, Thèse de doctorat d'Etat, Université de Technologie de Compiègne.,1982

10 J.W. Hancock and A. C. Mackenzie, J. Mech. Phys. Solids, 24 (1976) 201-217.

11 R.D. Thomson and J.W. Hancock, Ductile Failure by Void Nucleation,Growth and Coalescence,.Int. J. Fracture, 26 (1984) 99-112.

12 Y. Yamamoto, Conditions for Shear Localization in the Ductile Fracture of Void-Containing Materials, Int. J. Frac. ,14 (1978) 347-365

13 V. Tvergaard, Ductile Fracture by Cavity Nucleation Between Larger Voids, J. Mech. Phys. Solids, 30 (4) (1982) 265-286.

14 R.D. Thomson and J.W. Hancock, Deformation of Porous Imperfections Relevant to Ductile Fracture, Proc. Sixth Int. Conf. Fract., New Delhi,1984, pp1321-1327

15 G.C. Li and I.C. Howard ,The Effect of Strain Softening in the Matrix Material During Void Growth, J. Mech. Phys. Solids,31(1) (1983) 85-102.

Computational Methods for Predicting Material Processing Defects, edited by M. Predeleanu   181
Elsevier Science Publishers B.V., Amsterdam, 1987 — Printed in The Netherlands

A NOTE ON ADIABATIC FLOW LOCALIZATION IN VISCOPLASTIC SOLIDS

K. H. KIM and L. ANAND
Department of Mechanical Engineering, Massachusetts Institute of Technology
Cambridge, MA 02139, USA

SUMMARY
A history dependent dimensionless parameter λ which represents the integral in time of the ratio of the rate of flow softening to the rate of strain-rate hardening is identified as a possible flow localization parameter, and the attainment of a critically large value λ_c of λ is suggested as a simple criterion for monitoring the beginning of severe adiabatic flow localization. A full two-dimensional large deformation finite element simulation of a plane strain tension test on a thermo-elasto-viscoplastic material under adiabatic conditions is performed and the initiation and growth of a naturally appearing band-like region of localization is followed from slow early growth to severe localization. By simultaneously monitoring level contours of λ it is demonstrated that the time of formation of a band-like region in which $\lambda > \lambda_c$ and which extends across the specimen, *correlates* very well with the beginning of significant flow localization.

INTRODUCTION

Localization of plastic flow into shear bands is a widely observed phenomenon. Once such bands are formed they tend to persist and the strain inside the bands can become very large; therefore, the formation of shear bands is an important precursor to ductile fracture. Shear band formation is usually associated with a flow softening behavior of the material with increasing deformation. Various softening mechanisms are possible, but under high-rate plastic deformation conditions such as machining, shock impact loading, ballistic penetration, and certain metal forming processes such as explosive forming, the softening is due primarily to thermal effects, and the shear bands that form are called "adiabatic shear bands". For materials which exhibit thermal softening, the process of adiabatic shear banding is an autocatalytic one: an increase in the strain-rate in a soft zone causes a local increase in the temperature which causes a local increase in strain-rate which further increases the local temperature and so on. The precise details of this autocatalytic thermoplastic instability mechanism depend upon the strain hardening, thermal softening, and strain-rate hardening of the material. The basic phenomenon was first pointed out by Zener and Holloman (ref. 1), and in recent years there has been a considerable amount of analytical and numerical work that has been performed to better understand and follow the development of the localization process. As examples of the some of the work in the literature we cite references 2–15. We note that apart from the recent numerical work of

LeMonds and Needleman (ref. 15) most previous numerical work has been one-dimensional, and has focussed on studying shear-band type instabilities in simple shearing motions.

As background for the present paper we note that an analysis of the stability of two-dimensional plane homogeneous deformations has been recently presented by Anand, Kim and Shawki (ref. 9). The linear perturbation stability analysis of these authors is a generalization of a similar one-dimensional analysis of Clifton (ref. 5) and Bai (ref. 6) for simple-shearing motions. In their analysis Anand et al. assume isotropy, neglect elasticity and use a flow rule for the (plastic) stretching in which the direction of viscoplastic flow is in the direction of the deviator $\mathbf{T}'$ of the Cauchy stress $\mathbf{T}$, while the magnitude of plastic flow is proportional to an equivalent plastic strain rate $\dot{\bar{\gamma}}^p$ which is constitutively defined by a function $\dot{\bar{\gamma}}^p = f(\bar{\tau}, \bar{\gamma}^p, \theta)$. Here $\bar{\tau} \equiv \sqrt{(1/2)\mathbf{T}' \cdot \mathbf{T}'}$ is the equivalent shear stress; $\bar{\gamma}^p$, the time integral of $\dot{\bar{\gamma}}^p$, is the equivalent plastic shear strain; and θ is the absolute temperature. They assume that the strain rate function f is invertible such that one can write

$$\bar{\tau} = g(\bar{\gamma}^p, \dot{\bar{\gamma}}^p, \theta). \tag{1}$$

Corresponding to (1), the partial derivatives

$$S \equiv \partial g / \partial \bar{\gamma}^p, \tag{2}$$

$$R \equiv \partial g / \partial \dot{\bar{\gamma}}^p, \tag{3}$$

$$T \equiv -\partial g / \partial \theta, \tag{4}$$

denote the rates of strain hardening, strain-rate hardening and thermal softening, respectively. Further, for adiabatic deformations they take the energy balance equation to be

$$\dot{\theta} \approx (\omega / \rho c) \, \bar{\tau} \dot{\bar{\gamma}}^p, \tag{5}$$

where ρ, c, and $\omega (\approx 0.9)$ are the mass density, specific heat, and the fraction of plastic work converted to heat. Confining attention to materials for which $S > 0$, $R > 0$ and $T > 0$, they find that for dynamic (i.e., inclusion of inertial effects), adiabatic (i.e., neglect of heat conduction) deformations:

1. The *necessary condition* for the *initiation* of shear bands is

$$\left[S - \left(\frac{\omega \bar{\tau}}{\rho c} \right) T \right] < 0. \tag{6}$$

2. The maximum incipient rate of growth is for shear band perturbations at angles

$$\chi = \pm \pi / 4 \tag{7}$$

   relative to the maximum principal stress direction. Accordingly, the emergent shear bands are expected to form at orientations given by (7).

3. The *incipient rate of growth* of the *emergent* shear bands is given by (-P), where

$$P \equiv \left[\frac{S - (\omega \bar{\tau} / \rho c) T}{R} \right]. \tag{8}$$

These results clearly bring out the interactions of various material characteristics on adiabatic shear localization. They show the important interplay between the stabilizing effect of strain hardening and the destabilizing effect of thermal softening. High strength metallic materials generally exhibit relatively low strain hardening so that the resistance to adiabatic shear localization in these materials is low. Note that the effects of thermal softening increase with increasing flow stress $\bar{\tau}$ and decreasing density ρ. Also, the flow stress $\bar{\tau}$ increases, the specific heat c decreases, and the thermal softening T is enhanced as the temperature decreases. Thus, high strength materials which exhibit a low rate of strain hardening are very susceptible to adiabatic shear localization at low temperatures. Note also that the incipient rate of growth of the shear band is inversely proportional to the rate of strain-rate hardening R. Thus a weak strain rate sensitivity, that is a small value of R, promotes faster growth rates.

An inherent limitation of this (and all other) linear perturbation analyses is that it provides only (a) the *necessary* conditions for the *initiation* of shear bands, and (b) the *orientations* and the *incipient rate of growth* of the *emergent* shear bands. It does not provide any information regarding the more interesting stages of localization when the strain, strain rate, and temperature in the shear bands becomes much larger than elsewhere. To predict the beginning stage of *significant* flow localization with any reasonable accuracy, except perhaps for materials with a very low strain-rate sensitivity, it seems necessary to devise a new criterion. To this end, in what follows we attempt to formulate a simple new criterion for significant adiabatic flow localization in viscoplastic solids.

LOCALIZATION CRITERION

Here we do not neglect elasticty and thermal expansion effects, and we take the rate constitutive equation for the stress to be given by $\mathbf{T}^\nabla = \mathcal{L}[\mathbf{D} - \mathbf{D}^p] - \mathbf{\Pi}\dot{\theta}$, where with $\mathbf{W}$ denoting the spin tensor, $\mathbf{T}^\nabla \equiv \dot{\mathbf{T}} - \mathbf{WT} + \mathbf{TW}$, is the Jaumann derivative of the Cauchy stress; with μ and κ the elastic shear and bulk moduli and $\mathcal{I}$ and $\mathbf{I}$ the fourth and second order identity tensors, $\mathcal{L} \equiv 2\mu\mathcal{I} + (\kappa - (2/3)\mu)\mathbf{I} \otimes \mathbf{I}$ is the fourth order isotropic elasticity tensor; with α the coefficient of thermal expansion, $\mathbf{\Pi} \equiv 3\kappa\alpha\mathbf{I}$ is the second order isotropic stress-temperature tensor; $\mathbf{D}$ is the streching tensor, and the flow rule is taken as $\mathbf{D}^p = \dot{\bar{\gamma}}(\mathbf{T}'/2\bar{\tau})$, where $\mathbf{T}'$ is the deviatoric part of Cauchy stress tensor $\mathbf{T}$, $\bar{\tau}$ is the equivalent shear stress, and $\dot{\bar{\gamma}}^p = f(\bar{\tau}, \bar{\gamma}^p, \theta)$ is the equivalent plastic shear strain-rate. As before, it is assumed that the strain-rate function f can be inverted to give $\bar{\tau}$ in terms of $\bar{\gamma}^p$, $\dot{\bar{\gamma}}^p$ and θ; see equation (1).

Differentiation of (1) with respect to time gives

$$\dot{\bar{\tau}} = S\dot{\bar{\gamma}}^p + R\frac{d}{dt}(\dot{\bar{\gamma}}^p) - T\dot{\theta}, \tag{9}$$

where S, R and T are the rates of strain hardening, strain-rate hardening and thermal softening defined in equations (2)–(4). For adiabatic deformations the energy balance equation is given by (5). Substituting for $\dot{\theta}$ from (5) into (9) and rearranging, one obtains

$$\frac{d}{dt}(\dot{\bar{\gamma}}^p) + P\dot{\bar{\gamma}}^p = Q, \tag{10}$$

where P is defined in (8) and $Q \equiv (\dot{\bar{\tau}}/R)$. This equation was first derived by Shawki (ref. 14) who points out that it may be viewed "locally" as a nonlinear ordinary differential equation for $\dot{\bar{\gamma}}^p$. This equation has the (implicit) "solution":

$$\dot{\bar{\gamma}}^p = \dot{\bar{\gamma}}_i^p \exp(\lambda) \left[1 + \left(\dot{\bar{\gamma}}_i^p \right)^{-1} \int_{t_i}^{t} Q \exp(-\lambda) dt \right], \tag{11}$$

where $\dot{\bar{\gamma}}_i^p$ is the value of $\dot{\bar{\gamma}}^p$ at some initial time t_i, and

$$\lambda \equiv \int_{t_i}^{t} (-P)\, dt = \int_{t_i}^{t} \left[- \left\{ \frac{S - (\omega\, \bar{\tau}/\rho\, c)\, T}{R} \right\} \right] dt. \tag{12}$$

Note that equations (11) and (12) hold for *arbitrary* three-dimensional adiabatic deformations of bodies obeying the generic form of the constitutive equations assumed here.

For $S > 0$, $R > 0$, $T > 0$, the linear perturbation stability analysis gave the result that in homogeneous plane adiabatic motions shear band instability becomes *possible* when the parameter P becomes negative. Semiatin et al. (see, e.g. refs. 10, 11) have previously deduced a "flow localization parameter" α (see their equation (15) in ref. 10) which is proportional to the flow softening rate and inversely proportional to the strain rate sensitivity. From their numerous studies they have concluded that *noticeable localization* usually does not occur until $\alpha \approx 5$. If we interpret[1] their α as $(-P/\dot{\bar{\gamma}}^p)$, then a possible criterion for noticeable flow localization is that P should be "sufficiently negative". Such a criterion has also been suggested by Shawki (ref. 14) who uses G (see his equation (2.172)) to denote the parameter that we have here called P.

Semiatin and Jonas (ref. 11, p. 75) remark, "the α parameter provides an insight into the tendency to form shear bands as well as the likely degree of localization or severity of shear banding. Although the $\alpha = 5$ criterion is principally a rule of thumb, process modeling using finite element methods has confirmed the usefulness of this parameter." A few pages later (p. 84) they remark, "Another feature illustrated by the process simulation results is the fact that flow localization is a process not an event. Strain and strain-rate concentrations do not occur instantaneously. For this reason, flow localization cannot be expected to appear fully developed when α reaches some critical value (such as 5) at some point in the flow field."

Based on these remarks, and as is clearly suggested by equations (11) and (12), we note that the occurrence of a large value of the equivalent plastic shear strain-rate *at a material point* depends not only on the instantaneous sign of P or an instantaneous negative value of P, but on the sign and value of the integrated history λ of $-P$. Inspection of equation (11) reveals that in general there is a complicated interaction between the term $\exp(\lambda)$ and the term in the square brackets. However, it can be argued that the term in the square bracket of the equation (11) is bounded between the numbers one and zero such that the dominant term which contributes to the high value of the equivalent plastic shear strain-rate is $\exp(\lambda)$. Thus a simple criterion ("rule of thumb") for the localization of plastic deformation is

---

[1] The considerations of Semiatin, Jonas and co-workers are tied very closely to a particular power-law type constitutive equation for the shear stress.

$$\lambda > \lambda_c. \tag{13}$$

The satisfaction of this criterion at a material point should indicate that the equivalent plastic shear strain-rate at that point is very high.

To make this criterion specific we need to specify the lower limit of integration t_i in equations (11) and (12), and specify the value of λ_c in (13). Two possible choices for t_i are:

1. $t_i = 0$.

2. $t_i =$ the time when P first changes sign from positive to negative.

Recall that as long as P is positive the linear perturbation analysis predicts that the material is stable and the necessary condition for the formation of shear bands is not satisfied. Thus for materials which exhibit some strain hardening, if choice 1 is made then λ is accumulating a negative contribution[2] until such time as P turns negative. Since, as graphically commented by R. J. Clifton[3], "You can't put stability in the bank!", this choice for the lower limit of integration is not attractive, and the choice for t_i to be prefered is the time when P first changes sign from positive to negative. In this case λ is always positive, and λ_c has to be "sufficiently positive". As with the α-criterion of Semiatin, Jonas and co-workers, there does not appear to be a rigorous way to precisely specify the value of λ_c. However, it should be possible to "calibrate" the value of λ_c by performing full non-linear finite-element analyses of representative numerical experiments such as plane-strain tension and compression, and axi-symmetric tension and compression for different constitutive functions.

Localization of deformation into a region (band-like or otherwise) is an initiation and growth phenomenon in which the strain, strain-rate and temperature in the region becomes much larger than elsewhere. In what follows we report on a numerical experiment which demonstrates a procedure for obtaining a calibration of the critically positive value of λ by performing a simulation of the plane strain tension test. We show that by monitoring the nucleation and growth of regions of λ, we can follow the initiation and development of regions of intense plastic deformation in the body. The appearance of a significant sized region of "significantly positive" λ in the body *correlates* very well with the beginning of significant flow localization as judged from the distortion of the finite-element mesh, the contours of the equivalent plastic shear strain, shear strain-rate and temperature, and also with the rapid drop in the load carrying capacity of the specimen. In the plane strain tension test simulation the shape of the region of localized plastic flow which evolves naturally is a band-like region.

NUMERICAL EXAMPLE

For the class of large deformation rate constitutive equations for isotropic thermo-elasto-viscoplasticity described in the previous section, Anand et al. (ref. 16) and Lush and Anand (ref. 17) have developed special semi-implicit and fully-implicit time-integration

---

[2] For materials which show no strain hardening λ is always positive.

[3] Private communication.

procedures, and they have incorporated these time-integration procedures into the general-purpose, non-linear finite element computer program ABAQUS (ref. 18). The numerical simulation reported here was performed by using this computer code.

The particular constitutive function for the equivalent tensile stress $\bar{\tau}$ used in the numerical analysis is an equation proposed by Lindholm and Johnson (ref. 19). These authors have reported dynamic torsion test data obtained from short gage-length, thin-walled tubular specimens made from several metals, and they have proposed a constitutive equation for the shear stress which accounts for strain hardening, strain-rate hardening and thermal softening. For small elastic strains and under the assumption of isotropy the constitutive equation proposed by these authors may be interpreted to have the following form:

$$\bar{\tau} = \left(A + B\left(\bar{\gamma}^p\right)^n\right)\left[1 + C\ln(\dot{\bar{\gamma}}^p/\dot{\gamma}_0)\right]\bar{f}(\theta), \tag{14}$$

where

$$\bar{f}(\theta) = \left[(\theta_m - \theta)/(\theta_m - \theta_0)\right]^a, \tag{15}$$

and A, B, C, n, $\dot{\gamma}_0$ and a are material constants, θ_0 is a reference temperature and θ_m is the melting temperature. In their experiments Lindholm and Johnson found the steel AMS 6418 to be very vulnerable to shear localization. For this steel they report the following values for the material constants in equation (14): $A = 869\,\text{MPa}$, $B = 200\,\text{MPa}$, $C = 0.01$, $n = 0.18$, $\dot{\gamma}_0 = 0.01\,\text{sec}^{-1}$, $\theta_0 = 300°\text{K}$ and $\theta_m = 1763°\text{K}$. Lindholm and Johnson have proposed the value $a = 1$ in equation (15), however, to accelerate the localization process in our numerical simulation, we have used the value $a = 2$. Further we set $\omega = 0.9$, and take $\rho = 7750\,\text{kg/m}^3$, $c = 0.477\,\text{kJ/(kg°K)}$ as representative values for the mass density and the specific heat for the steel.

The numerical example considered here is the simulation of a plane strain tension test (see Clausing ref. 20). The mesh shown in Fig. 1 has been chosen to model one-quarter of a typical specimen for such a test. The mesh consists of 456 ABAQUS continuum plane strain 4-node isoparametric quadrilateral (CPE4) elements. Finite element analyses of shear bands based on fine meshes of quadrilateral elements built up from four crossed triangles have been previously used by Needleman and co-workers (ref. 15) to numerically capture sharply localized shear bands. Accordingly, the mesh in the central region of the gage section of the specimen, where deformation localization is expected, has been refined. In this region two of the nodes of a typical quadrilateral element have been collapsed to produce a triangular element, and such triangular elements are arranged to build quadrilaterals made from four crossed triangles. The top boundary of the of the specimen is pulled upward at a constant speed which gives a nominal plastic shear strain rate of $\sim 1000\,\text{sec}^{-1}$ in the central gage-section of the specimen. In the numerical analysis *the effects of inertia are neglected* and using the static procedures of the finite element code ABAQUS the full non-linear solution to the problem has been obtained. Values of the equivalent plastic shear strain $\bar{\gamma}^p$, the equivalent plastic shear strain rate $\dot{\bar{\gamma}}^p$, the absolute temperature θ, the parameter P, and the parameter λ were calculated at every integration point at the end of every displacement increment, and level contours of these variables were obtained at numerous representative increments. A plot of the overall load versus displacement curve was also obtained.

Due to the geometry of the plane strain tension specimen, the deformation in the gage section is slightly inhomogeneous right from the beginning , and a mild neck forms and gradually grows in the central region of the specimen. Maximum load occurs at time t_1. Slightly after a maximum in the load-time curve, a region of negative P appears in the center of the neck at time t_2; see Fig. 2. Recall that according to the linear perturbation stability analysis a change in the sign of P from positive to negative was a necessary condition for the initiation of shear bands. At this stage the contours of $(\bar{\gamma}^p, \dot{\bar{\gamma}}^p, \theta)$ exhibit mild inhomogeneities in these field variables but there are no signs of any significant band-like localization. After P first turns negative the contours of $(\bar{\gamma}^p, \dot{\bar{\gamma}}^p, \theta)$ do show a band-like region of localized deformation but the localization is slow and the load is only slowly decreasing.

At time t_3, which is substantially greater than t_2, a narrow band-like region of $\lambda \approx 10$ extends from the center of the neck to the surface of the specimen, and localization as evidenced by the contours of $(\bar{\gamma}^p, \dot{\bar{\gamma}}^p, \theta)$ has intensified, the load is beginning to drop more rapidly, and the deformed mesh is starting to show the localization; see Fig. 3.

After time t_3 the total load decreases more rapidly, and as shown dramatically in Fig. 4, at time t_4 the deformation has essentially been concentrated in a narrow band. At this time the value of λ in the band is $\lambda \approx 30$.

DISCUSSION AND CONCLUSIONS

A finite element simulation of a plane strain tension test under adiabatic conditions has been successfully performed. To the best of our knowledge, ours is the first full two-dimensional analysis of a plane strain tension test in a rate and temperature dependent solid under adiabatic conditions (neglect of heat conduction). Elasticity, thermal expansion and large geometry changes are accounted for, but inertial effects have been neglected.

We have used the numerical simulation of the plane strain tension test to follow the initiation and growth of a band-like region of localization through slow early growth to severe localization. It is shown, as expected, that the significant stage of severe local- ization, i.e. when the load starts to drop rapidly, is poorly correlated with the instant when $\{P \equiv (S - (\omega \bar{\tau}/\rho c) T)/R\}$ turns negative. However, the beginning of severe local- ization correlates fairly well with the time when there first develops a band-like region of $\lambda \equiv \int_{t_i}^{t} P \, dt > \lambda_c$ across the specimen.

In order to follow the localization process in an arbitrary deformation history it ap- pears attractive to monitor the parameter λ and its contours in addition to the parameters $(\bar{\gamma}^p, \dot{\bar{\gamma}}^p, \theta)$ and thier contours. It is important to note that like $\bar{\gamma}^p$ and θ, the parameter λ depends upon the entire deformation history experienced by each material point. Because of the point-wise nature of the parameter λ, the attainment of a sufficiently positive value of λ does not by itself predict localization which is usually understood to occur when the strain, strain-rate and temperature in a region (band-like or otherwise) becomes much larger than the regions which surround it. However, by monitoring the nucleation and growth of of a region where λ is positive we would have automatically monitored the region where the strain, strain-rate and temperature are greater than in the regions which surround it. To determine the shape of the region of localized deformation in a given boundary value problem, the full non-linear solution to the problem has to be carried out. In the plane strain tension test simulation the shape of the region of localized flow which evolves naturally is a band-like region. Preliminary calculations show that this is also the case for plane strain compression. However, for axi-symmetric tension the development of a substantial region of positive λ correlates with significant diffuse necking rather than shear localization.

Apart from the interest in monitoring regions of λ during numerical studies conducted to understand the physical phenomenon of flow localization, the monitoring of this parameter may be useful for determining the stage in a finite element analysis when it may become necessary to start worrying about significant mesh distortion and the need for a re-zoning operation.

ACKNOWLEDGEMENTS.

Helpful discussions with A. M. Lush, T. G. Shawki, D. M. Parks and R. Abeyaratne are gratefully acknowledged. Partial support for this work was provided by the U. S. National Science Foundation (Contract No. MEA-8315117).

REFERENCES

1 C. Zener and J. H. Hollomon, Effect of Strain Rate Upon Plastic Flow of Steel, Journal of Applied Physics, 15 (1944) 22 -32.
2 R. F. Recht, Catastrophic Thermoplastic Shear, J. Appl. Mech., 86 (1964) 189-193.
3 R. S. Culver, Thermal Instability Strain In Dynamic Plastic Deformation, in: R. W. Rhode et al. (Eds.), Metallurgical Effects at High Strain Rates, Plenum Press, New York, 1973, pp. 519-530.
4 A. S. Argon, Chapter 7 in The Inhomogeneity of Plastic Deformation, Am. Soc. Metals, Metals Park, Oh., 1973, pp. 161-189.
5 R. J. Clifton, Chap. 8 in Materials Response to Ultra-High Loading Rates, NRC Report. NMAB-356, 1980, pp. 129-142.
6 Y. L. Bai, Thermo-plastic Instability in Simple Shear, J. Mech. Phys. Solids, 30 (1982) 195-207.
7 R. J. Clifton, J. Duffy, K. A. Hartley and T. G. Shawki, On Critical Conditions for Shear Band Formation at High Strain Rates, Scripta Metall., 18 (1984) 443-448.
8 T. J. Burns, Approximate Linear Stability Analysis of a Model of Adiabatic Shear Band Formation, Quart. Appl. Math. 43 (1985) 65-84.
9 L. Anand, K. H. Kim and T. G. Shawki, Onset of Shear Localization in Viscoplastic Solids, J. Mech. Phys. Solids, in press.
10 S. L. Semiatin, M. R. Staker and J. J. Jonas, Plastic Instability and Flow Localization in Shear at High Rates of Deformation, Acta Metall. 32 (1984) 1349-1354.
11 S. L. Semiatin and J. J. Jonas, Formability & Workability of Metals: Plastic Instability & Flow Localization, ASM series in metal processing, American Society for Metals, 1984.
12 A. M. Merzer, Modelling of Adiabatic Shear Band Development from Small Imperfections, J. Mech. Phys. Solids, 30 (1982), pp. 323-338.
13 T. W. Wright and R. C. Batra, The Initiation and Growth of Adiabatic Shear Bands, International Journal of Plasticity, 1 (1985), pp. 205-212.
14 T. G. Shawki, Analysis of Shear Band Formation at High Strain Rates and the Viscoplastic Response of Polycrystals, Ph. D. Thesis, Division of Engineering, Brown University, Providence, R.I., June 1985.
15 J. LeMonds and A. Needleman, Finite Element Analyses of Shear Localization in Rate

and Temperature Dependent Solids, Brown University/Army Research Office Report No. DAAG 29-85-K-0003/4, 1986.

16 L. Anand, A. Lush, M. Briceno and D. Parks, A Time-Integration Procedure for a Set of Internal Variable Type Elasto-Viscoplastic Constitutive Equations, Computers and Structures, accepted for publication.

17 A. Lush and L Anand, Implicit Time-Integration Procedures for a Set of Internal Variable Constitutive Equations for Hot-Working, Proceedings of the 2nd International Conference on Numerical Methods in Industrial Forming Processes, Gothenburg, 25-29 August 1986, A. A. Balkema Publishers, Rotterdam, pp. 131-137.

18 H. D. Hibbitt, ABAQUS/EPGEN, A General Purpose Finite Element Code With Emphasis on Nonlinear Applications, Nuclear Engineering and Design, 77 (1984), pp. 271-297; Reference Manuals (Sept. 1984), Hibbitt, Karlsson and Sorensen Inc., Providence R.I..

19 U. S. Lindholm, and G. R. Johnson, Strain-Rate Effects in Metals at Large Shear Strains, in: J. Mescall and V. Weiss (eds.), Material Behavior Under High Stress and Ultrahigh Loading Rates, Sagamore Army Materials Research Conference Proceedings, 29 (1983), Plenum Press, pp. 61-79.

20 D. P. Clausing, Effect of Plastic Strain Rate on Ductility and Toughness, International Journal of Fracture Mechanics, 6 (1970), pp. 71-85.

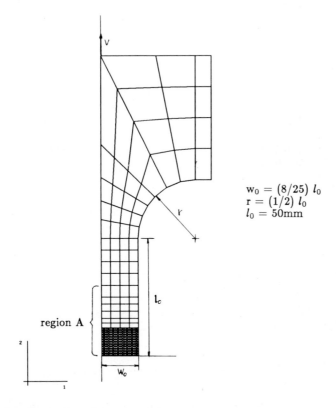

$$w_0 = (8/25)\, l_0$$
$$r = (1/2)\, l_0$$
$$l_0 = 50\text{mm}$$

Fig. 1. Finite element mesh for the simulation of a plane strain tension test. The 456 element mesh represents one quarter of the specimen. All subsequent figures show only the region A and the associated level contours of various quantities.

190

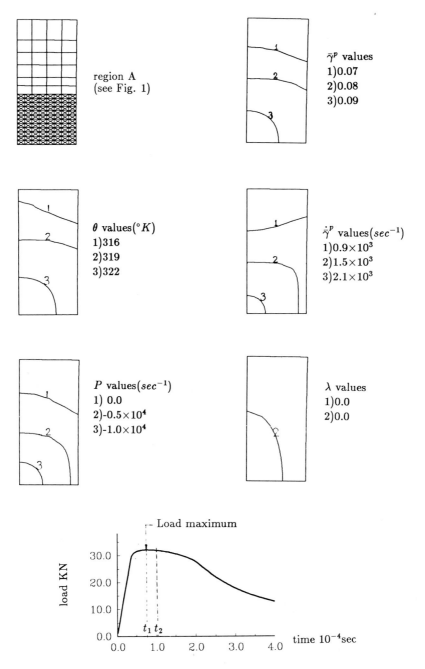

Fig. 2. Deformed mesh and the contour plots of the parameters $\bar{\gamma}^p$, θ, $\dot{\bar{\gamma}}^p$, P and λ at the time t_2 when there first forms a region of negative P across the neck of the plane strain tension specimen.

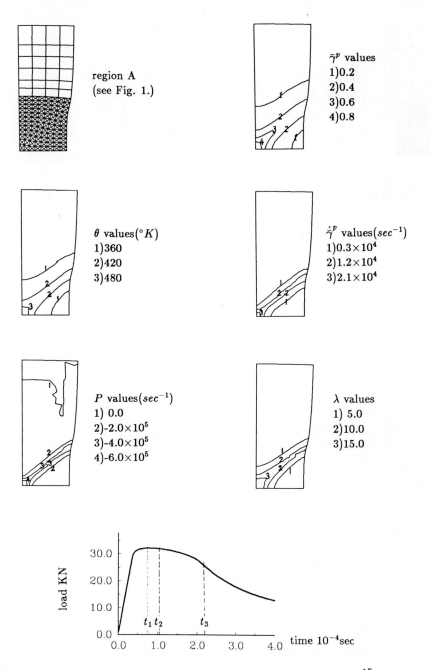

region A
(see Fig. 1.)

$\bar{\gamma}^p$ values
1)0.2
2)0.4
3)0.6
4)0.8

θ values($^\circ K$)
1)360
2)420
3)480

$\dot{\bar{\gamma}}^p$ values(sec^{-1})
1)0.3×10⁴
2)1.2×10⁴
3)2.1×10⁴

P values(sec^{-1})
1) 0.0
2)-2.0×10⁵
3)-4.0×10⁵
4)-6.0×10⁵

λ values
1) 5.0
2)10.0
3)15.0

Fig. 3. Deformed mesh and the contour plots of the parameters $\bar{\gamma}^p$, θ, $\dot{\bar{\gamma}}^p$, P and λ at time t_3. Note that a band-like region of $\lambda \approx 10$ extends across the neck of the specimen and at the same time contours of $(\bar{\gamma}^p, \theta, \lambda)$ show definite signs of localization.

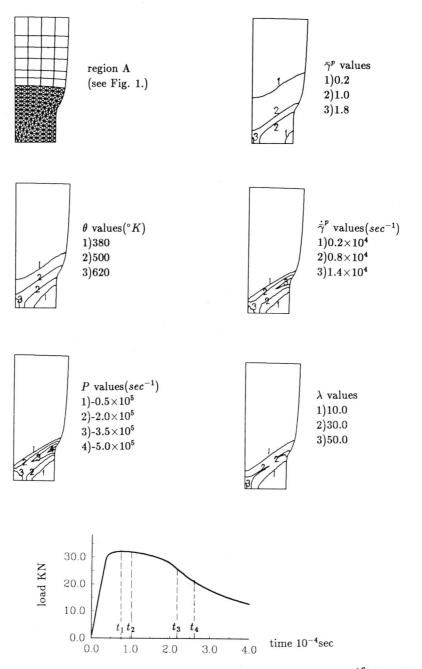

region A
(see Fig. 1.)

$\bar{\gamma}^p$ values
1)0.2
2)1.0
3)1.8

θ values($^{\circ}K$)
1)380
2)500
3)620

$\dot{\bar{\gamma}}^p$ values(sec^{-1})
1)0.2×10⁴
2)0.8×10⁴
3)1.4×10⁴

P values(sec^{-1})
1)-0.5×10⁵
2)-2.0×10⁵
3)-3.5×10⁵
4)-5.0×10⁵

λ values
1)10.0
2)30.0
3)50.0

Fig. 4. Deformed mesh and the contour plots of the parameters $\bar{\gamma}^p$, θ, $\dot{\bar{\gamma}}^p$, P and λ at the stage of fully developed shear localization inside the neck.

Computational Methods for Predicting Material Processing Defects, edited by M. Predeleanu   193
Elsevier Science Publishers B.V., Amsterdam, 1987 — Printed in The Netherlands

A COMPUTATIONAL METHOD TO ESTIMATE THE FORMABILITY OF COLD FORGING STEELS

S. Kivivuori[1], I. Lahti[2] and V. Ollilainen[2]

[1]Helsinki University of Technology, Lab. of Metal Working and Heat Treatment
Vuorimiehentie 2 A, 02150 Espoo, FINLAND
[2]Ovako Steel, Imatra Steel Works, 55100 Imatra, FINLAND

SUMMARY

   The formability of four low alloyed cold forging steels has been studied.
The deformation paths and the limit strains have been measured by testing
small cylindrical specimens under uniaxial compression. Various height to
diameter ratios and friction conditions were used.
   The stresses on the free surface of deformed specimens have been calculated
on the basis of the deformation paths. It has been assumed that the ductile
fracture occurs by rapid growth of voids and other inhomogeneities of material
when hydrostatic stress component to shear strength ratio reaches a critical
value as proposed by Oyane. The forming limit curves of the steels tested,
have been predicted by using this criterion.
   In the present paper it has been demonstrated the decissive effect of the
inclusion content on the forming limit curves of cold forging steels.

INTRODUCTION

   Cold forming processes are usually limited either by the high deformation
force or the ductile fracture. Several investigators (refs. 1-3) have reported
that the principal mechanism of ductile fracture includes three different
stages. At a first stage voids are formed around inclusions and other inhomogeneities in steel due to decohesion of the inhomogeneity matrix-interface or
cracking of the inclusions. If the deformation is continued the voids grow and
coalesce into a microcrack, which further grows and becomes visible when it
reaches the specimen surface at the later stage of deforming process.

   The stress and strain relations as well as the microstructure and strain
history have a great influence on the fracturing process (refs. 4-7). Large
positive strains and tensile stresses during metal working processes can cause
ductile fracture by rapid grow and coalescence of inhomogeneities. Proper tool
design and good lubrication minimize stress and strain concentrations caused
by the inhomogeneous deformation thus allowing higher amounts of reduction.
Lahti and Sulonen (ref. 8) have reported in their paper that a very important
factor affecting the upsetability of tested steels was the inclusion
properties including the size, shape and volume fraction of inclusions and
their distribution.

Kudo and Aoi (ref. 9) published the first analysis of cold forgeability by upsetting tests in 1967. More recently Thomason (ref. 10) and Kobayashi (ref. 11) applied essentially similar experimental technique in their analysis. In 1972 Lee (ref. 12) presented a fracture criterion for cold forming processes based on the linear relation between the total surface strains at fracture. Later on several investigators (refs. 6,8,13,14) have paid attention to this testing method suitable for estimating the cold formability of metals. A criterion for ductile fracture is presented by Oyane (ref. 15). This criterion has proved to be a very suitable to asses the ductility of porous materials.

The aim of the present paper is to demonstrate the effect of steel cleanliness on cold formability of low carbon boron steel and to estimate the formability by using the ductile fracture criterion proposed by Oyane.

EXPERIMENTAL MATERIALS AND PROCEDURES

Materials

The test materials were taken from industrially produced heats of low carbon boron steel. Steel making practices of the heats were varied in order to get different levels of cleanliness.

The chemical compositions of the test materials are shown in Table I.

TABLE I

The chemical compositions of (test) materials.

Heat code	% C	% Si	% Mn	% P	% S
A	0.17	0.28	0.93	0.021	0.014
B	0.20	0.25	0.90	0.017	0.018
C	0.17	0.10	0.70	0.014	0.029
D	0.18	0.19	1.00	0.011	0.016
E	0.18	0.21	1.00	0.019	0.019

After billet rolling the steels were rolled to 16 mm dia wire rod. The test materials were cut from the as rolled coils.

Determination of inclusion content

Inclusion contents were determined for each heat according to standard ASTM E 45-76 microscopical method D with the exception that both the bulk and the surface values of the volume fractions of inclusions were determined.

Upset tests

For the upset tests, cylindrical specimens, 14 mm in diam. with heights of 14,21 and 28 mm were machined out of the rods. Fracture strains and reductions in height at fracture were measured in the upset test for each steel making practice. The testing conditions were varied systematically by affecting the friction and using specimens having different height to diameter ratios. The pressing speed was maintained constant being about 1 mm/s. For the measurement of the equatorial strains, a grid of circles of 2 mm in diam. was etched on the cylindrical surface of the specimen.

EXPERIMENTAL RESULTS

Inclusion content

The inclusion contents and hardness of the materials are given in Table II. The bulk and surface figures are clearly different. This is partly due to the macroslag, which was detected near the surface of the specimens taken from heats D and E. The volume fraction of the macroinclusions was estimated and added to the figures of microinclusions.

TABLE II

Inclusions contents of the heats.

	A	B	C	D	E
Bulk content (%)	0.18	0.15	0.15	0.21	0.17
Surface content (%)	0.18	0.19	0.20	0.34	0.28
Hardness (HB)	153	150	140	145	160

Forming limit curves and strain paths

The forming limit curves corresponding the five steel making practices of the test material are displayed in Fig. 1. Straight lines have been drawn through 3-5 experimental values which all represent different forming conditions. Each points is an average of twenty test pieces.

The equatorial free surface strains were measured during deformation from the etched circles for several upsetting conditions. The best fitting polynomials were computed for each strain path as shown below (Eqs. (1)-(4))

$$\epsilon_\theta = -6.4 \, \epsilon_z^3 - 0.7 \, \epsilon_z^2 - 0.5 \, \epsilon_z \qquad (1)$$
(rough dies, $h_0/d_0 = 1.0$)

$$\epsilon_\theta = -3.4 \, \epsilon_z^3 - 0.7 \, \epsilon_z^2 - 0.5 \, \epsilon_z \qquad (2)$$
(rough dies, $h_0/d_0 = 1.5$)

$$\epsilon_\theta = -2.25\ \epsilon_z{}^3 - \epsilon_z{}^2 - 0.5\ \epsilon_z \tag{3}$$
(rough dies, $h_0/d_0 = 2.0$)

$$\epsilon_\theta = 0.7\ \epsilon_z{}^2 - 0.36\ \epsilon_z \tag{4}$$
(smooth dies, $h_0/d_0 = 1.5$)

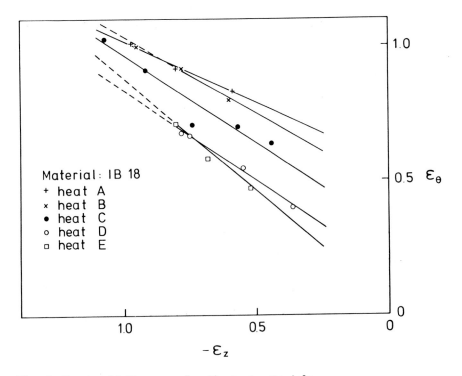

Fig. 1. Forming limit curves for the test materials.

The strain paths, as $\epsilon_\theta - \epsilon_z$-curves are given in Fig. 2. For comparison the results include also strain paths for two other steels in addition to the experimental material already mentioned. It seems that the steel grade has no effect on the course of the strain paths, probably because of the constrained mode of deformation.

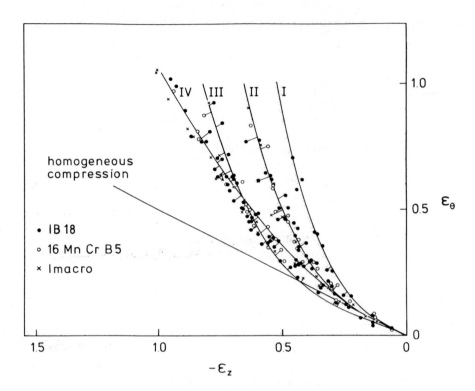

Fig. 2. Strain paths in various forming conditions. I: h_o/d_o = 1, sticking friction, II: h_o/d_o = 1.5, sticking friction, III: h_o/d_o = 2.0, sticking friction, IV: h_o/d_o = 1.5, smooth dies.

The strain ratio, $d\epsilon_\theta/d\epsilon_z$, can be calculated from the derivatives of the polynomials computed for the strain paths i.e. from Eqs. (5) to (8)

$$d\epsilon_\theta/d\epsilon_z = -19.2 \, \epsilon_z^2 - 1.4 \, \epsilon_z - 0.5 \tag{5}$$
(rough dies, h_o/d_o = 1.0)

$$d\epsilon_\theta/d\epsilon_z = -10.2 \, \epsilon_z^2 - 1.4 \, \epsilon_z - 0.5 \tag{6}$$
(rough dies, h_o/d_o = 1.5)

$$d\epsilon_\theta/d\epsilon_z = -6.75 \, \epsilon_z^2 - 2.0 \, \epsilon_z - 0.5 \tag{7}$$
(rough dies, h_o/d_o = 2.0)

$$d\epsilon_\theta/d\epsilon_z = 1.4 \, \epsilon_z - 0.36 \tag{8}$$
(smooth dies, h_o/d_o = 1.5)

In the case of the upset test the Levy-Mises equations have the following form

$$\frac{d\bar{\varepsilon}}{2\bar{\sigma}} = \frac{d\varepsilon_\theta}{2\sigma_\theta - \sigma_z} = -\frac{d\varepsilon_r}{\sigma_z + \sigma_\theta} = \frac{d\varepsilon_z}{2\sigma_z - \sigma_\theta} \quad . \tag{9}$$

The effective strain increment can be defined as

$$d\bar{\varepsilon} = \sqrt{\frac{2}{3}} (d\varepsilon_\theta^2 + d\varepsilon_r^2 + d\varepsilon_z^2)^{\frac{1}{2}} \quad . \tag{10}$$

By using equations (9), the definition of effective strain increment (10) and the constant volume condition, one can express the stress ratio $\sigma_M/\bar{\sigma}$ as a function of strain ratio $d\varepsilon_\theta/d\varepsilon_z$

$$\frac{\sigma_M}{\bar{\sigma}} = \frac{1}{\sqrt{3}} \frac{d\varepsilon_\theta/d\varepsilon_z + 1}{(d\varepsilon_\theta^2/d\varepsilon_z^2 + d\varepsilon_\theta/d\varepsilon_z + 1)^{\frac{1}{2}}} \quad . \tag{11}$$

Ductile fracture criterion

Oyane (ref. 15) has proposed the following criterion for ductile fracture:

$$\bar{\varepsilon} \ (\text{at fracture}) = -\frac{1}{A} \int_0^{\bar{\varepsilon}_f} \frac{\sigma_M}{\bar{\sigma}} \cdot d\bar{\varepsilon} + C \quad . \tag{12}$$

By using eqs. (10) and (11) it can be obtained

$$\frac{2}{\sqrt{3}} \int_0^{\varepsilon_{zf}} [(\frac{d\varepsilon_\theta}{d\varepsilon_z})^2 + \frac{d\varepsilon_\theta}{d\varepsilon_z} + 1]^{\frac{1}{2}} d\varepsilon_z = -\frac{1}{A'} \int_0^{\varepsilon_{zf}} (\frac{d\varepsilon_\theta}{d\varepsilon_z} + 1) \ d\varepsilon_z + C \quad , \tag{13}$$

where A' and C are material constants and A' = 2/3 A.

With the aid of eqs. (5) to (8) the eq. (13) can be integrated for the four upsetting condition studied in this work. The constants A and C have been determined for the five heats graphically as shown in figures 3 and 4. The values of the constants are given in Table III.

The forming limit curves shown in Fig. 1 were also calculated using Oyane's criterion, eq. (13). The comparison of the experimental results and the theory are presented in Fig. 5.

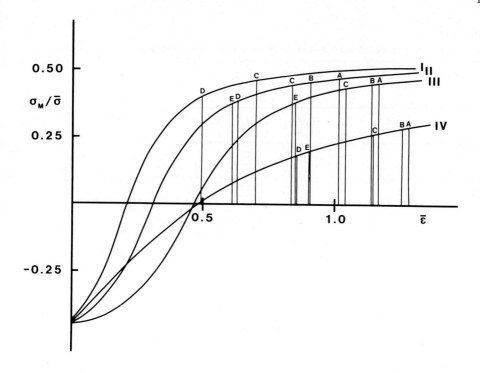

Fig. 3. The graphical determination of the integral.

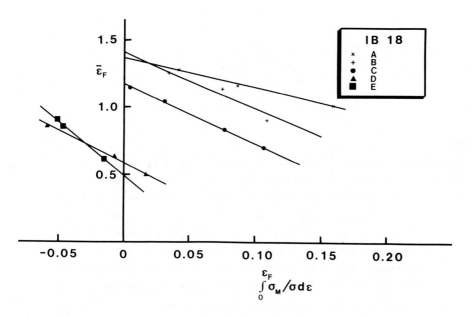

Fig. 4. The graphical determination of the constants A and C.

TABLE III

The values of the constants A and C in the Oyane criterion.

Constant	Heat				
	A	B	C	D	E
A	-0.44	-0.25	-0.23	-0.21	-0.15
C	1.37	1.40	1.17	0.58	0.50

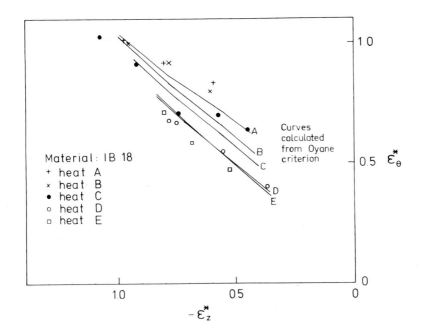

Fig. 5. The forming limit curves calculated using Oyanes criterion.

DISCUSSION

In this study the Oyane criterion for ductile fracture was used to estimate the forming limit curves in upsetting. The integration of the criterion was made possible by expressing the formula as a function of the slope of the deformation path, for which the polynomials including only ϵ_z were solved.

The theory and the test results are in a reasonable agreement the best fittings obtained for the heats D and E. In the case of heats A, B and C the theory underestimates the upsetability in sticking friction conditions but gives a good agreement in the tests with smooth dies. The difference between these two conditions is the magnitude of the hydrostatic stress, which is higher for the sticking friction conditions.

In Oyane's criterion the integral in eq. (12) takes into account the effect of hydrostatic stress component. When $\sigma_M = 0$ one obtains $\bar{\epsilon}_F = C$. When the integral has a negative value i.e. σ_M is negative (tensile) most of the time, $\bar{\epsilon}_F$ is smaller than C and when the integral has a positive value, i.e. σ_M is positive (compressive) most of the time, $\bar{\epsilon}_F$ is greater than C. This is in good agreement with the physical model of ductile fracture, which includes void initiation, growth and coalescence. The higher negative (tensile) value σ_M has the more severe is the void growth and the smaller becomes the strain at fracture. This has also been verified experimentally with the aid of outer hydrostatic pressure (ref. 15).

The results in Fig. 5 indicate that the Oyane criterion overestimates the effect of hydrostatic stress for heats A, B and C i.e. the most clean ones, while the agreement with the test results for the "unclean" heats D and E is exellent.

In spite of the discrepancy discussed here, it seems that the Oyane criterion can be used to estimate the upsetability of various materials. The criterion lists the heats in the right order from best to worst and the theoretically predicted forming limit curves lie roughly on the proper level.

The further formulation of the Oyane equation made in this study improves the useability of the criterion. Still, the determing of the constants in the formula is rather laborious. By using torsion tests with a possibility of varying the outer hydrostatic pressure could be better method to determine the constants.

CONCLUSIONS

The test results of this study show that the steel cleanliness is an important factor affecting cold upsetability.

For the test materials with about equal mechanical properties, chemical composition and surface quality but different inclusions contents, the Oyane criterion gave reasonable estimates for the forming limit curves in upsetting.

For the three cleanest heats the theoretically predicted forming limit curves overestimate the effect of hydrostatic stress component. For the two unclean heats the agreement with test results was exellent.

REFERENCES

1 Mc Clintock, F.A., A criterion for ductile fracture by the growth of holes, J. Appl. Mechanics 35 (1968) 363-371.
2 Kozasu, I., Shimizu, T., Kubota, H., The effect of nonmetallic inclusions on ductility and toughness of structural steels, Transactions ISIJ 13 (1973) 20-28.

3   Thomason, P.F., A theory for ductile fracture by interal necking of
    cavities, J. Inst. Metals 96 (1968) 360-365.
4   Edelson, B., Baldwin, W., The effect of second phases on the mechanical
    properties of alloys, Trans. ASM 55 (1962) 230-250.
5   Beachem, C.D., Yoder, G.R., Elastic-plastic fracture by homogeneous
    microvoid coalescence tearing along alternating shear planes, Met. Trans. 4
    (1973) 1145-1153.
6   Kivivuori, S., Sulonen, M., Formability limits and fracturing modes of
    uniaxial compression specimens, CIRP Ann. 27 (1978) 141-145.
7   Gill, F., Baldwin, W., Proper wiredrawing improves cold heading, Metal
    Progress 85 (1964) 83-85.
8   Lahti, I., Sulonen, M., The influence of non-metallic inclusions on cold
    upsetability of low carbon steel, Scand. J. of Metallurgy 11 (1982) 9-16.
9   Kudo, H., Aoi, K., Effect of compression test condition upon fracturing of
    a medium carbon steel, J. Japan soc. Tech. of Plasticity 8 (1967) 17-27.
10 Thomason, P.F., The use of pure aluminium as an analogue for the history of
    plastic flow, in studies of ductile fracture criteria in steel compression
    specimens, Int. J. mech. Sci. 10 (1968) 501-518.
11 Kobayashi, S., Deformation characteristics and ductile fracture of 1040
    steel in simple upsetting of solid cylinders and rings, Journal of
    Engineering for Industry, Trans. ASME 92 (1970) 391-399.
12 Lee, P.W., Fracture in cold forming of metals - a criterion and Model, PhD
    thesis, Drexel University, Philadelfia, Pa., 1972.
13 Oudin, J., Ravalard, Y., Stevenin, B., Some comments on instability,
    ductile.fracture and formability limit diagrams, Mech. Res. Comm. 6 (1979)
    51-60.
14 Sowerby, R., Chandrasekaran, N., The cold upsetting and free surface
    ductility of some commercial steels, J. Applied Metalworking 3 (1984)
    257-263.
15 Oyane, M., Criteria of ductile fracture strain, Bulletin of the JSME 15
    (1972) 1507-1513.

STRESSES AND DISLOCATION DENSITY DURING GROWTH OF SINGLE III-V

SEMICONDUCTOR CRYSTALS FROM THE MELT

J. C. LAMBROPOULOS

Department of Mechanical Engineering, University of Rochester,
Rochester, N.Y. 14627 (U.S.A.)

SUMMARY
    Finite element techniques are used in order to calculate the thermal
stresses and dislocation densities ($\{111\}<1\bar{1}0>$ slip systems) during the
Czochralski growth of single semiconductor crystals of III-V compounds with
cubic symmetry (such as GaAs or InP.) It is shown that the effect of elastic
anisotropy can not be neglected, and that it leads to higher stresses and
dislocation densities than modelling of the growing crystal as elastically
isotropic. Nevertheless, it is shown that, for crystals growing along the
<100>, <110> or <111> directions, it is possible to reduce the numerical
problem of calculating the thermal stresses from 3-D (independent variables
r, θ, and z; dependent variables the radial, circumferential and axial displacements u, v, and w) to 2-D (independent variables r,z) by properly modifying the dependence of the elastic stiffness matrix on the circumferential
coordinate θ.

INTRODUCTION
    The dislocations generated during the growth of semiconductor crystals of

GaAs or InP from the melt by the Czochralski method (Figure 1) can greatly

degrade the electronic and optical properties of the grown materials. A

recent overview has been presented by Jordan et al. (ref. 1). The presence

of dislocations has been attributed to excessive thermal stresses during

growth, following the pioneering work of Penning on Ge (ref. 2).

    The dislocation density in grown III-V crystals can greatly exceed the

density in Ge and Si. Jordan et al. (ref. 3) studied the temperature and

thermal stresses during Czochralski growth of GaAs and InP (refs. 3-5) and

showed that dislocation densities calculated from thermoelastic stresses

correlate with observed dislovation patterns. Nevertheless, the stress

calculation of Jordan et al. (refs. 1, 3-5) was approximate in that it

involved the plain strain assumption, thus being inapplicable to regions

within one crystal diameter from the crystal's ends.

    More recently, other precise stress calculations have been reported. Iwaki

and Kobayashi (refs. 6-7) investigated the effect of the Biot and Peclet

numbers,and also established the inadequacy of the plain strain approximation

near the solid-liquid interface. Duseaux (ref. 8) examined realistic thermal boundary conditions; Szabo (ref. 9) established the significance of the axial curvature of the temperature distribution in determining the thermal stresses; Lambropoulos (ref. 10) examined the attainment of quasi-steady state conditions during growth and showed that all significant stress variation, including extremum values, occurs within one radius from the solid-liquid interface. Motakef and Witt (ref. 11) established the significance of the encapsulant for the resulting dislocation density.

In all works cited above (refs. 1-11), it has been assumed that the crystal is elastically isotropic, although the material grown is a single crystal of cubic symmetry, and characterized by the elastic constants C_{11}, C_{12} and C_{44}. Some earlier work by Antonov and coworkers (refs. 12,13) established the approximate effect of elastic anisotropy. Their analysis was asymptotic in that it is valid only for long, thin cylinders, thus being inapplicable in the region within one diameter from the solid-liquid interface where all significant stress variation is known to occur (refs. 6-7, 9-11).

It is the objective of this paper to examine the accuracy of the isotropic assumption for crystals grown along the <100>, <110> or <111>directions. Complete account for anisotropy leads to a 3-D problem in which all six stress components depend on the radial, axial and circumferential coordinates. The problem is solved by using finite element techniques. Finally, it is shown that the problem can be reduced to 2-D (only radial and axial displacements which depend on r, z only) by properly modifying the elastic stiffness matrix. Such a reduction from 3-D to 2-D greatly simplifies the numerical calculation, while still accurately accounting for the effect of anisotropy and for the effect of crystal growth direction on the resulting thermal stresses and dislocation densities.

TEMPERATURE DISTRIBUTION

The axisymmetric temperature $T(r,z)$ in the growing crystal has been determined analytically under quasi-static conditions (ref. 3). The essential assumptions are that the solid-liquid interface is planar and maintained at the melting temperature T_m, and that the other sides of the crystal undergo Newtonian cooling into a medium maintained at the temperature T_a and characterized by the heat transfer coefficient h. The thermal parameters characterizing the temperature distribution are assumed to be independent of T (refs. 1-7, 9-13).

Data typical of the growth of GaAs are chosen (ref. 3) for the numerical examples to be presented in the sequel, thus

$$B = 1.0 \quad , \quad P = 0.025 \tag{1}$$

where $B=hr_0/k$ is the Biot number, and $P=p_0r_0/\kappa$ is the Peclet number. k,κ are thermal conductivity and diffusivity of the crystal, respectively. p_0 is the growth rate and r_0 is the radius (figure 1.) The aspect ratio of the crystal, L/r_0 is chosen as equal to 3, a value typical of GaAs or InP (refs. 3,10, 11).

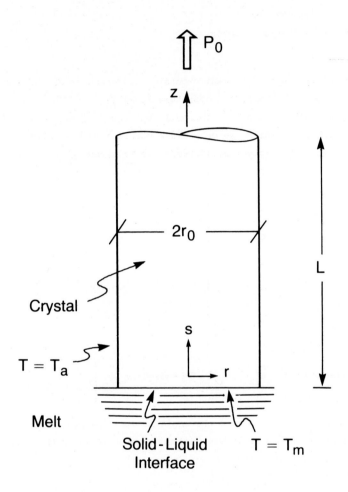

Fig. 1. The geometry of a single cylindrical crystal grown from the melt by the Czochralski method. The cooling coefficient of the surroundings is h.

ANISOTROPIC THERMOELASTIC STRESSES

Due to the cylindrical crystal geometry, the calculations are performed in the cylindrical coordinate system r, θ, z. The thermoelastic constitutive relation for the crystal is written as

$$\underline{\sigma} = \underline{C} \, (\underline{\varepsilon} - \underline{\varepsilon}^{th}) \tag{2}$$

where $\underline{\sigma}$ is the stress vector

$$\underline{\sigma} = (\, \sigma_{rr}\ \sigma_{\theta\theta}\ \sigma_{zz}\ \sigma_{rz}\ \sigma_{\theta z}\ \sigma_{\theta r}\,), \tag{3}$$

$\underline{\varepsilon}$ is the strain vector

$$\underline{\varepsilon} = (\, \varepsilon_{rr}\ \varepsilon_{\theta\theta}\ \varepsilon_{zz}\ \gamma_{rz}\ \gamma_{\theta z}\ \gamma_{\theta r}\,), \tag{4}$$

and $\underline{\varepsilon}^{th}$ is the thermal strain vector

$$\underline{\varepsilon} = \alpha\, T\, (\, 1\quad 1\quad 1\quad 0\quad 0\quad 0\,) \tag{5}$$

where α is the coefficient of thermal expansion (independent of T), and T is the temperature above the ambient temperature level T_a.

For $[001]$ growth, we denote by θ the angle between $[100]$ and an arbitrary position vector, both of which lie on a plane parallel to the solid-liquid interface. The matrix $\underline{C}$ is given by ($0<\theta<\pi/4$ due to symmetry)

$$\begin{bmatrix} C'_{11} & C'_{12} & C_{12} & 0 & 0 & C'_{16} \\ & C'_{11} & C_{12} & 0 & 0 & -C'_{16} \\ & & C_{11} & 0 & 0 & 0 \\ & \text{symm.} & & C_{44} & 0 & 0 \\ & & & & C_{44} & 0 \\ & & & & & C'_{44} \end{bmatrix} \tag{6a}$$

where

$$\left.\begin{array}{ll} C'_{11} = C_{11} + H(1-\cos4\theta)/4, & C'_{12} = C_{12} - H(1-\cos4\theta)/4 \\ C'_{44} = C_{44} - H(1-\cos4\theta)/4, & C'_{16} = H\sin4\theta/4 \end{array}\right\} \tag{6b}$$

For $[111]$ growth, we denote by θ the angle between the projection of $[110]$ onto the (111) plane and an arbitrary position vector lying in that plane ($0<\theta<\pi/3$). The matrix $\underline{C}$ is now given by

$$\begin{bmatrix} C'_{11} & C'_{12} & C''_{12} & H_c & -H_s & 0 \\ & C'_{11} & C''_{12} & -H_c & H_s & 0 \\ & & C'_{33} & 0 & 0 & 0 \\ & \text{symm.} & & C'_{13} & 0 & -H_s \\ & & & & C'_{13} & -H_c \\ & & & & & C''_{13} \end{bmatrix} \tag{7a}$$

where

$$C'_{11} = C_{11} + H/2, \quad C'_{12} = C_{12} - H/6, \quad C''_{12} = C_{12} - H/3$$

$$C'_{33} = C_{11} + 2H/3, \quad C'_{13} = C_{44} - H/3, \quad C''_{13} = C_{44} - H/6 \tag{7b}$$

$$H_c = \sqrt{2}\,H\cos3\theta/6, \quad H_s = \sqrt{2}\,H\sin3\theta/6.$$

For $[011]$ growth, θ is the angle between $[100]$ and an arbitrary vector both of which are parallel to the solid-liquid interface. Now $0<\theta<\pi/2$, and $\underline{C}$ is given by

$$\begin{bmatrix} C'_{11} & C'_{12} & C'_{13} & 0 & 0 & C'_{16} \\ & C''_{11} & C''_{13} & 0 & 0 & C''_{16} \\ & & C'_{33} & 0 & 0 & C'_{36} \\ & \text{symm.} & & C'_{44} & C'_{45} & 0 \\ & & & & C''_{44} & 0 \\ & & & & & C'''_{44} \end{bmatrix} \tag{8a}$$

where

$$\begin{aligned}
&C'_{11} = C_{11} - H(-7+4\cos2\theta+3\cos4\theta)/16, \quad C''_{11} = C_{11} - H(-7-4\cos2\theta+3\cos4\theta)/16 \\
&C'_{12} = C_{12} - 3H(1-\cos4\theta)/16, \quad C'_{13} = C_{12} - H(1-\cos2\theta)/4 \\
&C''_{13} = C_{12} - H(1+\cos2\theta)/4, \quad C'_{33} = C_{11} + H/2, \quad C'_{36} = -H\sin2\theta/4 \\
&C'_{16} = H\sin2\theta(1+3\cos2\theta)/8, \quad C''_{16} = H\sin2\theta(1-3\cos2\theta)/8 \\
&C'_{44} = C_{44} - H(1-\cos2\theta)/4, \quad C''_{44} = C_{44} - H(1+\cos2\theta)/4 \\
&C'''_{44} = C_{44} - 3H(1-\cos4\theta)/16, \quad C'_{45} = -H\sin2\theta/4
\end{aligned} \tag{8b}$$

In eqs. (6)-(8) H denotes the anisotropic parameter defined by (ref. 14)

$$H = 2C_{44} - C_{11} + C_{12} \tag{9}$$

where C_{11}, C_{12} and C_{44} are the cubic elastic constants.

It is convenient to also define the anisotropic parameter A (ref. 14)

$$A = 2C_{44} / (C_{11} - C_{12}) \tag{10}$$

and the equivalent Poisson's ratio ν

$$\nu = C_{12} / (C_{11} + C_{12}) \tag{11}$$

The parameters H, A, ν are related by

$$H = C \ (A-1) \ (1-\nu) \ / \ (1+\nu) \tag{12}$$

where C is equal to

$$C = C_{11} + C_{12} - 2C_{12}^2/C_{11} \tag{13}$$

C reduces to $E/(1-\nu)$ for isotropic materials (A=1, H=0).

It is obvious that a constitutive law such as the ones presented in eqs. (2), (6)-(8) requires a 3-D numerical calculation. Nevertheless, following a qualitative remark by Schimke et al. (ref. 15), we simplify the matrix $\underline{C}$ of eqs. (6)-(8) by assuming that the dependence on the angle θ shown in eqs. (6b), (7b), (8b) can be neglected, or equivalently, that the circumferential displacement vanishes and that the stresses are independent of θ.

Under these assumptions, for [001] growth $(0<\theta<\pi/4)$ the constitutive law becomes

$$
\begin{bmatrix} \sigma_{rr} \\ \sigma_{\theta\theta} \\ \sigma_{zz} \\ \sigma_{rz} \end{bmatrix}
=
\begin{bmatrix}
C_{11}+H/4 & C_{12}-H/4 & C_{12} & 0 \\
& C_{11}+H/4 & C_{12} & 0 \\
\text{symm.} & & C_{11} & 0 \\
& & & C_{44}
\end{bmatrix}
\begin{bmatrix} \varepsilon_{rr} - \alpha T \\ \varepsilon_{\theta\theta} - \alpha T \\ \varepsilon_{zz} - \alpha T \\ \gamma_{rz} \end{bmatrix}
\tag{14}
$$

For growth along [111] $(0<\theta<\pi/3)$

$$
\begin{bmatrix} \sigma_{rr} \\ \sigma_{\theta\theta} \\ \sigma_{zz} \\ \sigma_{rz} \end{bmatrix}
=
\begin{bmatrix}
C_{11}+H/2 & C_{12}-H/6 & C_{12}-H/3 & 0 \\
& C_{11}+H/2 & C_{12}-H/3 & 0 \\
\text{symm.} & & C_{11}+2H/3 & 0 \\
& & & C_{44}-H/3
\end{bmatrix}
\begin{bmatrix} \varepsilon_{rr} - \alpha T \\ \varepsilon_{\theta\theta} - \alpha T \\ \varepsilon_{zz} - \alpha T \\ \gamma_{rz} \end{bmatrix}
\tag{15}
$$

For growth along [011] $(0<\theta<\pi/2)$

$$
\begin{bmatrix} \sigma_{rr} \\ \sigma_{\theta\theta} \\ \sigma_{zz} \\ \sigma_{rz} \end{bmatrix}
=
\begin{bmatrix}
C_{11}+7H/16 & C_{12}-3H/16 & C_{12}-H/4 & 0 \\
& C_{11}+7H/16 & C_{12}-H/4 & 0 \\
\text{symm.} & & C_{11}+H/2 & 0 \\
& & & C_{44}-H/4
\end{bmatrix}
\begin{bmatrix} \varepsilon - \alpha T \\ \varepsilon_{\theta\theta} - \alpha T \\ \varepsilon_{zz} - \alpha T \\ \gamma_{rz} \end{bmatrix}
\tag{16}
$$

It is obvious that the constitutive laws of equations (14)-(16) represent 2-D problems, and as such are much easier to solve.

RESULTS

Both eqs. (6)-(8) and (14)-(16) were used for the 3-D finite element solution on a grid consisting of bilinear isoparametric "brick" elements. The material parameters used correspond to III-V compounds (A=2, ν=0.33). The equivalent shear stress τ_e defined by

$$\tau_e^2 = s_{ij}s_{ij}/2 \tag{17}$$

where s_{ij} is the stress deviator, was used in order to compare the thermoelastic stresses of the isotropic case (A=1), to the exact 3-D anisotropic case (A=2) of eqs. (6)-(8) and to the approximate anisotropic case (A=2) of eqs. (14)-(16). In all cases, the stress τ_e was measured in units of $C\alpha\Delta T$, where C is given by eq. (13) and $\Delta T=T_m-T_a$.

It was found that the isotropic τ_e underestimates the exact anisotropic stress τ_e, derived via eqs. (6)-(8), with a maximum error of about 40%. The approximate anisotropic stress τ_e, derived via eqs. (14)-(16), overestimates the exact anisotropic stress τ_e with a maximum error of about 6%. It is clear that the 2-D problems of eqs. (14)-(16) provide an adequate description of the exact 3-D anisotropic problems.

To estimate the effect of the anisotropic constant A on the resulting dislocation density ρ, we have used a measure for ρ that has been used extensively in the crystal growth literature (refs. 1, 3-7, 11), i.e.

$$\rho = \sum_R \tau_R^e \tag{18}$$

where the sum is over all 12 slip systems of the type $\{111\}<1\bar{1}0>$ available for the plastic deformation of crystals with the zinc-blende structure (such as GaAs and InP). The excess resolved shear stress τ_R^e is defined for each slip system by

$$\tau_R^e = \begin{cases} |\tau_R| - \tau_{CRSS} & \text{for } |\tau_R| > \tau_{CRSS} \\ \\ 0, \text{ otherwise.} \end{cases} \tag{19}$$

where τ_R is the resolved shear stress calculated by

$$\tau_R = \sigma_{ij} \, n_i \, m_j \tag{20}$$

where σ_{ij}, n_i and m_j are the components of the stress tensor, slip plane normal and slip direction, respectively (all referred to the crystallographic axes). In eq. (19), τ_{CRSS} is the critical resolved shear stress of the crystal, a level above which dislocations flow and multiply. τ_{CRSS} has a strong temperature dependence, but we have chosen the value at the melting point T_m since it is near the solid-liquid interface that all significant stressing occurs.

In units of $C\alpha\Delta T$, a typical value of τ_{CRSS} is 0.004 for GaAs or InP (ref. 1).

Figure 2 shows the dislocation density ρ for growth along <100>, <111> or <110>. The density ρ is shown near the outer periphery ($r=0.95r_0$) of the growing crystal, since it is at the outer periphery that the maximum dislocation density is observed. The plots are for that value of θ at which the maximum dislocation is observed (refs. 1,3,7).

The isotropic calculation (A=1) underestimates the exact anisotropic value of ρ (A=2), derived via eqs. (6)-(8), with a maximum error of 20-30% for <100>growth, and with a maximum error of 30-60% for <111> or <110> growth. The approximate anisotropic analysis (A=2), via. eqs. (14)-(16), overestimates the dislocation density with a maximum error of about 6%. It is concluded that, since the density of dislocations must be minimized during growth, the isotropic analysis greatly underestimates ρ. A more detailed analysis of the effect of anisotropy is forthcoming (ref. 16).

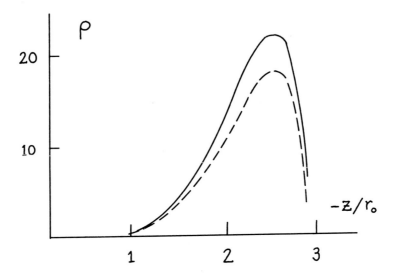

Fig. 2A . Dislocation density ρ, measured in units of $C\alpha\Delta T/100$, near the outer periphery of a cylindrical crystal grown from the melt. The growth direction is <100>. The dislocation density is shown at $\theta=0$, where the maximum value is observed. The solid line is the exact 3-D analysis (A=2). The dashed line corresponds to the isotropic case (A=1). B=1.0, P=0.025, $L/r_0=3$. The solid-liquid interface is at $z=-3r_0$.

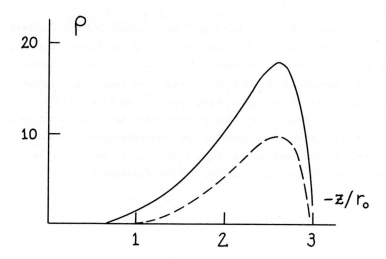

Fig. 2B. Dislocation density ρ, measured in units of $C_\alpha \Delta T/100$, near the outer periphery of a cylindrical crystal grown from the melt. The growth direction is <111>. The dislocation density is shown at $\theta = \pi/6$, where the maximum value is observed. The solid line is the exact 3-D anisotropic result (A=2). The dashed line corresponds to the isotropic analysis (A=1). B=1.0, P=0.025, L/r_0=3. The solid liquid interface is at $z = -3r_0$.

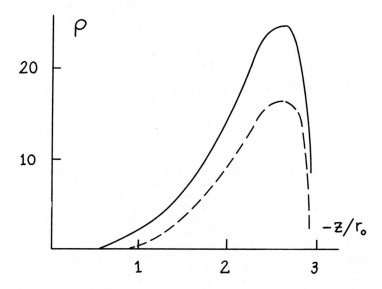

Fig. 2C. Dislocation density ρ, measured in units of $C_\alpha \Delta T/100$, near the outer periphery (r/r_0=0.95) and at the location of maximum ρ ($\theta = \pi/2$), for a cylindrical crystal grown from the melt. The solid line is the exact 3-D result (A=2). The dashed line is the isotropic case (A=1). B=1.0, P=0.025, L/r_0=3. The crystal growth direction is along <110>. The solid-liquid interface is at $z = -3r_0$.

DISCUSSION

From the previous analysis it is clear that isotropic stress calculations can underestimate the stresses by as much as 40% and the dislocation density by as much as 60% during the Czochralski growth of single semiconductor crystals of III-V compounds from the melt. Inclusion of anisotropy leads to 3-D numerical calculations, with constitutive laws such as eqs. (6)-(8). It has been shown that it is possible to reduce the problem to 2-D by essentially neglecting the dependence of stresses on the circumferential coordinate θ, leading to constitutive laws such as eqs. (14)-(16). It was shown that such simplification overestimate the stresses and the dislocation density by about 6%.

REFERENCES

1 A.S. Jordan, A.R. Von Neida and R. Caruso, The theoretical and experimental fundamentals of decreasing dislocations in melt grown GaAs and InP, J. Crystal Growth, 76 (1986) 243-262.
2 P. Penning, Generation of imperfections in Ge crystals by thermal strain, Philips Research Reports, 13 (1958) 79-97.
3 A.S. Jordan, R. Caruso and A.R. Von Neida, A thermoelastic analysis of dislocation generation in pulled GaAs crystals, Bell Sys. Tech. J., 59 (1980) 593-637.
4 A.S. Jordan, A.R. Von Neida and R. Caruso, The theory and practice of dislocation reduction in GaAs and InP, J. Crystal Growth, 70 (1984) 555-573.
5 A.S. Jordan, A.R. Von Neida, R. Caruso and J.W. Nielsen, A comparative study of thermal stress induced dislocation generation in pulled GaAs, InP, and Si crystals, J. Appl. Phys., 52(5) (1981) 3331-3336.
6 T. Iwaki and N. Kobayashi, Residual stresses in Czochralski grown crystal, J. Appl. Mech., 48 (1981) 866-870.
7 N. Kobayashi and T. Iwaki, A thermoelastic analysis of the thermal stress produced in a semi-infinite cylindrical single crystal during the Czochralski growth, 73 (1985) 96-110.
8 M. Duseaux, Temperature profile and thermal stress calculation in GaAs crystals growing from the melt, J. Crystal Growth, 61 (1983) 576-590.
9 G. Szabo, Thermal strain during Czochralski growth, J. Crystal Growth, 73 (1985) 131-141.
10 J.C. Lambropoulos, Stresses near the solid-liquid interface during the growth of a Czochralski crystal, J. Crystal Growth, 80 (1987) 245-256.
11 S. Motakef and A.F. Witt, Thermoelastic analysis of GaAs in LEC growth configuration, J. Crystal Growth, 80 (1987) 37-50.
12 P.I. Antonov, S.I. Bakholdin, E.V. Galaktionov, E.V. Tropp and S.P. Nikanorov, Anisotropy of thermoelastic stresses in shaped sapphire single crystals, J. Crystal Growth, 52 (1981) 404-410.
13 P.I. Antonov, S.I. Bakholdin, E.V. Galaktionov and E.A. Tropp, Effect of anisotropy on thermoelastic stresses appearing during the growth of shaped single crystals, Izv. Ak. Nauk SSSR, Ser. Fiz., 44(2) (1980) 255-268.
14 J.P. Hirth and J. Lothe, Theory of Dislocations, 2nd edn., Wiley, N.Y., 1982.
15 J. Schimke, K. Thomas, and J. Garrison, Approximate solution of plane orthotropic elasticity problems, Management Information Services, Detroit.
16 J.C. Lambropoulos, The isotropic assumption during the Czochralski growth of single semiconductor crystals, submitted to J. Crystal Growth (1987).

Computational Methods for Predicting Material Processing Defects, edited by M. Predeleanu   213
Elsevier Science Publishers B.V., Amsterdam, 1987 — Printed in The Netherlands

NUMERICAL ANALYSIS OF THE INFLUENCE OF VARIOUS DEFECTS ON SUPERPLASTIC FRACTURE
UNDER UNIAXIAL TENSION

Jianshe LIAN, Bernard BAUDELET and Michel SUERY
Institut National Polytechnique de Grenoble, E.N.S. de Physique de Grenoble
Génie Physique et Mécanique des Matériaux, Unité Associée au CNRS 793
Bp 46-38402 Saint Martin d'Hères, Cedex France

SUMMARY
   Necking development and fracture during superplastic tensile deformation are
analysed by considering various initial imperfections in the deformed specimen
and its environment, i.e. geometrical defect, localized initial grain size
inhomogeneity, temperature difference and cavitation. For each of the three
first cases analytic relationships giving limit strain with m-value are
presented which allow comparison between these initial defects as limiting
factors for superplastic strain. Numerical simulations performed with
reasonable values for these defects are able to explain the scatter in the m-
elongation relation for various superplastic alloys.

I. INTRODUCTION
   There has been considerable interest in the subject of plastic instability
and necking development under uniaxial tension since the analysis by
Hart(ref.1). Hart's analysis dealt with load and flow instability under
uniaxial tension, which took into account both strain hardening and strain
rate sensitivity of the material. This work have been extended in ref.2-4 that
gave a good understanding of the plastic and superplastic instability and the
development of macroscopic necking.

   It is recently recognized that the existence of internal cavities and their
growth during deformation have an important influence on superplastic necking
development and fracture(ref.5-8). Cavitation is a phenomenon common to hot
plastic deformation of many alloys and it is not uniquely confined to
superplasticity. However, because of their very high strain rate sensitivity,
superplastic alloys are able to tolerate quite large volume fraction of
cavities before failing. As a consequence, necking development and fracture of
superplastic materials are considered as processes controlled by both strain
rate sensitivity and cavity growth.

   For superplastic tensile deformation, an initial geometrical or a mechanical
defect can induce a neck and the influence of such defects has been analysed
recently(ref.4). However, in general, the defect can be thought as either an
evoluating defect, i.e. internal cavitation, or a generalized initial
imperfection which localizes deformation. In this paper a simple analysis of
necking development and fracture for superplastic tensile deformation is

presented which includes cavitation as an evoluating defect and temperature difference and initial grain size inhomogeneity along the tensile axis as initial constant imperfections.

II. ANALYSIS

II-1. Basic analysis of necking development

For the phenomenological description of uniaxial tensile deformation, Hart(ref.9) proposed a differential constitutive equation:

$$\delta(\ln\sigma) = \gamma\delta\varepsilon + m\delta(\ln\dot{\varepsilon}) \tag{1}$$

where the strain rate sensitivity parameter (m) and the strain hardening (γ) are considered as material constant. On the assumption that there is an initial geometrical defect in the specimen and from the constant volume condition and load condition, an expression of necking development can be derived(ref.4):

$$(1-f)^{1/m}\exp(-\varepsilon_N(1-\gamma)/m)d\varepsilon_N = \exp(-\varepsilon_H(1-\gamma)/m)d\varepsilon_H \tag{2}$$

where f is the initial geometrical defect $(=(A_{HO}-A_{NO})/A_{HO})$. The subscripts N and H rerer to the necked and homogeneous region of the specimen respectively. For superplastic material, γ is usually taken as zero so that eqn.(2) can be integrated to give the strain in the homogeneous region at failure:

$$\varepsilon_f = -m\ln(1-(1-f)^{1/m}) \tag{3}$$

In order to consider other types of imperfections than a geometrical defect, eqn.(2) can be written as:

$$D_f\exp(-\varepsilon_N(1-\gamma)/m)d\varepsilon_N = \exp(-\varepsilon_H(1-\gamma)/m)d\varepsilon_H \tag{4}$$

where D_f is a generalized function of defect. This function will be derived in the following for various imperfections.

II-2. D_f as a localized temperature difference

The effect on superplastic deformation of a temperature difference along the gauge length of the specimen has been considered by Hamilton(ref.10) who showed that such inhomogeneity may play a major role in observed elongation. In the present work, the neck is considered to be induced by a temperature difference which is assumed to be constant along the specimen during superplastic deformation. To describe this defect the following constitutive law is used(ref.11):

$$\dot{\varepsilon}kTexp(Q/RT) = K(b/d)^P(\sigma/G)^n \tag{5}$$

where $n=1/m$, d is grain size, p grain size index, b Burgers vertor, k Boltzmann's constant, R gas constant and Q is activation energy; G is shear modulus and K is a constant. As the other parameters in this equation are considered constant except temperature (T), it can be rewritten as:

$$\dot{\varepsilon}Texp(Q/RT) = A'\sigma^n \tag{6}$$

The condition of load equilibrium allows calculation of D_f in eqn.(4):

$$D_f(\delta T/T) = (1+\delta T/T)exp(-C\delta T/T) \tag{7}$$

with $\quad \delta T/T = (T_N - T_H)/T_H \quad ; \quad T_N > T_H \quad$ and $\quad C = Q/RT$

For a superplastic mechanism accommodated by grain boundary diffusion, C equals 10-15 and for one accommodated by lattice diffusion, C equals 23-27(ref.11). Equation (4) then becomes:

$$\varepsilon_{fT} = -mln(1-(1+\delta T/T)exp(-C\delta T/T)) \tag{8}$$

This relation gives the limit strain for a temperature difference as an initial imperfection.

II-3. D_f as an initial grain size inhomogeneity

Neck development is assumed to be induced by an initial grain size difference along the specimen. In this case from eqn.(5) we have:

$$\dot{\varepsilon} = A''\sigma^n d^{-p} \tag{9}$$

The same type of derivation as for a temperature difference yields:

$$D_f(\delta d/d) = (1-\delta d/d)^P \tag{10}$$

If the grain size remains constant during deformation (although this is not true in most cases) the corresponding expression for limit strain is:

$$\varepsilon_{fd} = -m\ln(1-(1-\delta d/d)^P)$$ (11)

with $\qquad \delta d/d = (d_H-d_N)/d_H; \qquad d_H > d_N$

II-4. D_f as cavity growth

Cavitation during superplastic deformation is a significant feature of many alloys, which was well recognized in recent years. Cavity growth mechanism is usually considered to be plasticity controlled, which gives for the evolution of cavity volume fraction with strain(ref.7,8):

$$C_v = C_{v0}\exp(\beta\varepsilon)$$ (12)

where C_{v0} is the initial cavity volume fraction and β is the cavity growth parameter. This equation was found to be in good agreement with experimental results of several superplastic alloys(ref.5).

In this case, the function D_f in eqn.(4) taking account the cavity growth is:

$$D_f = D_f(g)D_f(C_v)=(1-f)^{1/m}D_f(C_v)$$ (13)

$$D_f(C_v) =((1-C_{v0}\exp(\beta\varepsilon_N))/(1-C_{v0}\exp(\beta\varepsilon_H)))^{1/m}$$ (14)

where the geometrical part is also included for inducing a beginning of neck development. It is to be noted, however, that f can be produced by a difference of the initial cavity volume fraction along the specimen.

Equation (4) with D_f given by eqn.(13) and (14) can not be integrated analytically so that numerical calculations are needed to derive the limit strain in the case of cavity growth.

III RESULTS AND DISCUSSION

III-1 Comparison between different initial imperfections

In order to compare the effects of different kind of initial imperfections on limit strain, the geometrical defect will be considered as the reference. The geometrical defect f_{eqT} equivalent to a temperature difference $\delta T/T$ for the same limit strain is then:

$$f_{efT} = 1-(1+\delta T/T)^m\exp(-mC\delta T/T)$$ (15)

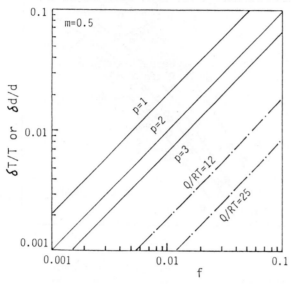

Figure 1 Comparison between initial grain size inhomogeneity
or temperature difference and initial geometrical
imperfection for their influences on limit strain

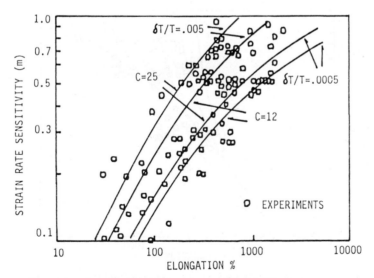

Figure 2 Experimental plot of m-elongation relationship for
various materials (ref.12,13) in comparison with
theoretical prediction in the case of a temperature
difference as an imperfection along the tensile axis

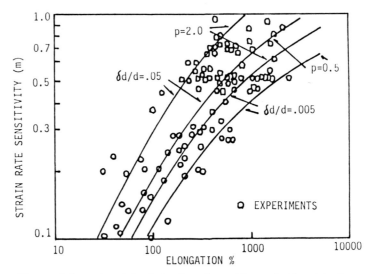

Figure 3 Experimental plot of m-elongation relationship for
various material (ref.12,13) in comparison with
theoretical prediction in the case of a grain size
inhomogeneity as an imperfection along the tensile axis

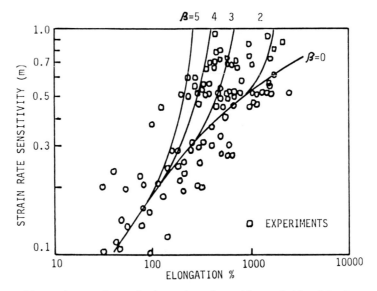

Figure 4 Experimental plot of m-elongation relationship for
various materials (ref.12,13) in comparison with
theoretical prediction in the case of cavity growth
with different cavity growth rate (β)

Similarly for the grain size inhomogeneity we have:

$$f_{efd} = 1-(1-\delta d/d)^{mp} \qquad\qquad (16)$$

The comparison between these defects is given in figure 1 for m=0.5. It is shown that $\delta d/d$ (with p=1, 2 and 3) has approximately the same effect on limit strain as the geometrical defect f. Higher limit strains are, however, obtained with smaller p-values. For p=1/m, the effects of $\delta d/d$ and f are exactly equivalent. This result has to be discussed in the light of the grain size values for a superplastic material and the scatter encountered in its determination. Indeed a geometrical defect f of 5% which can be considered as very damaging corresponds to a difference in grain size of only 0.5 micron for an average grain size of 10 microns, which is usually within the scatter of the measurement. This result leads to the conclusion that very precise determination of grain size is necessary to detect any difference along the specimen which would have an influence on the limit strain. However, the real situation is more complex since 0.5 micron is generally much smaller than the widthness of the distribution of the grain size in a specimen. Another analysis has to be performed to take into account such a distribution and any difference from the average distribution. Nevertheless the calculation shows that a grain size inhomogeneity, if it is unambiguously detectable, will greatly affect the limit strain of a superplastic material.

A temperature change $\delta T/T$ is also a dramatic imperfection in comparison with a geometrical defect, as can be seen in fig.1. Indeed, a small local increase of temperature will result in a large decrease of limit strain, since a change of temperature of 0.1% has the same effect as a geometrical defect of about 1%. This result leads to the conclusion that precise control of temperature along the gauge length is necessary to get uniform elongation. From the present analysis, the limit strain also depends on the material. For the same temperature difference, a material deforming by a mechanism controlled by grain boundary diffusion will exhibit a higher limit strain than a material with a lattice diffusion mechanism.

III-2. Prediction of m-ε_f relationship

Experimental data of tensile elongation versus strain rate sensitivity (using Woodford's original data(ref.12) with additional results from ref.13) are presented in figure 2, 3 and 4. In the figures are also plotted the theoretical curves of the limit strain for imperfections being: temperature change, grain size inhomogeneity and cavitation. In each case, reasonable values are considered, that is: $\delta T/T$ between 5×10^{-4} and 5×10^{-3} (fig.2) (which corresponds to 0.5k to 5k at 1000k) and $\delta d/d$ between 0.5% to 5%, with p of 0.5 and 2(fig.3). For the effect of cavity growth, C_{vo} is taken as 0.001 and f as 0.005 with values of β in the range of 0 to 5(fig.4). Comparison between theoretical plots and experimental data shows that the discrepancy in the observed elongation for a given m-value can easily be explained by considering at least one of the previously mentioned defect. This result thus suggests that elongation to fracture of superplastic material is a very sensitive characteristic, a high strain rate sensitivity (m) being a necessary but not sufficient condition for high elongation.

IV CONCLUSION

Superplastic fracture is a process of necking development. High values of the strain rate sensitivity parameter (m) are essential for providing a large uniform deformation. However early fracture can be induced by various defects. A generalized defect function is introduced into a differential equation which expresses necking development during superplastic deformation under uniaxial tension. This defect function can characterize a geometrical defect, a temperature difference, an initial grain size inhomogeneity both along the tensile-axis or an internal evoluating defect such as cavitation. Equivalence between these defects has been establised.

REFERENCES

1 E.W.Hart, Theory of the tension test, Acta Metall., 15 (1967) pp351-355
2 J.W.Hutchinson and K.W.Neale, Influence of strain rate sensitivity on necking under uniaxial tension, Acta Metall., 25 (1977) pp839-846
3 F.A.Nichols, Plastic instability and uniaxial tensile ductilities, Acta Metall., 28 (1980) pp663-673
4 J.Lian and B.Baudelet, Necking development and strain to fracture under uniaxial tension, Mater.Sci.Eng., 84 (1986) pp157-162
5 M.J.Stowell, Cavity growth and failure in superplastic alloys, Metals Science 17(1983) pp1-11
6 H.M.Shang and M.Suery, Modelling of cavitation in two phase alloys under uniaxial stress systems, Metals Science, 18(1984) pp143-152
7 M.Suery, Cavitation in superplastic alloys theoretical, in B.Baudelet and M.Suery (Ed.), Superplasticity, Paris, 1985 pp9.1-19
8 J.Lian and S.Suery, Effect of strain rate sensitivity and cavity growth rate on failure of superplastic material, Mater.Sci.Tech., 2(1986) pp1093-1098
9 E.W.Hart, A phenomenological theory for plastic deformation of polycrystalline metals, Acta Metall., 18(1970) PP599-610
10 C.H.Hamilton, Superplasticity, Strength of metals and alloys (Ed) H.J.Mcqueen, J.P.Bailon, J.J.Dickson, J.J.Jonas and M.G.Akben, Montreal, Canada, v3 (1985) PP1831-1857
11 A.Arieli and A.K.Mukherjee, The rate-controlling deformation mechanisms in superplasticity--Acritical assessment, Metall.Trans., A13(1982) pp717-732
12 D.A.Woodford, Strain rate sensitivity as a measure of ductility, Am.Soc.Metall.Trans.Q 62(1969) p291
13 J.W.Edington, K.N.Melton and C.P.Culter, Superplasticity, Prog.Mater.Sci., 21 (1976) pp61-170

Computational Methods for Predicting Material Processing Defects, edited by M. Predeleanu
Elsevier Science Publishers B.V., Amsterdam, 1987 — Printed in The Netherlands

Finite Element Simulation of Bending Process of Steel-Plastic Laminate Sheets

A.MAKINOUCHI[1], H.OGAWA[2] and K.HASHIMOTO[3]

[1]Deformation Processing Laboratory, RIKEN (The Institute of Physical and Chemical Research), Wako, Saitama, 351-01 (Japan)

[2]The Institute of Vocational Training, Sagamihara, Kanagawa, 229 (Japan)

[3]R & D Laboratories.II, Nippon Steel Corporation, Sagamihara, Kanagawa, 229 (Japan)

SUMMARY

A steel/polymer/steel three layer laminate is a new composite sheet which recently starts to be used in automobile parts and panels for the purpose of vibration damping and weight reduction. Bending process of the laminate sheets is simulated by an incremental elastic-plastic finite element method under plane strain condition with taking special care for description of boundary conditions. The simulation clearly demonstrates the process of generation of folding defect in the laminate sheet at the bent flange, and calculated sheet geometries agree very well with experiment.

INTRODUCTION

A steel/polymer/steel three layer laminate is a new composite sheet which is recently used in several automobile parts and panels. Two types of steel/polymer/steel laminate are available on the market. One type is " a light weight laminate steel sheet" and the other is "a vibration damping steel sheet", and the structures are illustrated in Fig.1. The light weight laminate steel sheet is composed of rather thin skin steel layers and a thick polymer core layer, so that it can maintain almost equal bending rigidity as of a simple steel sheet of same total thickness in spite of the weight reduction by 30% to 50%. On the other hand, the vibration damping steel sheet has a very thin polymer layer between two steel sheets, and has excellent noise-insulating and vibration-damping properties.

In spite of these superior service properties, the steel/polymer/steel laminates do not gain big usage in automobile industry so far. A main reason for this exists in the fact that several geometrical defects take place in forming processes. The defects are typically observed in bending operation. One example of the defects is the folding of the sheet at V-bent flange, which never arises in bending of the simple steel sheet. Another example is a localized necking which sometimes occurs in the outer skin steel layer of bent portion, even at non-severe deformation stage.

222

The purpose of this work is to provide an elastic-plastic finite element computer program to simulate the entire bending process and to predict such defects precisely. The computer program developed here is based on the updated Lagrangian formulation (ref.1), and deals with the bending problem as a plane strain deformation (ref.2). Therefore, no special assumption such as the Kirchhoff-Love assumption, which is commonly adopted in the sheet analysis, is assumed. Mechanically exact boundary conditions are formulated to deal with the highly non-linear sheet-tool contact problem in incremental way, and in order to maintain the boundary conditions unchanged within one step, an extended r-minimum method (ref.3) is utilized to limit the size of each incremental step.

The V-bending process of four different laminate sheets cited in Table 1 are simulated and the results are compared with experiments.

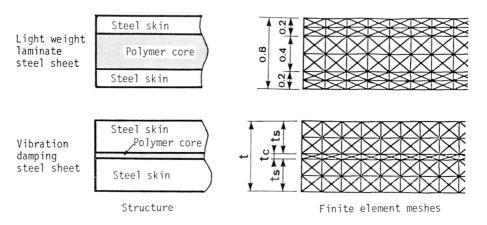

Structure   Finite element meshes

Fig.1   Structure of two types of steel/polymer/steel laminate sheet.

TABLE 1

Four different laminate sheets used in the experiment and the calculation.

Type of laminate sheet		Thickness of each layer (mm) ts / tc / ts	Polymer in intermediate layer
Light weight laminate steel sheet	Nylon-core	0.2 / 0.4 / 0.2	Nylon 6
	Polypropylene-core	0.2 / 0.4 / 0.2	Polypropylene
Vibration damping steel sheet	For high temperature use	0.4 / 0.07 / 0.4	Acrylic acid copolymer
	For normal temperature use	0.6 / 0.04 / 0.6	Polyester

FINITE ELEMENT MODELING

Material model

Two important problems must be considered in order to model the deformation process of the steel/polymer/steel laminate sheet. One is description of the visco-elastic-plastic constitutive equation of polymers used as core material and the other is prediction of the adhesive state of steel-plastic interface.

The J_2-flow law

$$\overset{\circ}{\tau}_{ij} = \frac{E}{1+\nu}\left[\delta_{ik}\delta_{jl}+\frac{\nu}{1-2\nu}\delta_{ij}\delta_{kl}-\frac{3\alpha(\frac{E}{1+\nu})\sigma'_{ij}\sigma'_{kl}}{2\bar{\sigma}^2(\frac{2}{3}H'+\frac{E}{1+\nu})}\right]D_{kl} . \tag{1}$$

is employed as a constitutive equation of polymers as well as of steels. In this equation $\overset{\circ}{\tau}_{ij}$ is the Jaumann rate of Kirchhoff stress, σ_{ij} is the Cauchy stress, D_{ij} is the rate of deformation which is the symmetric part of the

TABLE 2
Material constants used in the calculation.

	Material	Stress-plastic strain relation	Yield stress $(kgf \cdot mm^{-2})$	Young's modulus $(kgf \cdot mm^{-2})$	Poisson's ratio
1	Steel	$\bar{\sigma}= 55.9(0.02+\bar{\varepsilon}^p)^{0.193}$	26.4	17160	0.3
2	Nylon 6	$\bar{\sigma}= 13.5(0.1+\bar{\varepsilon}^p)^{1.6} + 5.46$	5.8	280	0.33
3	Polypropylene	$\bar{\sigma}= 3.0(0.1+\bar{\varepsilon}^p)^{2.0} + 2.37$	2.4	85	0.33
4	Acrylic acid copolymer	$\bar{\sigma}= 0.58(0.6+\bar{\varepsilon}^p)^{1.3} + 1.2$	1.5	20	0.33
5	Polyester	$\bar{\sigma}= 0.12(0.5+\bar{\varepsilon}^p)^{1.2} + 0.25$	0.3	10	0.33

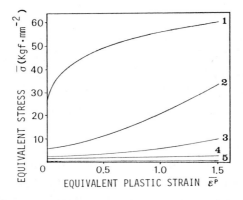

Fig.2 Stress-strain relations of a steel and polymers which compose the laminate sheets.

velocity gradient $L_{ij}(=\partial v_i/\partial x_j)$, H' is the plastic hardening ratio, E is the Young's modulus, ν is the Poisson's ratio, and α takes 1 for the plastic state and 0 for the elastic state or the unloading.

The J_2-flow law expresses behaviour of a rate and temperature independent isotropic material, so that is not accurate enough to model visco-elastic-plastic deformation of polymers which is the strongly affected by the strain rate, the temperature and the strain-induced anisotropy. However, in this study we dare employ (1) as the constitutive equation of four different polymers, because this equation is simple and all the material constants can be determined from only the uniaxial tension test.

The equivalent stress-equivalent plastic strain relation of polymer is represented by a modified power law of the form

$$\overline{\sigma}=c(\varepsilon_0+\overline{\varepsilon}^p)^n+b \tag{2}$$

The constants in this equation are determined from uniaxial tension test of sheet polymers used as core material as well as of the steel sheet used for skin layer, and are given in Table 2 together with the elastic constants. These stress-strain relations are plotted in Fig.2.

Complete adhesion between steel and polymer is assumed in the calculation, which means that no peeling and no sliding exist at the interface. In calculation two different finite element meshes shown in Fig.1 are used to model the structure of the vibration damping steel sheets and the light weight laminate steel sheets.

Basic equations for computer program

The variational principle is written (ref.1) by

$$\int_v \{ (\overset{\circ}{\tau}_{ij}-2\sigma_{ik}D_{kj})\delta D_{ij}+\sigma_{jk}L_{ik}\delta L_{ij}\} \ dV =\int_s \overset{\cdot}{t}_i\delta v_i dS. \tag{3}$$

where x_i and v_i represent the position and the velocity of particle respectively and $\overset{\cdot}{t}_i$ denotes the prescribed surface traction rate.

The constitutive law (1) and the work-hardening equation (2) are incorporated into the variational principle (3), and provide bases for the incremental elastic-plastic finite element computer program.

Boundary conditions

The sheet bending process is analyzed under the plane-strain condition. Dimension of tools used in the experiment is shown in Fig.3(a). Because of the symmetry with respect to the center line AB, only the right half of the tool and workpiece are modeled as in Fig.3(b). Components of vectors and tensors are referred to the global coordinate system (x,y,z) or a local coordinate system (l,n,z). The l-axis of the local coordinate system is taken along a longitudinal fiber of the sheet and the n-axis is normal to l.

Consider that at a certain stage of the bending process, the nodes P_1 and P_2 on the upper surface BD of the sheet are in contact with the punch as shown in Fig.3(b). The boundary conditions at P_1 and P_2 are given by

$$\dot{f}_l = 0 \ and \ v_n = 0, \tag{4}$$

assuming zero friction force on the punch-sheet interface. Here, $\dot{f}_l$ is the l-component of the rate of the nodal force vector and v_n is the n-component of the nodal velocity vector. The nodal forces f_n at P_1 and P_2 are examined in the subsequent step, and if any of such forces becomes zero, the condition at the corresponding point are changed to those of a free boundary

$$\dot{f}_x = 0 \ and \ \dot{f}_y = 0 \tag{5}$$

for the next step. The positions of the nodes other than P_1 and P_2 on the upper surface are also examined in the subsequent step, and if a node, say P_3, reaches the punch surface, the boundary conditions at P_3 are changed to (4).

Since the die radius used in the experiments is very small ($r_d = 0.3mm$), it is assumed in the calculation that only one node D_1, which is nearest to the die edge, is in contact with the die. Care must be taken to describe the boundary condition for D_1, because they are referred to the local coordinate system which rotates with a material element. Consider the nodal force $\underset{\sim}{f}$ acting to the node D_1, which has only n-component, $\underset{\sim}{f} = f_n \underset{\sim}{n}$, because of no friction and then rate of $\underset{\sim}{f}$ is written in the form

$$\underset{\sim}{\dot{f}} = \dot{f}_n \underset{\sim}{n} + f_n \underset{\sim}{\dot{n}}. \tag{6}$$

Since the rate of the base vector $\underset{\sim}{\dot{n}}$ is given by $\underset{\sim}{\dot{n}} = -\dot{\theta} \underset{\sim}{l}$ using the angle of the

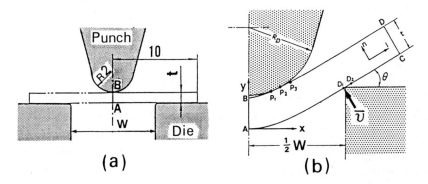

Fig.3 Geometry and dimension of tools and a specimen. (a) Set-up. (b) Boundary conditions for the calculation.

local coordinate system θ in Fig.3(b), substitution of this equation into (6) yields

$$\dot{f} = \dot{f}_n \underset{\sim}{n} - \dot{\theta} f_n \underset{\sim}{l} = -\frac{\overline{v}}{R} f_n \underset{\sim}{l} + \dot{f}_n \underset{\sim}{n} .$$

Here we used a relation

$$\dot{\theta} = v_n/R = \overline{v}/R$$

in which $\overline{v}$ and R represent the prescribed velocity and the rate of rotation at the point D_1 respectively. Therefore, the boundary conditions

$$\dot{f}_l = -\overline{v} f_n/R \qquad \text{and} \qquad v_n = \overline{v} \tag{7}$$

are given for D_1. The radius R is determined in the previous step using the relation $R = v_n/\dot{\theta}$. When the next node D_2 becomes the nearest to the die edge instead of D_1 at some later stage, several transitional steps satisfying the conditions

$$f_l = 0 \text{ and } f_n = -f_n \quad \text{for } D_1, \text{ and}$$

$$f_l = 0 \text{ and } v_n = 0 \qquad \text{for } D_2 \tag{8}$$

are used until f_n vanishes at the node D_1.

Note that in the computer program the step size is limited not only by yielding of one element but also by the boundary conditions. The step size is therefore the smallest load or displacement increment among the following three: (i) the increment required just to allow one element in the elastic state to reach the yield point; (ii) the increment which allows one free node P_3 to reach the punch surface; and (iii) the step sizes which make the nodal forces at P_1, P_2, and D_1 become just zero. By adopting this method, linearity of the equations is maintained within each step. Moreover, the nodes in contact with the tools are chosen automatically and the boundary conditions at each step are determined by the computation itself throughout the entire bending process.

EXPERIMENT

The four laminate sheets given in Table 1 are bent by using tools illustrated in Fig.3(a). In order to study deformation of each layer of the laminates, a striped pattern is printed on the edge surface of the sheet specimen and the deformed pattern is photographed at certain stages of the bending process. The relation between punch travel and forming force is recorded during the process. Teflon film lubricant is used to realize satisfactory lubrication between the tool and the sheet, corresponding to the no friction condition which is assumed in the calculation.

RESULTS

Fig.4 shows deformed geometries of the light weight laminate steel sheet with polypropylene core at four different bending stages. The calculated geometries which are shown in the right of the figure are compared with photographs taken at similar bent stages in the experiment. Both geometries at each stage agree very well each other. Folding of the sheet in bend flange is clearly observed both in the calculation and the experiment. The folding starts already at the bent angle $\theta=15°$, and increasing its degree as the process proceeds. It is clearly seen that strong shear deformation is produced in the polypropylene layer in the die cavity but no shear deformation exists outside the die edge. The sudden change of shear strain in the polypropylene layer along the sheet causes the folding defect in the bent flange. The shear between two skin steel layers ΔS is plotted in Fig.5 to clarify the magnitude of shear

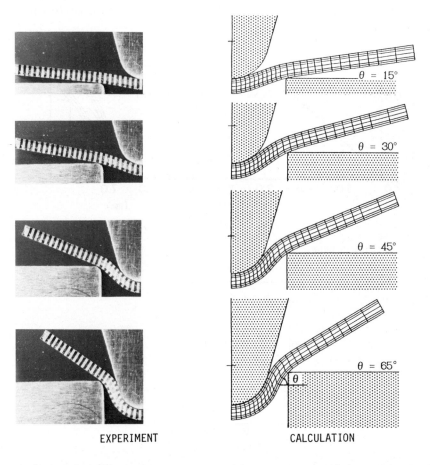

EXPERIMENT   CALCULATION

Fig.4 Deformed geometries in bending process of the light weight laminate steel sheet with polypropylene core. (W = 6 mm)

deformation in polypropylene core. The distribution of ΔS indicates the characteristic nature of the bending deformation of the laminate sheet, and the calculated distribution agrees very well with the experiment.

Fig.6 shows the punch travel – forming force relation. Good agreement between the calculation and the experiment is obtained also in this relation. The sudden decreases in the forming force in the calculated curve correspond to the calculation stage in which the finite element node in contact with the die edge shifts one to another under the boundary conditions (8).

Fig.7, Fig.8 and Fig.9 are deformed geometries of three different laminate sheets. The shear between two skin steels is observed along entire sheet length in bending of the vibration damping steel sheet for normal temperature use (Fig.9) in which very soft core material ,polyester, is used.

CONCLUSIONS

A finite element computer program is developed to simulate bending process of the steel/ polymer/steel laminate sheets.

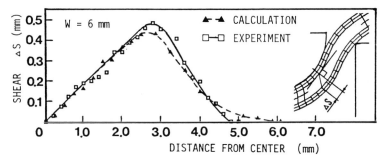

Fig.5 Distributions of shear between two skin steel layers along sheet length at the stage of θ=65° for the light weight laminate steel sheet with polypropylene core.

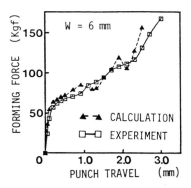

Fig.6 Punch travel-forming force relations in bending process of the light weight laminate steel sheet with polypropylene core.

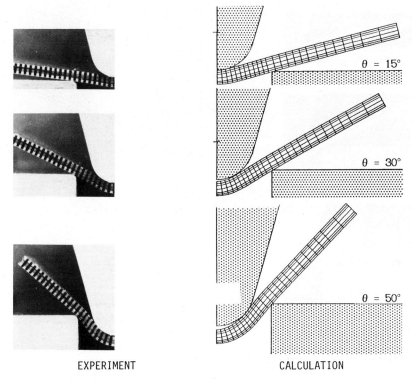

EXPERIMENT   CALCULATION

Fig.7 Deformed geometries in bending process of the light weight laminate
      steel sheet with nylon-6 core. (W = 6 mm)

The calculation clearly demonstrates the folding of sheet at the bent
flange caused by the concentrated shear deformation in polymer layer, and the
calculated results agree very well with the experiments.

ACKNOWLEDGEMENT
      The authors would like to thank Dr.Y.Nagai of Press Kogyo Co. for his help
in making the experimental apparatus, and also appreciate Mr.S.Yoshida's help
in the calculation and preparation of the paper.

REFERENCES

1 R.M.McMeeking and J.R.Rice, Finite element formulations for problems of
   large elastic plastic deformation. Int.J.Solids Struct., 11 (1975) 601-606.
2 A.Makinouchi, Elastic-plastic stress analysis of bending and hemming of
   sheet metal. Computer modeling of sheet metal forming process, in: N.M.Wang
   and S.C.Tang (Ed.), The Metallurgical Society of AIME, 1985, pp.161-176.
3 Y.Yamada, N.Yoshimura and T.Sakurai, Plastic stress-strain matrix and its
   application for the solution of elastic- plastic problems by the finite
   element method, Int.J.Mech.Sci., 10 (1968) 343-354.

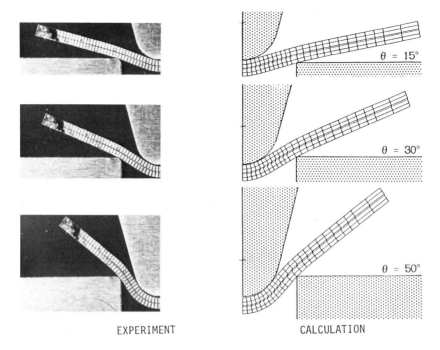

EXPERIMENT   CALCULATION

Fig.8 Deformed geometries in bending process of the vibration damping steel
       sheet for high temperature use. (W = 6 mm)

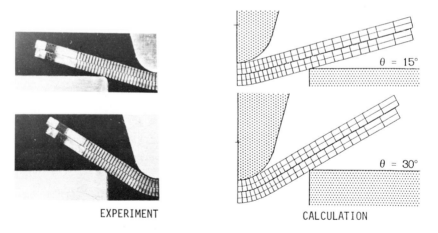

EXPERIMENT   CALCULATION

Fig.9 Deformed geometries in bending process of the vibration damping steel
       sheet for normal temperature use. (W = 8 mm)

Computational Methods for Predicting Material Processing Defects, edited by M. Predeleanu   231
Elsevier Science Publishers B.V., Amsterdam, 1987 — Printed in The Netherlands

DEFECTS IN THE PROCESSING OF METALS AND COMPOSITES

A.G. MAMALIS[1] and W. JOHNSON[2]
[1] Department of Mechanical Engineering, National Technical University of Athens (Greece)
[2] Faculty of Engineering, Purdue University, Indiana (USA)

SUMMARY
   Physical defects which occur during massive and sheet metal
forming are surveyed. The defects are treated descriptively but
metal plasticity theory is employed to explain some of them.
Defects arising during the processing of composite materials
(e.g. fibre-reinforced materials, metal-matrix composites, cera-
mic-matrix composites and bonded materials) are also documented.

INTRODUCTION
   Defects and limitations to material processing which can and
do arise in products are deserving of attention because their
economic consequences can be great. The spectrum of material
working and fabrication defectiveness embraces (ref. 1) the following:
(a) The occurrence of defects due to interaction between the
workpiece material, the tooling and the friction between the
latter and the process geometry;
(b) some forms of metallurgical structure which result from
purely mechanical action;
(c) the limits of performance imposed by the material properties
themselves with a given tooling and stressing system;
(d) elastic springback and generated residual stresses.
   Interaction of the above-mentioned features during material
processing makes it difficult to account precisely for the defects encountered in terms of the mechanics;   certain defects are
associated with particular processes whilst some defects are
peculiar to some materials. Until recently defects have tended to
draw little attention from academic researches into the mechanics
material processing because are mostly devoted to considering
ideal materials and processes which work successfully.
   Below, a wide range of defects are discussed which occur

during massive and during sheet metal processing; also, defects
arising during the processing of composite materials is now able
to be fairly well documented. The defects are treated descriptively but metal plasticity theory is employed to explain the
former; on the whole not much analysis has been applied to quantify their onset. The list of defects described is by no means
exhaustive and readers are referred to the other articles (refs.
1-3) for some detailed studies and an extensive list of references to particular defects.

Metal working defects of metallurgical origin, i.e. pipes,
seams and segregation, etc., in ingots, are not reviewed in this
paper but an extensive treatment of such can be found in the book
by Engel and Klingele (ref. 4).

PROCESSING OF METALS

Table 1, mainly reproduced from refs. (1) and (2), lists the
major defects that arise in both massive and sheet metal forming
operations, whilst in Table 2, taken from ref. (1), some surface
defects that are encountered in either one particular process or
a number of them are summarised. Some characteristic common
defects are outlined below.

Massive forming

Failures occurring during such compression- like massive working processes (rolling, forging, extrusion, piercing, drawing of
rod, sheet, wire and tube) are predominantly caused by so-called
"secondary" tensile stresses. Among other factors governing
surface and internal cracking are the ductility of the materials,
the temperature and residual stress field (and the magnitude of
the local hydrostatic pressure) around the position of the crack.

(i) Fracturing is observed in the rolling of slabs at the
edges (edge-cracking) where longitudinal tensions (or strains)
may develop under certain rolling conditions, or along the cen-
tre-plane leading to alligatoring (about mid-thickness) or in-
zippering (through thickness fracture) (ref. 5). As pointed out
elsewhere (ref. 2), the formation of internal or surface cracks
during the rolling of heavy ingots in a blooming mill, when the
reduction per pass is small basically is encouraged by inhomogeneous deformation.

Edge-cracking in rolling was examined by Schey (ref. 6) and

three causes adduced for its occurrence, viz. limited ductility, variation of the stresses across the width of the rolled material and uneven deformation at the edges. In subsequent passes, "overhanging" material is not directly compressed but forced to elongate and therefore subjected to longitudinal tensile stresses leading to cracking; see also the work on edge-cracking in cold-rolling (ref. 7), hot-rolling (ref. 8), sandwich-rolling (ref. 9) and ring-rolling (ref. 1).

Longitudinal cracking on the curved surface of a billet when upsetting is due to the presence of secondary tensions. Thomason's work (ref. 2) studying longitudinal surface defect influence on ductility in cold-heading and upsetting is noteworthy; very usefully an index of surface quality of cold-heading of wire was proposed, (see also ref. 10).

(ii) Central cracking is a common defect in light-pass rolling, wedge-rolling, helical- or transverse-rolling and in rotary-forming operations; an axial fissure is observed in two-roll transverse-rolling or an annular one in the three-roll process. Large inclusions in the central regions of billets strongly predispose material to cracking. Another cause of cracking is believed to be excessive "kneading" of a section after forming, due to excess metal being trapped in this section (ref. 2).

A similar centre cavity may be formed also in a forged round billet because of the development of tensile stresses. The introduction of reliable criteria for the onset of void formation or crack initiation in ductile metals, in regions of complex stress, should soon make it possible to calculate or anticipate the site of creation of these costly internal fractures in metal forming processes; (see Fig. 1(a) and ref. 11).

The reasons for axial cavity formation or annular fissure development may, to some extent, be understood with the help of plane strain slip-line field solutions for forging. The approach of two opposed rigid flat-ended indenters will tend to open-up a billet at its centre due to heavy shearing. Any existing or incipient crack may lead to central voids being formed when the cohesive strength of the structure is exceeded in the presence of plastic straining, especially if the ratio of the die width to distance apart is sufficiently small; this geometry will cause tensile stresses towards the middle of the product. Nasmyth introduced his vee-anvil to avoid this defect in effect by using three intenters or dies. To combat therefore centre-bursting in open-die forging, flat dies may be replaced by curved or so-called swaging dies which introduce favourable lateral compression. The extrusion-forging of materials possessing a limited ductility is, therefore, more likely to be successful than closed-die forging, which in turn is superior to open-die forging (ref. 2).

When using dies of appropriate shape to develop a system of compressive stresses, it is possible to close existing internal cuvities (e.g. shrinkage cavities in ingots) during upsetting operations and to produce good material properties at healed positions; dies with dished working faces have been suggested (refs. 2,12).

Central-bursts (or chevrons), i.e. internal arrow-shaped defects, are occasionally encountered in the cold-extrusion and wire-drawing of round bars of steel (see Fig. 1(b)). An upper-bound approach for formulating a central-burst fracture criterion has been attempted (see refs. 13 and 14).

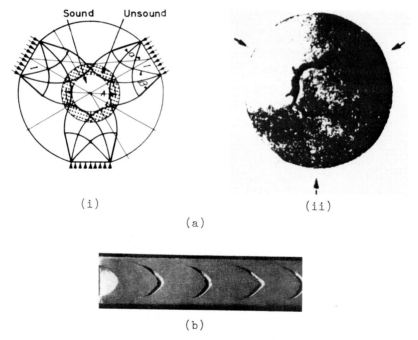

(i)

(ii)

(a)

(b)

Fig. 1. (a) Indentation by three equal size dies spaced at 120° to each other. (i) Slip-line field. (ii) Central circular cavity. (b) "Cup and cone" defect in copper wire drawing.

(iii) Unwanted end shapes can be developed particularly in the rolling of slabs and blooms and eradicating them by cutting them off can have important consequences for production costs. The form encountered at the front end of an initially square-ended ingot or slab after rolling is termed overhang due to folding over of the ingot head and that at the rear when well developed combines overlap and fish-tail; this is due to tail-end folding in the direction of its thickness for overlap whilst the width effect produces the fish-tail (see Fig. 2). These ends, which are cut off, constitute a crop-loss and may account for about 5% of a total throughput of bloom or slab weight. Even a small average

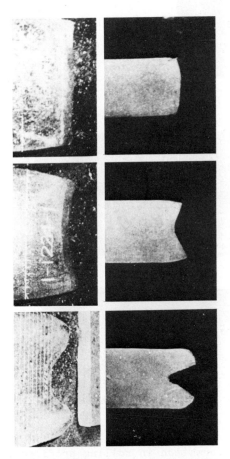

Fig. 2. Overlap development with reduction in actual slabs (left) and Plasticine models (right).

236

percentage reduction in the crop-loss of the annual volume of production of a mill leads to substantial material and monetary savings. Some experimental results and useful conclusions about these end defects as a result of experiments with Plasticine are available (refs. 1,2); N.K.K. of Japan have reported reducing their crop-losses substantially after modelling the phenomenon with Plasticine. Upper-bounds solutions for overlap developments have been suggested, (see refs. 1 and 2).

Fish-tail formation, i.e. an irregular and non-rectangular spread profile with concave edges is a defect also developed in ring-rolling, (see ref. 15 and Fig. 3(a)).

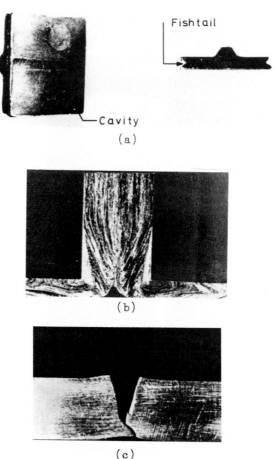

Fig. 3. (a) Cavity formation in rolling profiled rings.(b) Piping defect in extrusion-forging. (c) Depression formed in the rear surface of a strip during cutting.

(iv) In rolling a round or a section a type of fold developes if a flash or fin is formed in one pass (due to metal expressing itself between rolls outside of the section) and pressed into the metal during the subsequent pass (ref. 2). The promotion of these defects is aided by <u>overfilling</u> or <u>underfilling</u> passes. As just implied, in overfilling - "too much metal enters the pass" - and the result of a rolled-in lap occurs when the bar is turned through 90° on the next pass. Underfilled passes may lead to later passes turning sideways ("drunken" passes) (ref. 1).

<u>Folding</u> due to buckling is a common mode of failure in upsetting and heading operations, e.g. in upsetting the head of a cartridge case. However, buckling is not the only reason for folds and laminations. Folds at the edges of hammer-forged products originate frequently from too small reductions per pass when local deformation occurs near the surface; compare this with the rolling of ingots mentioned above. With subsequent passes involving heavy reductions, the metal may spread from both surfaces over the centre part and form folds or laminations. Folds or laps are also formed whenever metal folds over itself during die-forging, i.e. a flash or fin is formed in one stroke and pressed into the metal during a subsequent one (ref. 2).

(v) A well-known defect occurring at the rear end of extruded products is that of a <u>sinking-in</u> of the material, i.e. the beginning of piping or cavity formation on the bottom of a slug; it occurs during the final unsteady state phase of extrusion when the slug thickness has become greatly reduced. Upper-bounds for the prediction of this defect have been suggested (ref. 16).

<u>The end-extrusion defect</u> was recognised very early on, and on this occasion it is perhaps very appropriate (in France) that we should represent it by drawings given in the works of Henri Tresca (ref. 17), (see Figs. 4 and 5). The corresponding defect with side-extrusion is shown in Figs. 5(a), (b), (c), (e), (f) and (h). In a manner of speaking the curvature of the two side-extruded products, (see Figs. 5(e), (f), (g) and (h)), are noteworthy in that being curved rods then can be of no commercial value. This, therefore, poses the problem of how to determine the position of die orifices so that rotation is not caused. A problem with end asymmetric single rod extrusion or end multi-extrusions (where curvature may also occur here) is that of the product is not always normal to the orifice. Not merely would the

238

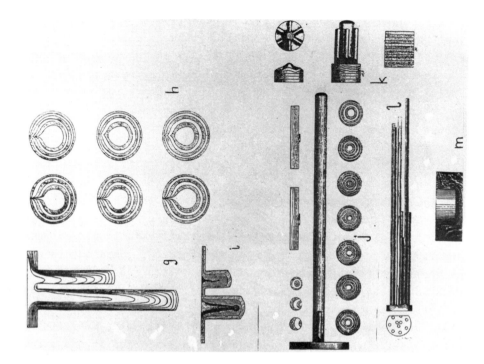

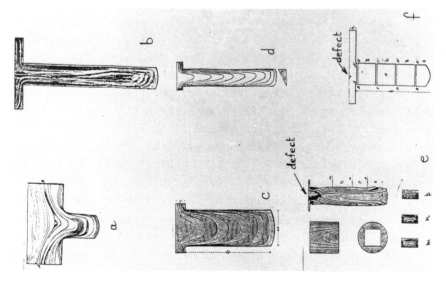

Fig. 4

product cross-section (if the orifice was circular) not be circular (see the exaggerated representation in Fig. 5(h)) but elliptical, the collecting of it for cutting to length would be difficult (of course long lands are used on dies - at the price of increased extrusion pressure - to ensure normal extrusion). Problems raised by different rates of extrusion through many orifices is one which needs more attention that the little it has received; to the best of the authors knowledge theoretical prediction of the relative volume flow and its direction or curvature through many holes simultaneously in extrusion is not available. Progress has only been made for plane strain flows using slip-line fields (ref. 18).

When rolling profiled T-shaped rings at large total reductions of ring wall thickness, it was found for certain rolling conditions a specific form of cavity formation arises; separation of the material from the main roll occurs where initially it was in contact, (see Fig. 3(a)). Accounting for this defect-producing ring-rolling phenomenon has been attempted together with some quantitative assessments by considering it to be an analogy to the piping defect well-known in axisymmetric or plane strain extrusion-forging situations, (see Fig. 3(b) and ref. 19).

It was shown by Hill (ref. 20) that a defect can be created by indenting with wedge-shaped or flat-ended dies strip resting on a rigid foundation. He found that the metal immediately below the tool, adjacent to the surface on which it was supported, was caused to rise towards the descending tool; in fact a pipe or gap was created between the supporting surface and the cut material, (see Fig. 3(c)); the essential feature is that no tension can be sustained between the cut material and the foundation.

Fig. 4. (a) The much published extrusion of two thick discs won under the action of a hammer. (b) An often published extrusion of a cylinder formed of two hemi-cylinders. (c) The end-extrusion of ten discs of lead through a square die. (d) Ten plates of lead end-extruded through a die which is a right angle triangle. (e) The flow of a block composed of concentric tubes end-extruded through a square orifice. (f) The end-extrusion of six discs of lead through an eccentric die. (g) The flow of ten discs of lead through two circular discs of different diameter. (h) The extrusion of six discs of lead through two circular dies. (i) The flow of two discs of lead through two circular dies of different size. (j) The end-extrusion of ten discs of lead through two dies. (k) The end-extrusion of twelve discs of lead through a large central die and six small surrounding dies. (l) The end-extrusion of six discs through ten equal dies. (m) A block pierced by two oppositely moving axial punches.

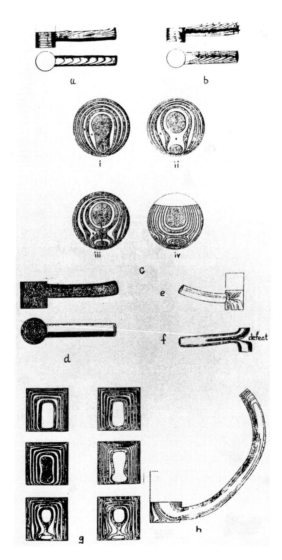

Fig. 5. (a) Side-extrusion of sixteen discs of lead through a circular side orifice. (b) As in (a) but size orifice square with a diagonal horizontal. (c) Cylindrical block composed of concentric tubes extruded through a circular side orifice; punch advanced from (i) to (iv). (d) As for (c). (e) Curving side-extrusion. (f) Test specimen of concentric tubes sideways extruded; note the large defect. (g) The side-extrusion through a square die of a block formed of concentric layers (tubes); the side of a square was horizontal. (h) A cylindrical block of concentric tubes sideways extruded through a rectangular orifice.

Sheet metal forming

   There are three types of end point or limiting condition for
sheet metal transverse loaded over some large fraction of its
area   and transmitting biaxial tensile stress in its own plane
in some regions in deep-drawing and pressing, mainly (ref. 2):
(a) Wrinkling or compressive instability in regions of high
compressive stress;
(b) local tensile instability in regions where a line of zero
extension can exist;
(c) failure in regions of biaxial tensile strain.
   Because sheet forming is largely a plane stress operation, the
above three conditions may be associated with a specific portion
of a yield surface, (see Fig. 15 of ref. 2). Forming limit diagrams based on the dimensions of the initial blank and the tool
have been suggested by various investigators.
   The defects which arise during various sheet forming processes
range from such common principal defects as wrinkling, buckling,
necking and fracture, springback and surface marks to ones of
specialised nature such as intergranular cavitation and void
formation, shape inaccuracy defects in large automotive panels,
decohesion, (see refs. 1 and 2 and Tables 1 and 2), and localised
shear-band formation, (see ref. 17); the latter seems first to
have been noticed and reported by Tresca.
   Springback when press-forming shallow automotive panels of
high strength steel sheets is of the utmost importance. Restri-
king is a method of correcting or compensating for springback
carried out by strongly "bottoming" the punch in the die so as to
produce a coining action. Components significantly restruck
suffer substantially lower magnitudes of springback than do lig-
htly treated ones. Heavy bottoming also confers closer adherence
to die shape; (see the work reported in ref. 2). As shown (ref.
22) springback is reduced with increasing blank-holder force in
stretch-bending tests; (see also ref. 23).
   As in sheet metal forming operations, springback is one of the
most common problems of metal spinning (ref. 24). Spinning at
ambient temperature gives rise to parts with greater curvatures
than specified ones and desired tolerances may not be achieved
due to excessive springback. The amount of springback varies
directly with the ratio of the radius to which the sheet is to be
bent to workpiece thickness and inversely with the forming tempe-

TABLE 1

Physical defects in metal processing (ref. 2)

ROLLING	EXTRUSION, PIERCING
Flat- and section-rolling	Christmas tree (fir tree)
Edge cracking	Hot and cold shortness
Transverse-fire cracking	Radial and circumferential
Alligatoring (crocodiling)	cracking
Fish-tail	Internal cracking
Folds, laps	Central burst (chevrons)
Flash, fins	Piping (cavity formation)
Laminations	Sucking-in
Ridges-spouty material	Corner lifting
Ribbing	Skin inclusions (side and
Sinusoidal fracture	bottom of the billet)
Zippering	Longitudinal streaks
Cross-,transverse- and helical-	Laps
rolling	Laminated fractures
Central cavity (axial or ann-	Mottled appearance
ular fissure)	Extrusion defect
Overheated ball bearing	Impact-extrusion
Roll mark	Multiple tensile "necks"
Folding (or seaming), laps,	Thermal break off
fringes	DRAWING OF ROD, SHEET, WIRE
Squaring	AND TUBE
Necking	Internal bursts (cup and
Triangulation and triangular	cone chevron)
fish-tail	Transverse surface cracking
Ring-rolling	Edge cracking
Cavities	Chips of metal, bulging
Fish-tail	Poor surface finish
Edge cracking	Folding and buckling
Straight-sided forms	Fins, laps, spills
FORGING	Chatter (vibration) marks
Open- and closed-die forging,	Season cracking
Upsetting, Indentation	Island-like welding
Longitudinal cracking	Pulling out
Hot tears and tears	Run out
Edge cracking	DEEP DRAWING
Central cavity	Wrinkling
Centre bursts	Puckering
Cracks due to t.v.ds and	Tearing (necking)
thermal cracks	Edge cracking
Folds, laps	Orange peel
Flash, fins	Stretcher-strains (Lüders
Laminations	lines)
Orange peel	Earing
Shearing fracture	BENDING AND CONTOUR-FORMING
Piping	Cracking
Rotary-forging	Wrinkling
Mushrooming	Springback
Central fracture	HOLE FLANGING
Flaking	Lip formation
High energy rate forging	Petal formation
Piping	Plug formation
Dead metal region	BLANKING AND CROPPING
Laps	Distortion of the part
Turbulent metal flow	(doming and dishing)

(cont.)

TABLE 1 (cont.)

Cracking	Back extrusion(over reduction)
Martensitic lines	Under reduction
Eyes, ears, warts, beards, tongues	PEEN FORMING, BALL-DROP FORMING
SPINNING, FLOW TURNING, SHEAR FORMING	Overlapping dimples
	Orange peel
Springback	Surface tearing
Wall fracture (shear and cir- cumferential splitting)	Break-up of surface grains
	Intergranular cracking
Wrinkling	Wrinkling
Buckling	Microfissures
	Folds

TABLE 2

Surface defects in massive and sheet metal forming (ref. 1)

Alligator skin	Mottled surface
Blisters	Overlapping (laps)
Burnt surface	Peeling (orange peel, pebbles)
Check marks	Pickle pitting
Cold laps	Pinchers
Cold shut	Pinheads (blisters)
Crowfeet (check marks)	Pitting
Dark areas	Ragging marks
Die lines	Roakes
Edge cheks	Rolling mill laps
Fishhooks (check marks)	Rough surface
Folds	Scratches
Galls	Seams
Ghost lines	Shearing defects
Grease spots	Shivers
Greasy surface	Snakes
Guide marks	Spangled surface (mottled surface)
Hair lines (cracks)	Stretcher-strains
Hair seams (seams)	White spots on steel ingots

rature. Hot-spinning and spinning at higher pressures eliminate most of the springback observed in cold-spinning operations. Intermediate annealing and deliberate "overforming" also reduce springback. The mechanical strength parameter, for practical purposes, which governs the amount of springback is the ratio of yield strength to ultimate tensile strength; higher ratios gene- rally result in greater springback.

Relating to the shape inaccuracy of large panels possessing complex contours, such geometrical surface defects as surface warp, wrinkling and surface deflection have been reported, (see Table 2). These defects are attributed to non-uniform stress distribution in the plane of the sheet; in addition, the inherent difficulties encountered in bending a sheet of metal under tension were pointed to as major factors giving rise to shape fixing defects (ref. 22).

Failure modes obtained in the piercing and hole-flanging of sheet metals are listed in Table 1, (see also ref. 25). These occur either by petalling, plugging or lip-fracture, (see Fig. 6). In the context of the hole-expanding limit test, it has been pointed out that the limit is generally lowered by the presence of microcracks at the hole-edge produced during the punching of the hole; strain concentration obtains and causes early failure (ref. 22).

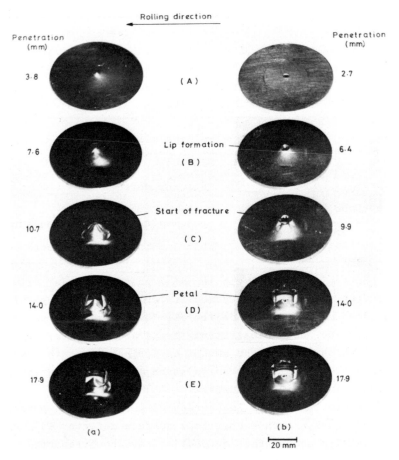

Fig. 6. Progressive perforation of thick aluminium plates with a square punch of 45° semi-angle for specimens (a) without and (b) with an initial hole.

PROCESSING OF COMPOSITE MATERIALS

A wide range of common defects that arise in composite material fabrication and subsequent processing is presented in Table
3, reproduced from refs. (1) and (3). The kinds of composites
treated are:

(a) Fibre-reinforced plastics;

(b) metal-matrix composites;

(c) clad/bonded materials;

(d) coated materials.

Buckling phenomena and fracture modes associated with the
plastic collapse of thin wall structures of fibre-reinforced
composites are examined, (see ref. 27). A photograph, taken from
this work, showing the splitting of tube and the cracking of the
fibres of a thin wall square tube subjected to axial loading is
shown in Fig. 7.

Damage regions occurring in bands parallel to the direction of
maximum shear stress during the forging of thorium-coated
tungsten wires embedded in a fine-grain nickel-base super-alloy
composite are shown in Fig. 8, (see ref. 26).

The principal defect encountered in deep-drawing and stretch-
forming of coated sheets has been recognised to be the flaking or
peeling off of the coat material. Flaking limit strain diagrams

Fig. 7. Axial collapse of a square thin-walled tube of fibre-
reinforced composite showing splitting of tube and cracking of
fibres.

TABLE 3

Defects in the manufacturing of composite materials (refs. 1,3)

FIBRE-REINFORCED PLASTICS	
Incomplete impregnation of fibre	
Incomplete cure of resin	
Poor wetting and subsequent poor adhesion of fibre to matrix	
Bubbles	
Voids	Manufacture
Delaminations	
Briken strands	
Loose ends of fibres	
Knotted strands	
Wrinkled strands and crevices	
Crazing cracks	
Local resin-rich areas	
Severe delamination	
Concealed cuts	Forming
Rupture of resin starved layers	
Fibre pull-out	Tensile loading
Fibre-matrix debonding	
Splitting	Compressive
Buckling	loading
Transverse cracking of fibres	Static-, dynamic-
Parallel splitting of laminates	piercing
Delamination	
Translamination cracking perpendicular to fibres	Impact
Spalling	
Surface flaws (step, hole, ripple, branch, fissure, crack) of whiskers	
METAL-MATRIX COMPOSITES	
Incompatibility of fibre and matrix	
Poor wettability	
Reaction between fibre and matrix	
Inadequate percolation of the matrix material to properly surrounding the fibres	Manufacture
Voids	
Porosity	
Matrix-filament debonding	
Axial densification	
Density gradients	Powder
Delaminations	compaction
Filaments to break	
Filament buckles	
Formation and flattening of ribbons	
Flaky area of carbon fibre	Fibre-coating
Void formation at the poles of fibres	
Transmatrix cracks	Forging
Decohesion of a matrix-fibre interface	
Brooming or crushing of component ends	
Transverse cracking	
Rough "tree-bark" surface finish	
Delaminations	Rolling
Fibre rupture as multiple necking	

(cont.)

TABLE 3 (cont.)

Transverse cracking Rough "tree-bark" surface finish Delaminations Burst Split into bundles of metal coated fibres Microcracks	Extrusion
Bamboo shape defect Breakage of continuous fibres or filament Voids Debonding	Drawing of bar, rod and wire
Filament breakage Splitting Brittle fracture	Bending

SANDWICH, CLAD/BONDED MATERIALS

Poor bonding Bond breakage Warping at high temperatures Edge delamination Destruction of the flyer plate	Manufacture
Wrinkling Earing Wall fracture Delamination or bowing	Deep-drawing, bending
Surface defects	Abrasive wear

COATED MATERIALS

Poor coating
Thinning and chipping at the corners
   and edges
Bubbles and specks due to excessive
   pickling
Fish scale, ruptures, flakes, peeling off
   due to absorbed hydrogen between coating
   and base material
Severe oxidation
Substrate
Voids
Porosity
Overspraying
	Manufacture

Surface marks Wrinkling Earing Springback	Sheet forming

have been suggested (ref. 22). Additionally, in forming coated steels, especially into box-shaped panels, there is the possibility of a loss of coat material at the corners. Deformation modes associated with the ball-drop forming of coated (electrolytic tin-plated and hot-dipped galvanised) steel plates are reported (ref. 28). Testing procedures to evaluate the material properties and coating adherence of coated materials are given (refs. 29 and 30, but see also ref. 22).

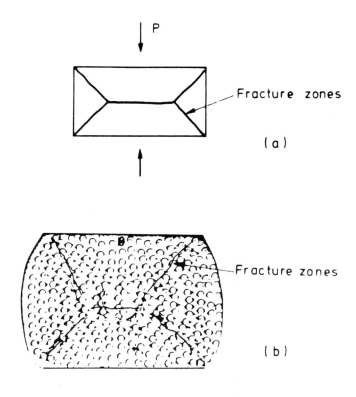

Fig. 8. Zones of severe damage and spread profile in a section normal to the direction of fibre alignment for a forged composite. (a) A schematic diagram of the fracture zones. (b) A micrograph of a polished specimen.

CONCLUDING REMARKS

The present paper is written to give an intepretive and consolidated view of material processing defects for the general interest of the manufacture engineer and for improving the academic teaching of the subject of material processing mechanics at a practical level. It is hoped that directing attention to a wide spectrum of defects will suggest to those concerned with developing new processes principles for avoiding them. We also hope that the catalogue of defects described will quicken interest in creative and successful material processing practice and promote better interpretation through theoretical analyses.

ACKNOWLEDGEMENTS

We are grateful to Ms. M. Bosklavitou, Mr. M.B. Ioannides, Dr. D.E. Manolakos and Mr. E. Papaioannou for helping with the preparation of our manuscript.

REFERENCES

1   W. Johnson and A.G. Mamalis, Common defects in the processing of metals and composite materials, in: A. Sawczuk and G. Bianchi (Eds.), Plasticity Today: Modelling, Methods and Applications, Elsevier, Amsterdam, 1984, Ch. 40, pp. 745-766.
2   W. Johnson and A.G. Mamalis, A survey of some physical defects arising in metal working processes, in: Proc. 17th Int. M.T.D.R. Conference, Birmingham, U.K., September, 1976, Macmillan, London, 1977, pp. 607-621.
3   W. Johnson and S.K. Ghosh, Some physical defects arising in composite material fabrication, J. Materials Sci., 16 (1981) 285-301.
4   L. Engel and H. Klingele, Rasterelektronmikroskopische Untersuchungen von Metallschäden, Carl Hanser Verlag, München, 1974.
5   O.D. Sherby, S. de Jesus, T. Oyama and E. Miller, Metal forming defects at large strain deformation at intermediate temperatures, in: E.H. Lee and R.L. Mallet (Eds.), Plasticity of Metals at Finite Strain, Division of Applied Mechanics, Stanford University, pp. 3-15.
6   J.A. Schey, Prevention of edge-cracking in rolling by means of edge restraint, J. Inst. Metals, 94 (1966) 193-200.
7   B. Dodd and P. Boddington, The causes of edge-cracking in cold-rolling, J. Mech. Working Technol., 3 (1980) 239-252.
8   N.M. Burman and P.F. Thomson, Edge-cracking and profile of hot-rolled aluminium slabs, J. Mech. Working Technol., 13 (1986) 205-217.
9   A.A. Afonja and D.H. Sansome, Edge-cracking in sandwich rolling, J. Mech. Working Technol., 3 (1979) 77-83.
10 A. Jenner and B. Dodd, Cold upsetting and free surface ductility, J. Mech. Working Technol., 5 (1981) 31-43.
11 H.A. Kuhn, Forming limit criteria-bulk deformation processes, in: J.J. Burke and V. Weiss (Eds.), Advances in Deformation Processing, Plenum Press, New York, 1978, pp. 159-186.
12 A.A. Chaaban and J.M. Alexander, A study of the closure of cavities in swing forging, in: Proc. 17th M.T.D.R. Conference, Birmingham, U.K., September, 1976, Macmillan, 1977, pp. 633-646.
13 B. Avitzur, Analysis of central bursting defects in extrusion and wire drawing, Trans. ASME, J. Engng. Industry, 90 (1968) 79.
14 E.H. Lee and R.M. McMeeking, Concerning analysis of central burst in metal forming, Trans. ASME, J. Engng. Industry, 100 (1978) 386-387.
15 A.G. Mamalis, J.B. Hawkyard and W. Johnson, Spread and flow patterns in ring-rolling, Int. J. Mech. Sci., 18 (1976) 11-16.
16 W. Johnson and H. Kudo, The Mechanics of Metal Extrusion, Manchester University Press, Manchester 1962.

250

17 W. Johnson, Some neglected researchers in the field of metal working, in: Proc. 3rd Int. Conf. on Structural Analysis Systems; CAD/CAM and FEM in Metal Working, North Staffordshire Polytechnic, Stafford, U.K., April, 1987.

18 W. Johnson, R. Sowerby and R.D. Venter, Plane Strain Slip-Line Fields, Pergamon Press, Oxford, 1982.

19 A.G. Mamalis, J.B. Hawkyard and W. Johnson, Cavity formation in rolling profiled rings, Int. J. Mech. Sci., 17 (1975) 669-672.

20 R. Hill, On the mechanics of cutting metal strips with knife-edged tools, J. Mech. Phys. Solids, 1 (1953) 265-270.

21 W. Johnson and T.X. Yu, The press brake bending of rigid linear work hardening plates, Int. J. Mech. Sci., 23 (1981) 307-318.

22 S.K. Ghosh, Principally on sheet metal forming defects as described in the 11th Biennial Congress of the IDDRG, Int. J. Mech. Sci., 23 (1981) 195-211.

23 W. Johnson and A.N. Singh, Elastic springback in circular blanks formed by bending with a hemi-spherical punch and die, Metallurgia and Metal Forming,47 (1980) 275-280.

24 ASM, Metals Handbook, 8th edn., Forming, Vol. 4, Metals Park, Ohio, USA, 1973.

25 W. Johnson and A.G. Mamalis, The perforation of circular plates with four-sided pyramidally-headed square-section punches, Int. J. Mech. Sci., 20 (1978) 849-866.

26 A.G. Mamalis, W. Wallace, A. Kandeil, M.C. de Malherbe and J.-P.A. Immarigeon, Spread and fracture patterns in forging fibre-reinforced composites, J. Mech. Working Technol., 5 (1981) 15-30.

27 A.G. Mamalis, On the axial collapse of thin-walled structures of composite materials, (to be published).

28 A.G. Mamalis, G.C. Vosniakos and A.J. Zavaliangos, Ball-drop forming of coated steel sheets, J. Mech. Working Technol., 15 (1987).

29 A.G. Mamalis and L.P. Hatzikonstantis, On the Yoshida buckling test of high strength and coated low-carbon steels, Annals of the CIRP, 36(1) (1987).

30 K.G. Lawrenz, Untersuchungen über das Tiefziehen kunststoffbeschichteter Stahlbleche, Industrie Anzeiger, 99(29) (1977) 515-516.

Computational Methods for Predicting Material Processing Defects, edited by M. Predeleanu   251
Elsevier Science Publishers B.V., Amsterdam, 1987 — Printed in The Netherlands

Damage Evolution Modeling in Bulk Forming Processes

Kapil K. Mathur and Paul R. Dawson

Sibley School of Mechanical and Aerospace Engineering
Cornell University, Ithaca, New York 14853

Abstract

A mathematical formulation is presented for the analysis of metal forming operations which lead to accumulation of material damage by the nucleation and growth of voids. Two different approaches are introduced to model the accumulation of damage within the context of this formulation. Emphasis of the formulation is on the application of two state variable models to describe the strain hardening, rate-dependent plastic flow with plastic dilatancy. The numerical predictions have been compared with reported sheet drawing experiments. Good agreement between the numerical predictions and the experimental data demonstrates the success of the formulation. The application of the mathematical formulation in the design of manufacturing operations has been further demonstrated by studying the effect of die angle geometry, for a fixed thickness reduction, on the accumulation of material damage and on the evolution of the hardness.

Notation

a_0	- Parameter in Hart's model	$\dot{Q}$	- Heat generation rate
C	- Parameter in Hart's model	R	- Universal gas constant
$\mathbf{d}$	- Rate of deformation	S_σ	- Surface with imposed traction
d'_{II}	- Effective rate of deformation $= \left[\frac{2}{3}\mathbf{d}'\mathbf{d}'\right]^{\frac{1}{2}}$	t	- Time
e	- Internal energy	$\mathbf{T}$	- Traction
f	- Internal porosity	β	- Function in Cocks and Ashby model
f_0	- Parameter in Hart's model	κ	- Lamé constant $= K_b - \frac{2}{3}\mu$
f_i	- Initial value of internal porosity	λ	- Parameter in Hart's model
G	- Elastic shear modulus	μ	- Shear viscosity
δJ	- Virtual rate of work	ρ	- Apparent density of workpiece
K	- Parameter in Hart's model	ρ_0	- True density of workpiece material
K_b	- Bulk viscosity	σ	- Stress tensor
m	- Parameter in Hart's model	σ_m	- Mean stress $= \frac{1}{3}\mathrm{Tr}(\sigma)$
m'	- Parameter in Hart's model	σ^*	- Hardness
M	- Parameter in Hart's model	σ'_{II}	- Effective stress $= \left[\frac{3}{2}\sigma'\sigma'\right]^{\frac{1}{2}}$
n	- Parameter in Cocks and Ashby model	τ	- Flow stress
Q, Q'	- Activation energies	θ	- Temperature

Introduction

Microstructural changes that may occur during a bulk forming operation can lead to internal damage in the workpiece. In ductile materials almost all material damage occurs as a consequence of nucleation and growth of microvoids. Damage is described in terms of microvoid formation at inclusions, grain boundaries or in regions of large deformation gradients. Material damage accumulated in the workpiece has been quantitatively measured as a decrease in the density of the workpiece (which locally becomes porous), relative void volume, or by the largest void size (ref. 1).

Several mechanisms have been identified for void growth. Voids may grow as a result of grain boundary diffusion, surface diffusion, creep around voids which lie within a grain or on a grain boundary, or a combination of these mechanisms. Microvoids generally nucleate at small strains. These microvoids then grow by ductile deformation of the metal. They eventually coalesce with other microvoids to form an elongated void of observable size. These elongated voids are commonly referred to as "arrowhead cracks". Propagation of such elongated voids results in ductile fracture.

Material damage leads to degradation of mechanical properties of the workpiece related to elasticity, yielding and plastic flow. This damage is responsible for phenomena such as central bursts in extrusion and alligator cracks in rolling. Alternately, the "damaged" workpiece may cause failure during the fabrication process or may effect the performance of the final product.

In this article two different approaches are discussed for modeling the accumulation of material damage in a bulk forming process. The first scheme uses two constitutive models to define Lamé constants which describe the shearing and volumetric behavior of the workpiece material. The second approach imposes mass continuity as a constraint equation thereby introducing pressures as additional unknowns which are necessary for integrating the evolution equation for damage. Both the techniques have been compared in detail. This formulation provides an effective mean to choose from possible alternatives the forming conditions that minimize material damage to a product. The accuracy of the numerical predictions have been assessed by direct comparison with reported experimental data.

A Metal Forming Model with Damage

The metal forming model presented in this article is for workpiece materials in which the internal porosity forms as the workpiece deforms. Balance laws govern the thermo-mechanical behavior of the workpiece. These laws must be solved along with the kinematic equations and boundary conditions to obtain the motion and the temperature of the workpiece due to deformation. For modeling the workability of metals during steady state deformation processing operations, an Eulerian reference frame is useful. In this reference frame, the metal deforms as it passes through a control volume which is fixed in space.

The governing equations of the metal forming model in the presence of internal porosity are briefly summarized below. Details are available in (refs. 2-3). For a homogeneous material, conservation of mass assuming no mass production can be expressed as

$$\dot{\rho} + \rho \nabla \cdot \mathbf{u} = 0. \tag{1}$$

In the above equation, the apparent density of the workpiece material is the product of the true density of the metal and the solid fraction,

$$\rho = \rho_0 (1 - f). \tag{2}$$

Simple linear constitutive relations have been assumed for the thermal response of the workpiece material. The internal energy and the heat flux are related to the spatial gradient of temperature via the specific heat and the conductivity. The change in the area/volume due to the presense of internal porosity has been explicitly included in the constitutive equation for the heat flux so that the thermal conductivity of the "undamaged" workpiece material (fully dense) can be used. As the temperature changes in most steady-state forming operations are small the temperature dependence of the specific heat and the thermal conductivity has been neglected.

For most forming operations large irrecoverable plastic strains occur; therefore, it is convenient to neglect the elastic behavior of the workpiece material. By assuming isotropy, the constitutive equations representing the mechanical response of the metal have been decomposed into the volumetric and deviatoric responses. The volumetric response relates the development of the internal porosity ("damage") to the change in the volume of the workpiece. The metal forming model assumes that the dominant mechanism responsible for changes in the volume of the workpiece is the accumulation of damage. In other words, the model neglects the volume changes due to elasticity and temperature changes. Consequently, the true density of the workpiece material is a constant and the volumetric response is expressed as

$$\mathrm{Tr}(\mathbf{d}) = \frac{\dot{f}}{1-f}. \tag{3}$$

The internal porosity has been treated as a state variable which is a measure of the current state of "damage" in the material. It is represented by the void area (volume) fraction in the workpiece. A variety of damage models have been published recently which either use a continuum approach or a mechanistic approach to describe the accumulation of material damage. These include the phenomenological damage model proposed by Davison et al. (ref. 4) and the mechanistic models of Chaboche (ref. 5), Cocks and Ashby (ref. 6), and Dragon (ref. 7). The reader is referred to the article by Mathur and Dawson (ref. 3) for a discussion on these damage models.

In the simulations that are presented later the mechanistic damage model proposed by Cocks and Ashby (ref. 6) has been used to model the accumulation of material damage due to void growth during the drawing of strips. This model is based on a single scalar state variable f which represents the internal porosity. The model, in the form implemented for the drawing simulations, assumes that material damage occurs by the growth and coalescence of voids which lie within a grain and on grain boundaries. This material damage accumulates according to

$$\dot{f} = \beta(n, \sigma_m, \sigma'_{II}) \left[\frac{1}{(1-f)^n} - (1-f) \right] d'_{II}. \tag{4}$$

The function β accounts for the influence of the stress state, the changing shape of voids, and the localization of the stress field because of grain boundary sliding and is defined as

$$\beta = \mathrm{Sinh}\left(2\frac{2n-1}{2n+1}\frac{\sigma_m}{\sigma'_{II}}\right)_{\mathrm{local}}. \tag{5}$$

Material damage is defined by a positive monotonically increasing function because it is more difficult to close voids once they are opened. Therefore,

$$\dot{f} \geq 0. \tag{6}$$

Since the rate of deformation in Equation (31) can be computed by the deviatoric response of the metal (represented by Hart's model (ref. 8) for drawing simulations), only one additional material parameter (n) needs to be evaluated. The effect of n on the simulations are been discussed in (ref. 3). The value of n has been taken to be 5.0 in the results presented in this article.

The deviatoric response of the metal is represented by a yield condition which relates the second invariant of the stress deviator as a function of the rate of deformation, temperature, and a set of state variables which represent the effect of the development of microstructure within the metal matrix on the mechanical properties of the metal. This behavior has been modeled by a version of Hart's model (ref. 8) appropriate for conditions when a single microstructural transient dominates. In its complete form Hart's model contains a scalar state variable known as the hardness and a tensor state variable called the anelastic strain. The anelastic strain has been neglected because the anelastic rate of deformation is assumed to be negligibly small compared to the plastic rate of deformation. For the same reason the elastic terms of the model have also been neglected. The resulting constitutive model reduces to a scalar state variable model. The mechanical analogy of this version of Hart's model is a parallel combination of an element which represents the dislocation glide on slip planes ("the viscous element") and another element representing the plastic flow controlled by dislocation motion ("the plastic element"). The constitutive equations which describe this version of the model are: Compatibility Equations:

$$d'_{II} = d'^p_{II} = d'^v_{II}; \qquad \sigma'_{II} = \sigma'^p_{II} + \sigma'^v_{II} \tag{7}$$

Invariant Relationships:

$$\sigma'^p_{II} = \sigma^* \exp[-(\tfrac{d^*}{d'_{II}})^\lambda]; \qquad \sigma'^v_{II} = G\left(\tfrac{d'_{II}}{a}\right)^{\frac{1}{M}} \tag{8}$$

$$d^* = f_0\left(\tfrac{\sigma^*}{G}\right)^m \exp(\tfrac{-Q}{RT}); \qquad a = a_0 \exp(\tfrac{-Q'}{RT}). \tag{9}$$

The evolution equation for the hardness is given by

$$\sigma^* = C\sigma^*\left(\tfrac{\sigma_{II}'^p}{\sigma^*}\right)^K \left(\tfrac{\sigma^*}{G}\right)^{-m'} d'_{II}. \tag{10}$$

This version of Hart's model has eleven material parameters that must be determined for the temperature and the rate of deformation ranges applicable to the forming conditions being simulated. Some material parameters such as the shear modulus (G), and the activation energies (Q and Q') were taken from handbook values. Other material parameters have been determined from a combination of relaxation data and constant rate of deformation compression data. The material parameters used in the simulations are listed in Table (1). Articles by Dewhurst and Dawson (ref. 9), Eggert and Dawson (ref. 10), and Dawson (ref. 11) discuss this version of Hart's model in greater detail.

A Levy-Mises type flow law is used to relate the stress deviator tensor to the rate of deformation tensor

$$\mathbf{d'} = \frac{3d'_{II}}{2\sigma'_{II}}\sigma'. \tag{11}$$

The effect of the internal porosity on the shearing behavior of the metal has been neglected because of small internal porosity levels.

Boundary conditions are necessary to complete the mathematical formulation of the metal forming model. For determining the motion of the metal through the control volume, velocity/traction boundary conditions must be specified over the entire surface. To compute the temperature distribution within the workpiece temperature/heat flux boundary conditions must be imposed. Initial conditions for the state variables must be specified on regions where the metal enters the control volume. Friction boundary conditions at the workpiece-tool interface can be easily implemented in this model. Details are available in (ref. 2).

Numerical Solution of the Model Equations

The motion of the metal through the control volume is computed from a variational principle which represents the virtual rate-of-work. Two different approaches are possible to incorporate the volumetric response of the metal into this variational statement.

The first approach uses the constitutive equations to define the Lamé constants. The volumetric response of the metal is used to define a bulk viscosity

$$K_b = \frac{\sigma_m}{\mathrm{Tr}(\mathbf{d})}, \tag{12}$$

and the deviatoric response of the metal is expressed by a shear viscosity

$$\mu = \frac{\sigma'_{II}}{3d'_{II}}. \tag{13}$$

The flow law representing the total response of the metal can then be written as

$$\sigma_{ij} = 2\mu d_{ij} + \kappa d_{kk}\delta_{ij}, \tag{14}$$

where κ is a Lamé constant and is related to the bulk and shear viscosities (see Notation). The variational statement representing the virtual rate-of-work is expressed as

$$\delta J = -\int_V \mathrm{Tr}(\sigma \cdot \delta\mathbf{d})dV + \int_V \rho\mathbf{g} \cdot \delta\mathbf{u}dV + \int_{S_\sigma} \mathbf{T} \cdot \delta\mathbf{u}dS = 0. \tag{15}$$

The Euler equations obtained from the above variational statement are the balance of linear momentum equation and the traction boundary conditions. By substituting interpolation functions for the velocity and the flow law for the stress a matrix equation is obtained. The nonlinearity in the volumetric and the deviatoric response results in the matrix equation being nonlinear.

The streamline integration method is used to integrate the coupled non-linear evolution equations associated steady-state metal forming simulations (ref. 12). The internal porosity is integrated using the conservation of mass equation along a streamline path through the flow field

$$\int df = \int [\mathrm{Tr}(\mathbf{d})(1-f)] \frac{dx_s}{u_s}. \tag{16}$$

The second approach imposes mass continuity as a constraint to the variational statement using a Lagrangé multiplier. Variational statements which introduce Lagrangé multipliers as additional unknowns lead to mixed formulations involving both velocity and stress as primary unknown variables. The resulting mixed formulation may be written as

$$\delta J = -\int_V \mathrm{Tr}(\sigma' \cdot \delta \mathbf{d}') dV + \int_V \rho \mathbf{g} \cdot \delta \mathbf{u} dV + \int_{S_\sigma} \mathbf{T} \cdot \delta \mathbf{u} dS + \delta \int_V \lambda \left[Tr(\mathbf{d}) - \frac{\dot{f}}{1-f} \right] dV = 0. \tag{17}$$

The Lagrangé multiplier obtained from the constraint is the hydrostatic state of stress that develops in the workpiece. In addition to the equations which represent the balance of linear momentum and the traction boundary conditions the conservation of mass equation is also one of the Euler equations for this statement. The final system of equations for the velocities and the pressures is of the form

$$\begin{pmatrix} K & G^T \\ G & 0 \end{pmatrix} \begin{pmatrix} U \\ -P \end{pmatrix} = \begin{pmatrix} F \\ X \end{pmatrix}. \tag{18}$$

The submatrix $[K]$ depends on the velocities at the nodal points because of the nonlinear yield criterion. Moreover, the force vector $\{X\}$ depends on the nodal point velocity as well as the nodal point pressure because the rate equation governing the accumulation of material damage is nonlinear. Additional caution is required in the solution of the above system of equations because of possible zero diagonal elements. Ill-conditioning due to zero diagonal elements can be avoided by an ordering scheme which ensures that for a nodal point columns representing pressures are reduced after the columns which represent velocity.

The internal porosity is now integrated using the evolution equation proposed by the constitutive model which describes the volumetric response of the workpiece material

$$\int df = \int [\dot{f}(\sigma_m, \sigma'_{II}, d'_{II}, f)] \frac{dx_s}{u_s}. \tag{19}$$

The hydrostatic state of stress at each point on the streamline is also necessary apart from other kinematic quantities such as the rate of deformation and the velocities to integrate the evolution equation. The pressure fields computed by finite element techniques can be quite noisy for nearly incompressible flow fields because, in general, the interpolation functions for the velocity cannot ensure that the conservation of mass is enforced within an element exactly. Continuity is satisfied either in an average sense within an element or at each sampling point. Several element types, pressure approximations, and integration orders were examined to determine the combination which gave optimal results for nearly incompressible flow situations. Special care is also needed to filter out the spurious pressure modes exhibited by some element types. Mathur and Dawson (ref. 3) discuss these concepts in detail.

Typically damage accumulation takes place in those regions within the workpiece where the hydrostatic state of stress is tensile. The flow field in these regions is compressible. In the remaining portion of the workpiece the material acts incompressible. Within the context of modeling the accumulation of material damage, the mathematical formulation must be able to treat both incompressible and compressible flow fields and must make smooth transitions between the two types of flow fields. This represented a major difficulty in the first approach because as the material approaches incompressible behavior the bulk viscosity (K_b) approaches infinity. For modeling fully-dense materials by this technique a large but finite value for the bulk viscosity is sufficient to ensure incompressibility. However, for modeling the accumulation of damage the computation of reliable pressure distributions is essential because the effect of hydrostatic stress on the evolution of the internal porosity is quite severe. Accurate and physically realistic pressure fields cannot be obtained by assigning a large value to the bulk viscosity whenever the flow field approaches the incompressible regime †. The transition from a compressible flow field to an incompressible flow field is trivial in the scheme which imposes the continuity equation as a constraint equation using Lagrangé multipliers. Whenever no accumulation of material damage occurs, $\dot{f} = 0$, and the variational statement reduces to the familiar expression used for fully-dense materials.

The two techniques for coupling the volumetric response of the workpiece material with the deviatoric response discussed above were compared for the drawing simulation discussed later. The penalty function approach computed unsatisfactory results. Although the trends predicted by this technique were consistent with experimental observations, the magnitude of damage predicted was very high. The computed flow field was very sensitive to the initial value of damage and was unrealistic for some initial values. Moreover, the convergence of the nonlinear problem for this scheme was extremely sluggish. The mixed formulation formulation was quite successful for coupling the two responses comprising the mechanical behavior of the metal. The predictions were well supported by experiments and convergence was quite rapid. The scheme proved to be very stable and was therefore chosen to compute the velocity field and the pressure distribution within the control volume for modeling the accumulation of damage in all the simulations discussed later. However caution is required against problems frequently encountered with mixed formulations (ref. 3).

The temperature distribution within the workpiece is obtained by using a Galerkin scheme wherein a is obtained from the energy balance equation. Using the constitutive equations for the thermal response of the metal, the heat flux and the internal energy are expressed as functions of temperature and temperature gradients. Standard finite element techniques then result in an unsymmetric matrix equation. This matrix equation is then solved for the temperature distribution.

The equations which must be solved to compute the flow field, the temperature distribution, and the evolution of the state variables are coupled. The iterative procedure uses previous solution for the temperature distribution and the state variables to set the system of equations for the flow field. Similarly, the current solution of the flow field is used while computing the temperature field or updating the state variables.

Application to Strip Drawing

The drawing operation involves pulling a sheet of metal through a pair of flat dies.

---

† The magnitude of the pressures is dependent on the magnitude of the bulk viscosity.

Strip drawing operation closely resembles a two-dimensional analogue of wire drawing operation. This operation has been analyzed in detail using the slip line method by Coffin and Rogers (ref. 1). Figure (1) shows the schematic of a strip drawing operation. This operation usually starts with workpiece materials which are initially fully-dense (or have a very small internal porosity). Forming conditions lead to a nucleation and growth of voids within the workpiece that can be observed as a decrease in the density of the workpiece.

The first three passes of this operation have been simulated by the metal forming model discussed above. Aluminum strips initially 5 mm thick were "drawn" through dies with die angles of 20° and a nominal reduction of 20 % per pass. These strips were pulled though frictionless dies at a constant velocity of 1 cm s^{-1}. The initial temperature of the workpiece and the dies was taken to be 373 K. These drawing conditions replicate one of the cases used by Coffin and Rogers for their experimental study on structural damage during drawing of metals.

The symmetric half of the deformation zone was discretized with 174 nine-node Lagrangian elements. The finite element mesh used in the analysis had 767 velocity nodes and 696 pressure nodes. Continuous piecewise quadratic velocity approximations and discontinuous piecewise bilinear pressure approximations were used. An integration step length of 0.02 mm was used to integrate the evolution equations for the state variables. To ensure that no mass entered or left the control volume the surface position upstream of the die was adjusted using a free surface algorithm. The corner of the workpiece as the metal emerges out of the die was rounded to prevent any discontinuities in the velocity. Each aspect of the numerical simulation has been discussed in greater detail by Mathur and Dawson (ref. 3).

Coffin and Rogers have measured the density of the drawn strips using a fluid displacement technique. The predictions made by simulations for the average "apparent" density of the workpiece have been compared with the experimental results reported by Coffin and Rogers in Figure (2). The computed decreases in the density of workpiece are in very good agreement with the reported density changes. The densities plotted in this figure are average values over the entire cross-section of the workpiece, although the void growth is concentrated near the centerline. The density of the central one-third of the thickness of the strip has been reported to be much less than the average density of the strip. This trend is also well captured by the multi-pass simulation.

Figure (3) shows the evolution of the hardness (the state variable in Hart's model) as the workpiece material passes through the first three sets of dies. The hardness increases with the overall reduction. Moreover, the surface hardens more than the interior of the workpiece. The rate of increase of the hardness constantly decreases with continued deformation. This is consistent with experimental observations.

The accumulation of material damage is shown in Figure (3). It is clear from the contour plots that almost all material damage occurs near the centerline of the strip. These predictions are in excellent agreement with the experimental observations of Coffin and Rogers who observed that damage was most severe in the central one-third of the thickness of the workpiece. The growth of damage is modest for small reductions but increases rapidly as the overall reduction increases. The simulations predict that material damage is cumulative and increases with successive drawing passes. This trend is also supported by experimental observations.

Effect of Geometry Changes on Material Damage

Die angles have known to effect the accumulation of material damage. Therefore, the effect of die angle geometry, for a fixed thickness reduction, on the accumulation of material damage and hardness has been studied in greater detail. The first pass of the strip drawing operation discussed above was simulated for a 30 % nominal reduction. The die angle geometries were taken as 10°, 20°, and 30°.

Figure (4) shows the hardness distribution for the three dies. The hardness changes are greater when the material is drawn through the 30° die because of a larger rate of deformation induced by a sharper geometry change. The surface of the strip hardens more than the centerline. The final hardness of the strip increases as the die angles increase. Almost all the evolution of the hardness occurs directly under the dies.

The accumulation of material damage as the metal emerges out of the three different dies is shown in Figure (4). The accumulation of material damage for the 10° die is minimal. However, appreciable damage occurs for the 30° die angle geometry indicating that larger geometry changes produce more material damage in the workpiece. Almost all damage accumulation occurs in the vicinity of the centerline of the strip where the hydrostatic component of the stress is tensile. Little or no damage accumulation occurs on the surface of the strip.

Conclusions

A metal forming model in which the internal porosity evolves has been presented. This model emphasizes the accurate description of the response of the material to the deformation and the microstructural changes that are taking place in the deforming material. Constitutive laws which use state variables have been used to describe the rate-dependent and temperature-dependent material behavior. Strain hardening effects and plastic dilatancy by void growth are included in the material description.

The trends predicted by the mathematical formulation for the hardness are consistent with trends observed experimentally. This indicates the success of the state variable model of Hart in capturing the essence of the strain hardening behavior during a large deformation process. The predictions for the accumulation of material damage also agree very well with the reported experimental data, demonstrating that the damage model of Cocks and Ashby fits in quite well within the framework of our viscoplastic formulation. More importantly, the two state variable models have coupled well, to reproduce both the effect of strain hardening and plastic dilatancy by the evolution of internal porosity.

Acknowledgements

This work has been supported by the National Science Foundation under Grant DMC-8352275 with matching support from ALCOA, the General Motor Corporation, and the XEROX Foundation. Computations supporting this research were performed on the Cornell National Supercomputer Facility, which is in part supported by the National Science Foundation, New York State, and IBM Corporation. The simulation results were postprocessed on a facility of the Mathematical Sciences Institute of Cornell University which is supported by the US Army Office of Basic Research.

REFERENCES

1 Coffin, Jr., L. F. and H. C. Rogers, "Influence of Pressure on the Structural Damage in Metal Forming Processes", *Trans. ASM 60*, 672, 1967.

2 Dawson, P. R., "A Model for the Hot and Warm Forming of Metals with Special Use of Deformation Mechanism Maps", *Int. J. Mech. Sci. 26*, 4, 227, 1984.

3 Mathur, K. K. and P. R. Dawson, "On Modeling Damage Evolution During the Drawing of Metals", *Mechanics of Materials*, in press.

4 Davison, L., A. L. Stevens and M. E. Kipp, "Theory of Spall Damage Accumulation in Ductile Materials", *J. Mech. Phys. Solids 25*, 11, 1977.

5 Chaboche, J. L. "Thermodynamic and Phenomenological Description of Cyclic Viscoplasticity with Damage", *ONERA Publication 3*, 156, 1978 (in French).

6 Cocks, A. F. C. and M. F. Ashby, "On Creep Fracture by Void Growth", *Prog. Mater. Sci.*, 189, 1982.

7 Dragon, A., "Plasticity and Ductile Fracture Damage Study of Void Growth in Metals", *Engineering Fracture Mechanics 21*, 4, 875., 1985

8 Hart, E. W., "Constitutive Relations for the Nonelastic Deformation of Metals", *ASME J. Eng. Mater. Tech.*, 193, 1976.

9 Dewhurst, T. B. and P. R. Dawson, "Analysis of large Plastic Deformations at Elevated Temperatures Using State Variable Models for Viscoplastic Flow", in: K. J. William, ed. *Proc. of Symposium on Constitutive Equations : Micro, Macro and Computational Aspects*, ASME WAM, 149, 1984.

10 Eggert, G. M. and P. R, Dawson, "On the Use of Internal Variable Constitutive Equations in Transient Forming Processes", *Int. J. Mech. Sci.*, **29**, 95, 1987.

11 Dawson, P. R., "On Modeling of Mechanical Property Changes During Flat Rolling of Aluminum", *Int. J. Solids Structures*, in press.

12 Agrawal, A. and P. R. Dawson), "A Comparison of Galerkin and Streamline Techniques for Integrating Strains from an Eulerian Flow Field", *Int. J. Numer. Methods Eng. 21*, 853, 1985.

TABLE 1

Model Parameters for Drawing Simulation

Parameter	Units	Value	Parameter	Units	Value
a_0	s^{-1}	9.64×10^{52}	Q'/R	K	1.45×10^4
f_0	s^{-1}	2.12×10^{19}	Q/R	K	1.45×10^4
M		7.80	m		5.00
λ		0.15	G	GPa	24.20
σ^*_i	MPa	64.00	m'		3.50
C		6.19×10^{-9}	K		4.50
n		5.0	f_i	%	0.01
ρ_0	$Kg\ m^{-3}$	2707	C_p	$J\ Kg^{-1}\ K^{-1}$	963
k	$W\ m^{-1}\ K^{-1}$	200			

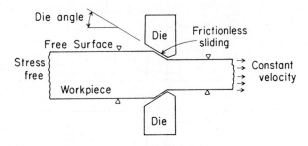

Figure 1(a). Strip Drawing Operation : Geometry and Boundary Conditions.

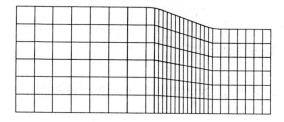

Figure 1(b). Strip Drawing Operation : Finite Element Mesh.

o   Reported by Coffin and Rogers (ref. 1).

· · · · · Simulation.

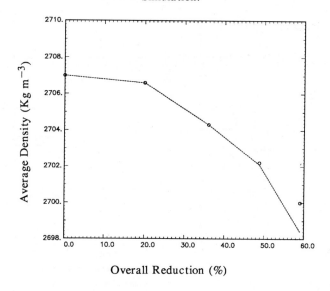

Figure 2. Density as a Funtion of Overall Reduction

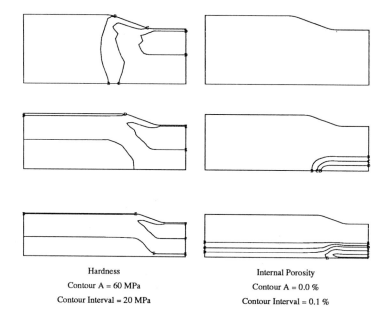

Figure 3. Evolution of State Variables with Overall Reduction

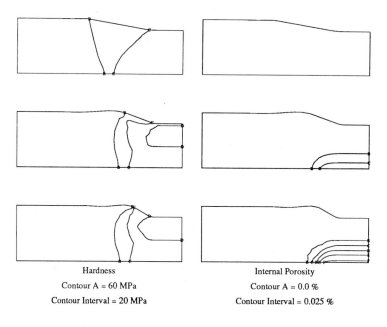

Figure 4. Effect of Die Angle on the Evolution of State Variables

Computational Methods for Predicting Material Processing Defects, edited by M. Predeleanu   263
Elsevier Science Publishers B.V., Amsterdam, 1987 — Printed in The Netherlands

PREDICTION OF THE WRINKLING INSTABILITIES BY LONG WAVELENGTH PERTURBATIONS IN
THIN SHEET METAL FORMING

P. MAZILU

Institut für Umformtechnik, Technische Hochschule Darmstadt, D-6100 Darmstadt/
FRG

SUMMARY
   A new instability criterion to predict wrinkling in free deformation areas
of industrial sheet metal stampings by long wavelength perturbations is derived.

INTRODUCTION
   The occurrence of wrinkling in the manufacturing of sheet metal parts can
mean failure of the final product. The avoidance of such wrinkles by a propre
design of sheet metal tools requires in advance a study of the stability. The
determination of general criteria for stability in metal sheet forming is a difficult task. More realistic is to look for criteria valid in certain areas of
the metal sheet and for particular type of perturbations.
   During a forming process a metal sheet presents two distinct type of areas:
the free areas and the areas in contact with the tools.
   The disturbances can be classified into following three categories:
   (i) perturbations with long wavelength implying no significative bending
moments;
   (ii) short wavelength perturbation involving bending moments which depend
only on the local geometry (local curvature) of the metal sheet;
   (iii) perturbations of intermediate range involving both bending moments and
dependence on the geometrical form of the metal sheet piece.

   In the following we shall restrict ourselves to the study of stability to
long wavelength perturbations in the free areas of the metal sheet forming.
Because, be definition, the long wavelength perturbations involve no bending
moments the stability must be studied in the frame work of a membrane state of
stress theory. Concerning the application such a membrane theory arises the
following question: It is the membrane theory able to supply realistic stability criteria or will such a theory predict erroneously instabilities at any compressive forces?
   A general mathematical reasoning shows that the stability criterion derived
within membrane theory will approximate so good the "true criterion" as the

equations of the membrane theory approximate the complete system of equations involving membrane forces and bending moments. The following example will illustrate this assertion.

Let us consider a circular cylindrical shell of thickness d subjected to an axial pressure P on the edges, Fig. 1. The shell, assumed to be in elastical equilibrium, is subjected to the displacement disturbances having the form

$$u = A \cos m \phi \cos \frac{\lambda x}{a} \quad ,$$

$$v = B \sin m \phi \sin \frac{\lambda x}{a} \quad ,$$

$$w = C \cos m \phi \sin \frac{\lambda x}{a} \quad .$$

These disturbances describe a buckling mode with m half waves around the circumference and $n = \frac{\lambda \ell}{\pi a}$ half waves along the length of the cylinder.

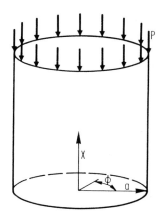

Fig. 1. Cylindircal sheel; coordinates and loads.

To determine is the critical pressure P corresponding to the bulking condition.

The critical value of P was determined within a shell theory involving both membrane forces and bending moments (see [1]) to be

$$P_s = \frac{E\,d}{1-\nu^2} \left\{ (1-\nu^2)\,\lambda^4 + \frac{d^2}{12a^2} \left[(\lambda^2+m^2)^4 - 2\,(\nu\lambda^6+3\lambda^4m^2+(4-\nu)\lambda^2m^4+m^6) + \right. \right.$$

$$\left. \left. + 2\,(2-\nu)\,\lambda^2\,m^2 + m^4 \right] \right\} \left[\lambda^2(\lambda^2+m^2)^2 + \lambda^2\,m^2 \right]^{-1} \quad .$$

. If one neglects the contribution of the bending moments to this critical
pressure one obtains

$$P_m = \frac{E\,d\,\lambda^4}{\lambda^2(\lambda^2+m^2) + \lambda^2\,m^2}\quad.$$

In Fig. 2 both critical pressures P_s and P_m have been plotted against
$\ell/na = \pi/\lambda$ as the abscisse. The shell theory was applied for $k = \frac{d^2}{12a^2} = 10^{-5}$.
In order to cover a wide range of values in both variables logarithmic scales
have been used.

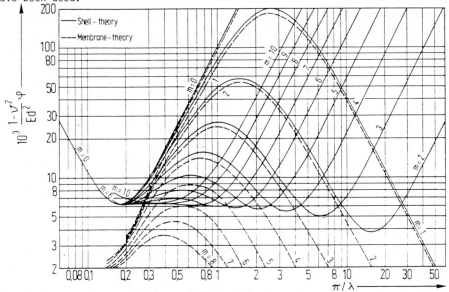

Fig. 2. Comparative evaluations of the critical pressures.

This example gives evidence of the possibilities and the limits of applica-
bility of the membrane theory. Table 1 shows how many harmonics can be calculated with good accuracy within the membrane theory.

TABLE 1

Buckling modes given back by the membrane theory.

ℓ/a	m	n
30	1	90
20	1	60
10	2	25
4	3	10
2	4	5
1.5	5	3

A second question is concerned with the practical importance of the stability to long wavelength disturbances. Actually in the sheet metal forming nearly all wrinkles in the final products are short wavelength buckles. As it was remarked by Geckeler [2] in 1928 these short wavelength buckles are caused by the subsequent evolution of the wrinkles. In the incipient state, as a rule, the wrinkles are long wavelength buckles. This fact was experimentally confirmed by other authors (see [3-4]).

The first step in deriving the stability criteria for long wavelength perturbations consists into choosing an appropriate form of the equilibrium equations for the membrane state of stress. Ore must remark that a shell theory based on Kirchhoff's hypothesis is not applicable to the sheet metal forming. The variable thickness and the distortion of the normal to the middle surface require other mathematical models than the classical shell theory.

The first section of the paper is concerning with the derivation of equilibrium equations of sheets under membrane state of stress. These equilibrium equations are derived by using the mathematical techniques of the minimal surface theory (for details see [5]).

The next section presents a mathematical definition of the stability with respect to wrinkling disturbances and establishes a stability criterion in agreement with the given definition. This criterion states the positive definitiveness of a certain quadratic form as a sufficient condition to ensure the stability to the long wavelength disturbances.

The paper ends with some considerations about the practical applications of the stability criterion.

EQUATIONS OF EQUILIBRIUM

Geometrical preliminaries

The geometrical shape of a thin sheet metal forming is defined by the equation of the middle surface $\Sigma(t)$ at any moment t during the forming process. Let $0\, x_1\, x_2\, x_3$ be a coordinate system such that $0x_3$-axis concides with the punch-direction. The middle surface equation will be

$$x_3 = z\,(x_1, x_2, t) \tag{1}$$

where, for any fixed t, the function $z\,(.,t)$ is defined over the domain D (projection of the metal sheet on $0x_1 x_2$-plane), Fig. 3.

Let us denote by ds an infinitesimal line element in the domain D. One the surface $\Sigma(t)$, ds corresponds to an infinitesimal line element

$$dS = \sqrt{1 + (z_{,1}\alpha_1 + z_{,2}\alpha_2)^2}\ ds \qquad ,$$

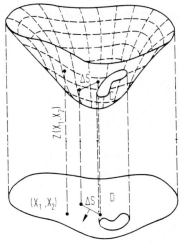

Fig. 3. Geometrical relations on the middle surface.

where α_1 and α_2 denote the direction cosinus of ds .

Let $\underset{\sim}{\tau}$ be the unit vector tangent to dS and $\underset{\sim}{\nu}$ the unit vector lying in the tangent plane and being normal to $\underset{\sim}{\tau}$. If $\underset{\sim}{N}$ denotes the unit vector normal to $\Sigma(t)$, then

$$\underset{\sim}{\nu} = \underset{\sim}{\tau} \times \underset{\sim}{N} \qquad (2)$$

holds.

Let us consider a point M on the sheet and a curve C on the surface through this point. Let $\underset{\sim}{\nu}$ and $\underset{\sim}{\tau}$ be the unit normal and unit tangent vector to this curve. We now consider a section through the sheet along the curve. Consider also an infinitesimal element of the section curve of length dS at the point M of the membrane. The stress state is entirely characterized by a force

$$\underset{\sim}{T} \nu \ ds \ ,$$

which acts on ds on those parts of the sheet for which $\underset{\sim}{\nu}$ is exterior normal. According to the hypothesis of the membrane state of stress $\underset{\sim\nu}{T}$ lies in the plane tangent to the sheet and no bending moments exist. In particular, for the case when C is a curve whose projection on $0x_1x_2$-plane is $x_1 = $ const. and $x_2 = $ const. correspond the stress vectors $\underset{\sim1}{T} \ dS_2$ and $\underset{\sim2}{T} \ dS_1$ respectively. Since $\underset{\sim\nu}{T}$, $\underset{\sim1}{T}$ and $\underset{\sim2}{T}$ are lying in the planes tangent to the sheet middle surface it is possible to break them down as follows

$$\begin{aligned}
\underset{\sim}{T} &= \sigma_\nu \, \underset{\sim}{\nu} + \sigma_\tau \, \underset{\sim}{\tau} \\
\underset{\sim1}{T} &= \sigma_{11} \, \underset{\sim}{\nu}^{(1)} + \sigma_{12} \, \underset{\sim}{\tau}^{(1)} \\
\underset{\sim2}{T} &= - \sigma_{21} \, \underset{\sim}{\tau}^{(2)} + \sigma_{22} \, \underset{\sim}{\nu}^{(2)}
\end{aligned} \qquad (3)$$

The formulation of the equilibrium conditions for the small triangular area element of the sheet-surface yields

$$T_\nu \, dS + T_1 \, dS_2 + T_2 \, dS_1 \; = \; 0 \quad ,$$

whence it follows

$$T_{\nu_1} \; = \; \left[\sigma_{11} \, \nu_1^{(1)} + \sigma_{12} \, \tau_1^{(1)} \right] \frac{dS_2}{dS_1} + \left[\sigma_{22} \, \nu_1^{(2)} + \sigma_{21} \, \tau_1^{(2)} \right] \frac{dS_1}{dS}$$

$$T_{\nu_2} \; = \; \left[\sigma_{11} \, \nu_2^{(1)} + \sigma_{12} \, \tau_2^{(1)} \right] \frac{dS_2}{dS} + \left[\sigma_{22} \, \nu_2^{(2)} - \sigma_{21} \, \tau_2^{(2)} \right] \frac{dS_1}{dS} \tag{4}$$

$$T_{\nu_3} \; = \; \left[\sigma_{11} \, \nu_3^{(1)} + \sigma_{12} \, \tau_3^{(1)} \right] \frac{dS_2}{dS} + \left[\sigma_{22} \, \nu_3^{(2)} - \sigma_{21} \, \tau_3^{(2)} \right] \frac{dS_1}{dS} \quad .$$

Let us denote

$$\tilde{n}_1 \; = \; \frac{dS_2}{dS} \quad \text{and} \quad \tilde{n}_2 \; = \; \frac{dS_1}{dS} \tag{5}$$

and

$$\nu_1^{11} = \nu_1^{(1)} \; ; \; \nu_2^{12} = \tau_2^{(1)} \; ; \; \nu_1^{21} = -\tau_1^{(2)} \; ; \; \nu_2^{22} = \nu_2^{(2)}$$

$$\nu_1^{12} = \tau_1^{(1)} \; ; \; \nu_3^{11} = \nu_3^{(1)} \; ; \; \nu_1^{22} = \nu_1^{(2)} \; ; \; \nu_3^{21} = -\tau_3^{(2)} \tag{6}$$

$$\nu_2^{11} = \nu_2^{(1)} \; ; \; \nu_3^{12} = \tau_3^{(1)} \; ; \; \nu_2^{21} = -\tau_2^{(2)} \; ; \; \nu_3^{22} = \nu_3^{(2)} \quad .$$

Then (4) can be written in more compact form

$$T_{\nu_i} \; = \; \sum_{\substack{j=1,2 \\ k=1,2}} \sigma_{jk} \, \nu_i^{jk} \, \tilde{n}_j \quad . \tag{7}$$

Let us consider a piece Δ of the sheet metal surface bounded by the curve C . The force and moments equilibrium conditions of this piece are

$$\oint_C T_{\nu_i} \, ds \; = \; 0 \quad , \quad i = 1, 2, 3 \quad ;$$

$$\oint_C (x_\ell \, T_{\nu_i} - x_i \, T_{\nu_\ell}) \, ds \; = \; 0 \quad , \quad i, j, \ell \; = \; 1, 2, 3 \quad .$$

Because Δ is arbitrar piece these conditions imply the following equilibrium equations:

$$\left[\sigma_{11}\frac{1+(z_{,2})^2}{\sqrt{1+(z_{,1})^2+(z_{,2})^2}}\right]_{,1}+\sigma_{21,2}-\left[\sigma_{22}\frac{z_{,1}z_{,2}}{\sqrt{1+(z_{,1})^2+(z_{,2})^2}}\right]_{,2}=0$$

$$\left[-\sigma_{11}\frac{z_{,1}z_{,2}}{\sqrt{1+(z_{,1})^2+(z_{,2})^2}}\right]_{,2}+\sigma_{12,1}+\left[\sigma_{22}\frac{1+(z_{,1})^2}{\sqrt{1+(z_{,1})^2+(z_{,2})^2}}\right]_{,2}=0$$

$$\left[\sigma_{11}\frac{z_{,1}}{\sqrt{1+(z_{,1})^2+(z_{,2})^2}}\right]_{,1}+\left[\sigma_{12}z_{,2}\right]_{,1}+\left[\sigma_{21}z_{,1}\right]_{,2}+\left[\sigma_{22}\frac{z_{,2}}{\sqrt{1+(z_{,1})^2+(z_{,2})^2}}\right]_{,2}$$
$$=0$$

$$(8)$$

and the symmetry conditions

$$-\sigma_{11}\frac{z_{,1}z_{,2}}{\sqrt{1+(z_{,1})^2+(z_{,2})^2}}+\sigma_{12}=\sigma_{21}-\sigma_{22}\frac{z_{,1}z_{,2}}{\sqrt{1+(z_{,1})^2+(z_{,2})^2}}\quad.\qquad(9)$$

Remark 1: Letting $\sigma_{11}=\sigma_{22}=p$ and $\sigma_{12}=\sigma_{21}=0$, then eq. (8) reduce to the equations of stabilized membranes

$$-z_{,1}\;1+(z_{,2})^2\;z_{,11}+2(z_{,1})\;z_{,2}z_{,12}-z_{,1}\;1+(z_{,1})^2\;z_{,22}=0$$

$$-z_{,2}\left(1+(z_{,2})^2\right)z_{,11}+2(z_{,2})\;z_{,12}-z_{,2}\left(1+(z_{,1})^2\right)z_{,22}=0$$

$$\left[\frac{z_{,1}}{\sqrt{1+(z_{,1})+(z_{,2})}}\right]_{,1}+\left[\frac{z_{,2}}{\sqrt{1+(z_{,1})+(z_{,2})}}\right]_{,2}=0\quad.$$

These three equations are equivalent to the well-known equation of minimal area (Plateau's problem).

Remark 2: The state of stress $\sigma_{11}=\sigma_{22}=p$, $\sigma_{12}=\sigma_{21}=0$ is encountered in soap-bubbles but not necessarily in the tent-membranes as the mathematiciens have recently stated (see [6]).

CRITERION OF STABILITY IN CASE OF WRINKLING PERTURBATIONS

Let us assume that in the undeformed state the sheet is plane and parallel to the (x_1,x_2)-plane. We denote the projection of the undeformed sheet on the (x_1,x_2)-plane with D_0. A punch acting in the x_3-direction deforms the sheet gradually into a surface described by the equation

$$x_3 = z(x_1,x_2,t)$$

$$(x_1,x_2) \in D(t), \quad D(0) = D_o, \quad t \in [0,T] \quad . \tag{10}$$

The function $z(x_1,x_2,t)$ together with the stress components $\sigma_{ij}(x_1,x_2,t)$; $\sigma_{12}(x_1,x_2,t)$; $\sigma_{21}(x_1,x_2,t)$ and $\sigma_{22}(x_1,x_2,t)$ must satisfy the equilibrium equation (8) and the symmetry conditon (9) at each moment t .

The inertia forces acting upon the sheet are neglected. In addition to the eqs. (8) and (9), constitutive relationships between the stresses and the strains must be considered. The strains are derived from the displacement vector field of components

$$u_1(X_1,X_2,t) , \quad i = 1, 2, 3 , \quad |(X_1,X_2) \in D_o , \quad t \in [0,T] ,$$

defined by the equation

$$u_i(X_1,X_2,t) = x_i - X_i \tag{11}$$

where X_i and x_i (i = 1, 2, 3) denote the coordinates of the same material point in the undeformed and deformed states respectively. Using these relations the function $z(x_1,x_2,t)$ is related with the displacement $u_i(X_1,X_2,t)$, i = 1, 2, 3 by the relation

$$z(X_1 + u_1(X_1,X_2,t) ; x_2 + u_2(X_1,X_2,t) ; t) = X_3 + u_3(X_1,X_2,t) . \tag{12}$$

From a mathematical point of view, wrinkling of a sheet can be taken as a deviation from (10) having the form

$$X_3 = z(x_1,x_2,t) + \delta z(x_1,x_2,t) . \tag{13}$$

In the general case, this deviation is accompanied by a modification of the stresses and displacements $\sigma_{ij}(x_1,x_2,t)$; i,j = 1, 2 ; $u_i(X_1,X_2,t)$; i = 1, 2, 3.

This leads to a complicated problem of stability, involving the solution of a complex system of equations, i.e., (8), (9) and constitutive equations.

In the following we shall restrict the procedure to the special case when disturbances of the form (13) are possible without modification of any other fields. This is the case when (13) is due to ideal rigid-plastic deformation, for example.

Let us first formulate the following definition of wrinkling stability.
Definition: The sheet metal forming process is stable with respect to wrinkling disturbances if for any perturbations occurring at any moments $t_o \in [0,T]$ having the form

$$(t-t_o) \; \delta z(x_1,x_2) \; , \; (x_1,x_2) \in D(t_o) \tag{14}$$

which satisfies the condition $\left. \partial z \right|_{\partial D} = 0$ on the boundary ∂D , there is an $\varepsilon > 0$ such that for $t - t_o \leq \varepsilon$ the variation of the work done by the external forces is always positive.

Since perturbations having the form $(t-t_o) \; \delta z(x_1,x_2)$ can be combined with those of the form $(t-t_o) \; (-\delta z(x_1,x_2))$, the above requirement of stability can be reformulated as follows:

"... there is an $\varepsilon > 0$ such that for $|t-t_o| \leq \varepsilon$ the variation of the work done by the external forces is always positive."

We shall proceed now with the derivation of a criterion of stability to wrinkling disturbances. First the expression of the variation of work done by the external forces is determined.

The existence of a perturbation in displacement having the form

$$(0, \; 0, \; (t-t_o) \; \delta z)$$

cannot occur without a corresponding change of the left-hand side term in the equilibrium equation (8). Due to the particular form of the disturbances (which are directed only along the Ox_3-axis) the changes of the first two equations (8) have no implication upon the work. For this reason we shall focus our attention only on the third equation (8). By replacing z with $z + \lambda \; \delta z$ where $\lambda = t-t_o$

$$\left[\sigma_{11} \frac{z_{,1} + \lambda \; \delta z_{,1}}{\sqrt{1 + (z_{,1}+\delta z_{,1})^2 + (z_{,2}+\lambda\delta z_{,2})^2}} \right]_{,1} + \left[\sigma_{12}(z_{,2}+\lambda\delta z_{,2}) \right]_{,1} + \left[\sigma_{21}(z_{,1}+\lambda\delta z_{,1}) \right]_{,2} +$$

$$+ \left[\sigma_{22} \frac{z_{,2} + \lambda \; \delta z_{,2}}{\sqrt{1 + (z_{,1}+\lambda\delta z_{,1})^2 + (z_{,2}+\lambda\delta z_{,2})^2}} \right]_{,2} = - \delta F_3(\lambda) \quad , \tag{15}$$

where $\delta F_3(\lambda)$ denotes the force component in the Ox_3-direction responsible for the disturbance. One observes that $\delta F_3(0) = 0$. By multiplying (15) with the velocity in the Ox_3-direction

$$v_3 = \frac{\partial [z(x_1,x_2,t_o) + (t-t_o) \; \delta z(x_1,x_2)]}{\partial t} = \delta z(x_1,x_2) \quad , \tag{16}$$

by integrating over $D(t_o)$, and by applying Green-Gauß formula, one obtains

$$\int_{D(t_o)} \left[\sigma_{11} \frac{(z_{,1}+\lambda\delta z_{,1}) \; \delta z_{,1}}{\sqrt{1 + (z_{,1}+\lambda\delta z_{,1})^2 + (z_{,2}+\lambda\delta z_{,2})^2}} + \sigma_{12}(z_{,2}+\lambda\delta z_{,2}) \; \delta z_{,1} + \right.$$
$$+ \sigma_{21}(z_{,1}+\lambda\delta z_{,1}) \; \delta z_{,2} +$$
$$\left. + \sigma_{22} \frac{(z_{,2}+\lambda\delta z_{,2}) \; z_{,1}}{\sqrt{1 + (z_{,1}+\lambda\delta z_{,1})^2 + (z_{,2}+\lambda\delta z_{,2})^2}} \right] dx_1 \; dx_2 = \int_{D(t_o)} \delta F_3(\lambda) \; \delta z dx_1 \; dx_2 \quad . \tag{17}$$

The power involved by the disturbances will than be

$$P(\lambda) = \int_{D(t_o)} \delta F_3(\lambda) \, \delta z(x_1, x_2) \, dx_1 \, dx_2 \qquad . \tag{18}$$

Consequently the variation $\Delta L(t, t_o)$ of work done by the external force will be

$$\Delta L(t, t_o) = \int_0^{t-t_o} P(\varsigma) \, d\varsigma \qquad . \tag{19}$$

Now we are able to prove the following criterion of stability to wrinkling disturbances.

Criterion: If the quadratic form

$$\phi(\delta z) = \int_{D(t)} \left[\sigma_{11} \frac{1 + (z_{,2})^2}{\sqrt{(1 + (z_{,1})^2 + (z_{,2})^2)^3}} \delta z_{,1} \, \delta z_{,1} + \left(\sigma_{12} + \sigma_{21} - \right. \right.$$

$$\left. - (\sigma_{11} + \sigma_{22}) \frac{z_{,1} z_{,2}}{\sqrt{(1 + (z_{,1})^2 + (z_{,2})^2)^3}} \right) \delta z_{,1} \, \delta z_{,2} + \sigma_{22} \frac{1 + (z_{,1})^2}{\sqrt{(1 + (z_{,1})^2 + (z_{,2})^2)^3}} \times$$

$$\left. \times \delta z_{,2} \, \delta z_{,2} \right] dx_1 \, dx_2 \tag{20}$$

is for any $t \in [0,T]$ positive definite in $\overset{o}{W}_2(D(t))^{\dagger)}$ (i.e., at any $t \in [0,T]$)

$$\phi(\delta z) \geq 0 \qquad .$$

For all $\delta z \in \overset{o}{W}_2(D(t))$ holds, then the sheet forming process is stable with respect to wrinkling disturbances.

Proof of Criterion: According to the definition, the stability with respect to wrinkling disturbances is ensured if for any disturbance δz there is an $\varepsilon > 0$ such that

$$\Delta L(t, t_o) \geq 0 \tag{21}$$

for any t provided the condition $|t - t_o| \leq \varepsilon$.

It is clear that (21) is satisfied for any $|t - t_o| \leq \varepsilon$ if and only if

---

$\dagger)$ $\overset{o}{W}_2(D)$ denotes the Hilbert space of functions defined over D vanishing on the boundary and having the gradients of integrable square.

$$\frac{\partial(\Delta L)}{\partial t}\bigg|_{t=t_0} = 0 \quad , \tag{22}$$

$$\frac{\partial^2(\Delta L)}{\partial t^2}\bigg|_{t=t_0} > 0 \quad . \tag{23}$$

Since $\delta F_3(0) = 0$, the condition (22) is automatically fulfilled. If one uses (19), (18) and (17) then (23) implies immediately the positive nature of (20). Conversely, if $\phi(\delta z)$ in a positive definite form, then (23) is satisfied this proves completely the criterion.

APPLICATIONS OF THE STABILITY CRITERION

There are two possibilities:

A.   The matrix

$$\left(\begin{array}{cc} \sigma_{11} \dfrac{1 + (z_{,2})^2}{\sqrt{(1 + (z_{,1})^2 + (z_{,2})^2)^3}} & -\dfrac{1}{2}\left[\sigma_{12} + \sigma_{21} - (\sigma_{11} + \sigma_{22})\right] \dfrac{z_{,1}\, z_{,2}}{\sqrt{(1 + (z_{,1})^2 + (z_{,2})^2)^3}} \\[4ex] \dfrac{1}{2}\left[\sigma_{12} + \sigma_{21} - (\sigma_{11} + \sigma_{22})\right] \dfrac{z_{,1}\, z_{,2}}{\sqrt{(1 + (z_{,1})^2 + (z_{,2})^2)^3}} & \sigma_{22} \dfrac{1 + (z_{,1})^2}{\sqrt{(1 + (z_{,1})^2 + (z_{,2})^2)^3}} \end{array} \right).$$

$$\tag{24}$$

is for all (x_1, x_2) positive definite. Then the functional (20) will be equally positive definite and consequently the stability to long wavelength disturbances holds.

B.   There are regions in D for which the matrix (24) is not positive definite. Then the positive definitivness of (20) must be investigated with respect to each particular type of disturbances. For example in the case of a rectangular domain D $(0 < x_1 < a , 0 < x_2 < b)$ one chooses the perturbations of the form

$$\delta z = \sin \frac{m \pi}{a} x_1 \sin \frac{n \pi}{b} x_2 \tag{25}$$

By setting (25) in (20) the functional $\phi(\delta z)$ will be transformed in a function $\phi(m,n)$. The maximal range in the plane (m,n) for which $\phi(m,n)$ is strong positive definite, defines the bounds for the wavelengths with respect to which the metal sheet forming is stable. For other type of domain D one must choose appropriate form of disturbances.

REFERENCES

[1] W. Flügge, Stresses in shells, Springer-Verlag, Berlin, Göttingen, Heidelberg 1962.

[2] J.W. Geckeler, Plastisches Knicken der Wandung von Hohlzylindern und einige andere Faltungserscheinungen an Schalen und Blechen, ZAMM 8 (1928) 341-352.

[3] E. Siebel, Der Niederhaltedruck beim Tiefziehen, Stahl und Eisen 74 (1954) 155-158.

[4] I. Romer, F. Fischer and F. Braun, Über den Einfluß eines variablen Niederhalterdrucks auf das Grenzziehverhältnis, Stahl und Eisen 21 (1984) 1065-1072.

[5] H. Gloeckl and P. Mazilu, Wrinkling stability to long wavelength perturbations, to be published.

[6] S. Hildebrand and A. Tromba, Panoptimum, Spektrum der Wissenschaft, 1987.

Computational Methods for Predicting Material Processing Defects, edited by M. Predeleanu
Elsevier Science Publishers B.V., Amsterdam, 1987 — Printed in The Netherlands   275

CRITICAL DAMAGE AND DUCTILE FRACTURE : QUANTITATIVE EXPERIMENTAL DETERMINATION

F.H. MOUSSY and P. LEFEBURE

Cold Forming Section. Institut de Recherches de la Sidérurgie Française -
185 rue du Président Roosevelt 78105 Saint-Germain-en-Laye (France)

SUMMARY
    Original metallographic observations are used to determine quantitati-
vely the porosity and porosity gradients near fracture surfaces. The techniques combining Ion Beam Polishing, Scanning Electron Microscopy, manual
drawings on tracing-paper and quantitative metallography lead to valid
porosity measurements. The main results are : porosity at fracture is not a
material constant, it decreases sharply with stress triaxiality; porosity at
fracture is low : 0.5 -10 %; the areas concerned by these porosities are very
small, 70-1000 μm². To predict fracture, models have to be able to describe
slightly porous materials at very fine scale. At this scale, the critical
damage value leading to fracture cannot be determined by classical mechanical
testing or physical investigations.

INTRODUCTION
    Deformation processing of metals and alloys is studied since a long time.
The goal of all these studies is to increase the formability of materials and
the service properties of the finished part. Presently, the development of
automatization particularly with the aid of computers and robots needs a
higher degree of knowledge concerning the material properties and their
evolution during forming. Among the factors limiting formability the most
important ones are :

- load levels necessary to deform the metal, leading to large capacity
  presses and tool wear
- localization of deformation by necking leading to unacceptable thickness variations in sheet for example
- fracture.

To understand and solve these problems, the behaviour of steels has to be
known from two different points of view :

- Deformation of the metal : stress-strain relationships, anisotropy,
  influence of strain history, flow-rule. This first aspect is
  represented by functions which can be modified during straining :
  induced anisotropy, memory effects for example.
- Limits to the deformation of metals : buckling, necking and fracture.
  This second aspect is represented by criteria, the application of which

requires the knowledge of the stress-strain relationships and specific limiting conditions.

For these two aspects, physical and mechanical approaches are developed. They are both individually not sufficient and are to be considered as complementary. This paper is relative to fracture criterion studies with a physical approach, but the link between mechanical and physical criteria is discussed. This analysis shows the necessity of determining the physical fracture mechanisms in these particular extreme conditions and the difficulty to model them correctly.

MECHANICAL AND PHYSICAL DESCRIPTION OF FRACTURE PROCESSES

From a macroscopical point of view, fracture occurs with or without the presence of an initial macrodefect : crack or notch. When an initial defect exists, Fracture Mechanics is applied to take into account the very high gradients of stresses and strains which are built up at the notch or crack root and which are described in terms of singularities. But in the more general case of a continuous uncracked specimen, fracture occurs either on the surface (upsetting test, torsion) or in the bulk material (tensile test, "chevrons" in wire drawing), apparently in a more progressive manner. In fact, these two fracture descriptions are relative to the same phenomenon, the scale at which the approach is conducted being the only difference. The unicity of these two descriptions of fracture is realized nowadays by the so called "Local Approaches" developed by "(refs. 1-2)".

The basic problem in studying and modeling fracture is to describe and understand how, starting from a continuous medium, a macrocrack can appear at a given point. Using the relevant analysis developed in (ref. 3)", it consists in the determination of the frontier between the areas of Damage and Fracture Mechanics.

Applied to solids, there is a discrepancy between the continuous description of the classical mechanics and the discrete constitution of real solids at atomic scale. It is however possible to overcome this contradiction by considering a sufficiently large volume of the material. Almost all of the treatises dealing with continuum media begin with the definition of the "elementary volume" which defines the limits of validity for the theories developed afterwards. Among them we can mention two of them :

♦ "Introduction to the Mechanics of a Continuous Medium" by L.E. MALVERN "(ref. 4)".

"The adjective _continuous_ refers to the simplifying concept underlying the analysis : we disregard the molecular structure of matter and picture it as being without gaps or empty spaces. We further suppose that all the mathematical functions entering the theory are continuous functions, except possibly at a finite number of interior surfaces separating regions of continuity".

♦ "Mécaniques des Matériaux Solides" by J. LEMAITRE and J.L. CHABOCHE "(ref. 5)".

"Par élément de volume au sens de la mécanique physique des solides, il faut entendre un volume suffisamment important par rapport aux hétérogénéités de la matière et suffisamment petit pour que les dérivées partielles des équations de la Mécanique des Milieux Continus aient un sens".

What is very important in these definitions is that the size of this "elementary representative volume" depends of microstructural or physical parameters "(refs. 3, 5)", in other words, it is not possible to define ex abrupto a universal "elementary volume".

These considerations can become important when we are looking for a modelization of the behaviour of metals integrating physical mechanisms (induced texture development leading to anisotropy) but are certainly determing when we are looking for a modelization of fracture.

Mechanical approach

♦ Continuum approach

This is a phenomenological one which consists in defining a fracture criterion of the type :

$$f (\sigma_{ij}, \epsilon_{ij}) = 0$$

or of an incremental type :

$$\int_{o}^{f} g (\sigma_{ij}, \epsilon_{ij}) \, d \epsilon_{ij} = 0$$

some examples are given below.

. RICHMOND and SPITZIG "(ref. 6)".

$$I_1 + a I_2 = b$$

where I_1 and I_2 are respectively the first and the second invariant of the stress tensor, a and b are material constants.

. $D = D_c$ "(ref. 7)".

D being damage defined in the most general form as a function of the point X and time t :

$$D = f (X,t).$$

. LATHAM and COCKCROFT "(ref. 8)".

$$\int_{0}^{\bar{\epsilon}_f} Max\ (\sigma_I,\ o)\ d\bar{\epsilon} = C$$

where σ_I is the highest principal strain.

All these formulations put forward the major role played by the tensile stresses and the stress triaxiality. So, without knowing the deformation and fracture mechanisms, it is possible to prevent fracture in choosing a judicious deformation mode which minimizes the tensile stresses. The validity of these criteria and the values of the constants are established by experiments leading to fracture for different strain paths. The mechanical values σ_{ij} and ϵ_{ij} have to be calculated at each point but are defined in the sense of macroscopical features belonging to continuum mechanics.

✦ Metal containing holes or porosity

Since earlier works on ductile fracture of steels, it is well known that deformation and fracture proceeds by nucleation, growth and coalescence of microvoids. Although the medium is no more continuous, a mechanical approach remains possible in modelling the different stages of the hole growth considering that the matrix between the holes is continuous. The cavities have to be considered regularly distributed and of simple geometric shape : spheres, cylinders, cubes, ellipsoïds. This modelization is also realized with finite element methods "(ref. 9)".

Some of these fracture criteria are mentioned here :
. Mc CLINTOCK (cylindric holes) "(ref. 10)".

$$\int_{0}^{\bar{\epsilon}_f} \left[\frac{2}{\sqrt{3}(1-n)}\ sinh\ (\frac{\sqrt{3}(1-n)}{2}\ .\ \frac{\sigma_a + \sigma_b}{\bar{\sigma}}) + \frac{\sigma_b - \sigma_a}{\bar{\sigma}} \right] d\bar{\epsilon} = K$$

σ_a and σ_b are the stresses at infinite perpendicular to the cylinder axis and n is the strain hardening coefficient.
. OYANE "(ref. 14)"

$$\int_{0}^{\bar{\epsilon}_f} \left(1 + \frac{\sigma_m}{A\ \bar{\sigma}} \right) d\bar{\epsilon} = K$$

or

$$\int_{0}^{\bar{\epsilon}_f} \left(1 + \frac{\sigma_m}{A\ \bar{\sigma}} \right) \bar{\epsilon}^c\ d\bar{\epsilon} = K$$

. HANCOCK and MACKENZIE "(ref. 12)"

$$\bar{\epsilon}_f = K \exp \left(\frac{-3\,\sigma_m}{2\,\bar{\sigma}} \right)$$

which results from an integration of the RICE and TRACEY model of hole growth.

In these criteria, the physical processes of fracture are taken into account. The fracture is defined in terms of a physical or more exactly a geometrical condition : critical porosity, coalescence when holes reach a given size or are sufficiently large to meet.

Physical approach

The basis of this approach consists in observing the deformation and damage mechanisms at microscopical scale and to follow them until fracture. It is thus possible to determine the microstructural components which are responsible for damage initiation, particularly the interfaces of low cohesion : particle-matrix interfaces, grain boundaries.... A physical definition of damage can be proposed and the growth of damage is followed quantitatively. The mechanisms of void coalescence leading to a macrodefect can only be qualitatively determined because fracture in a bulk material results from very localized and rapid phenomena. We find here the same difficulties presented before, relative to the definition of geometrical fracture criteria which are exactly of the same type as physical local fracture criteria. Different hypotheses have been proposed :

- critical value of $(\frac{L_2}{L_1})_c$ "(ref. 13)" where L_1 is the length of a void parallel to the tensile direction and L_2 the distance between two adjacent voids perpendicularly to the tensile direction.
- critical hole growth $(\frac{R}{R_0})_c$ "(ref. 1)".

One of the major difficulties of this physical approach is to obtain significant results and also to make observations of the hole growth which has proceeded under a real triaxial stress field, that is to say in the bulk material.

Link between mechanical and physical approaches

As shown here, mechanical approaches are the only ones to give general and quantitative results which can be applied at a macroscopic industrial scale, but they need some physical data which can only be obtained by observation. The fracture criterion even if formulated in a macroscopic mechanical form, should reflect processes at a microscopic level and take into account all the local physical phenomena : deformation, damage, microcracks, geometrical distribution of defects, behaviour of highly strained material.

Many theoretical approaches deal with porosity. Is a critical porosity a valuable fracture criterion or is the porosity, if sufficiently high, able to influence the plastic potential and so produce a decrease in the stress-strain curves leading to local instability and fracture ? The aim of this paper is to present some recent results allowing to answer partially these questions.

EXPERIMENTAL PROCEDURE

Steel

A plain carbon steel was selected for this study. Its chemical composition is given in table 1.

TABLE 1

Chemical composition of the steel (10^{-3} % weight).

C	Mn	Si	P	S	Cr	Mo	Al
420	720	230	26	31	150	10	45

The second phase particles which are considered to act on ductile fracture are cementite. Two heat treatments were realized to obtain globular cementite with an homogeneous (tempering of martensite) and heterogeneous (spheroïdization of ferrite-perlite) distribution. Inclusions, mainly Manganese Sulfides are globular.

The mechanical properties are given in table 2.

TABLE 2

Mechanical properties of the steel.

	σ_y MPa	Lüders elongation (%)	σ_u MPa	Elongation at fracture (%)	RA %
Homogeneous cementite	310	2	489	36	67
Heterogeneous cementite	468	2.4	586	28,1	63

Strain paths

Different strain paths were realized to compare the influence of stress and strain triaxiality on the fracture process (specimen dimensions in mm) :

- smooth tensile specimen (ST) : diameter 10. Length 50
- notched tensile specimen : diameter : mini 6, maxi 11, radius of curvature : 2.5 (AE 2.5) and 1.25 (AE 1.25)
- upsetting test on cylinders : diameter 20. Height 30.

Experimental quantification of local damage

This is the central part of this study. Among the numerous definitions of damage, we have chosen a physical and simple one. Damage is considered to be equal to the porosity, that is to say damage D, is a dimensionless scalar. This definition of damage is of course not sufficient to explain completely the steel properties, for example directional properties like anisotropy of ductility, but its knowledge is often necessary to establish the validity of theoretical approaches concerning fracture and flow of metals.

For D measurements, we have used the simple equation established firstly by DELESSE (1848).

$$L_L = A_A = V_V$$

where L_L, A_A and V_V are respectively the lineic, surfacic and volumic concentrations of a second phase in a matrix ; here, the second phase are the voids. The only hypothesis for the use of this equation are : the sample must be statistically representative. Since we are interested in the last stages of deformation, just before fracture, specimens were deformed until fracture and then cut perpendicularly to the fracture surface. The samples were mechanically polished and then Ion Beam Polished (IBP). This new method is actually the only one able to reveal properly microdefects down to .05 µm diameter or thickness. Numerous micrographs in Scanning Electron Microscopy (SEM) were used to cover relatively large areas in the vicinity of the fracture surface. Starting from these photographs, a drawing was manually made on tracing-paper, the holes being black and the matrix white. This very well contrasted image could be automatically analyzed with the Textur Analyse System (TAS) from Leitz (IRSID - ARMINES - LEITZ PATENT). Depending on the specimens, gradients of local damage have been established either parallel or perpendicular to the fracture surface. The D gradients being very localized and the fracture surface being highly irregular, it is necessary to describe the technique used and define the surface element on which the D measurement is done "(ref. 14)".

The D value corresponding to a distance d = hi from the fracture surface is measured by A_A determined on the area situated between two lines l_i and l_{i+1}, fig. 1, the line l_i being obtained by the translation of the fracture surface line l_o by a distance hi.

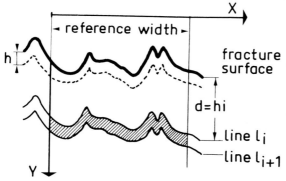

Fig. 1 -

Definition of the elementary area leading to the porosity histogramm perpendicular to the fracture surface.

For D gradients parallel to the fracture surface, the methodology is identical, fig. 2. The depth p is kept constant. For simplification we consider a set of axes OXY, O being on the fracture surface X parallel to this surface and Y perpendicular, fig. 1 and 2.

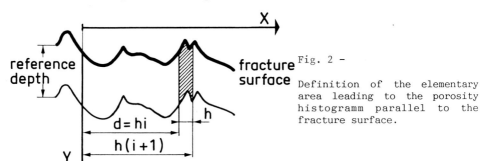

Fig. 2 -

Definition of the elementary area leading to the porosity histogramm parallel to the fracture surface.

Voids leading to dimples, some quantifications have been done on fracture surfaces to obtain additionnal informations on the final fracture deformation processes. SEM photographs were manually reproduced on a tracing-paper and then the size of the dimples was evaluated by automatic image analysis.

RESULTS AND DISCUSSION

Most of the observations were made on the steels heat treated so as to have homogeneous cementite structure because quantitative representative data are easier to obtain.

Although deformation and fracture mechanisms are at microscopic scale qualitatively independant of the stress and strain history, quantitavitely the behaviour is very different between the tensile (ST, AE 2.5 and AE 1.25) and the upsetting specimens; consequently, in the following the results will be presented in two different parts : tensile and compressive loading.

Tensile specimens

When fracture occurs in the bulk material, it starts usually at a unique place. This happens when the material is no longer able to resist to fracture for the local stress and strain states which are reached. As soon as the macrodefect is created, there is a redistribution of the stresses and strains which combines a stress decrease in certain areas away from the macrodefect and a concentration of stress at the extremities of this macrodefects. Ductile fracture in the bulk material, during the stage corresponding to the propagation of a defect, has to be described with Fracture Mechanics concepts. We are looking for critical fracture conditions leading to the macrodefect initiation but, on a fractured specimen it is not possible to locate this point. Metallographic observations do not permit to determine any singular area under the fracture surface located near the symmetry axis of the specimen; so we shall consider that any area observed near the central part of the fracture surface is representative of the conditions leading to local fracture.

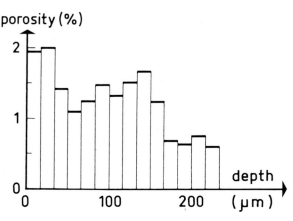

Fig. 3 - Porosity gradient perpendicular to the fracture surface.

To determine properly the porosity at fracture we have to select a measuring area sufficiently small relative to the porosity gradients and sufficiently large relative to the pore density. In each case, gradients were determined parallel and perpendicular to the fracture surface.

For AE 2.5 and AE 1.25 specimens the histogramms were realized parallel to X for a depth of 25 μm. The porosity varies from 0 to respectively 6 and 4 %. These high values are only obtained on very narrow areas: 5 or 10 μm which are not representative of the mean tendancy. The porosity value was so established on 160 μm in the X direction.

The porosity gradient in the Y direction was determined for the ST, fig. 3. There is an irregular but significant decrease from 2 to .5 % for Y going from 0 to 235 μm. For AE 2.5 and AE 1.25, the gradients are higher, the porosity decreases over distances of some tens of micrometers. Geostatistical considerations indicate that to determine a porosity of 1 % formed with holes of 0.5 μm diameter with an accuracy of 20 % it is necessary to observe 1500 μm². These statistical considerations validate our quantitative porosity determination.

The main results are presented in table 3.

TABLE 3 - Porosity quantification.

Cementite distribution	straining mode	V_V %	$\bar{S}$ μm²	Feret* //Fract. μm	Feret* ⊥Fract. μm	V_V grad //Fract %→% (over μm)	V_V grad ⊥ Fract. %→% (over μm)
Heterogeneous	ST	5.12	1.06	1.02	1.12	no	ND**
	ST	1.9	0.19	0.50	0.42	no	2 → 0.5 (230 μm)
	AE 2.5	0.97	0.26	0.57	0.49	no	1 →≈0.5 (10 μm)
Homogeneous	AE 1.25	0.48	0.21	0.53	0.42	no	0.5→≈0.3 (10 μm)
	UPSETTING	15	0.66	1.17	0.69	15→2 (120 μm)	16→1.5 7→0.7 (16 μm)

*Max Length //or ⊥ to fracture surface.

**not determined.

Triaxiality effects

It is well known that triaxiality plays a major role in damage growth. The theoretical and experimental works show clearly that for a given equivalent strain developed during two similar strain paths differing by stress triaxiality, hole growth and damage are higher for the higher triaxiality. Such a situation is realized in comparing the ST, AE 2.5 and AE 1.25 specimens. The triaxiality is increasing from ST to AE 1.25. The AE 2.5 and AE 1.25

were modelized by FEM (code ZEBULON) calculations. At fracture, the stress triaxiality $\sigma_m/\bar{\sigma}$ equals 1.1 for AE 2.5 and 1.45 for AE 1.25. Although not modelized, the stress triaxiality for ST is less than 1. To explain the decrease of ductility with increasing stress triaxiality, some authors consider that fracture occurs for a critical porosity, whatever the stress and strain history (for example, V_V^C = .06 for OYANE). Our measurements, which are statistically significant and representative, indicate clearly that the porosity at fracture is not a material constant. It decreases from 2 % for ST to 1 % for AE 2.5 and .5 % for AE 1.25 (table 3). These results, relative to the critical porosity at fracture, are not contradictory with the influence of stress triaxiality on porosity growth.

Flow mechanisms are governed by shearing or equivalent stress or second stress tensor invariant.

When the stress triaxiality or the first invariant increases, the shearing is less easy, the matrix becomes progressively more brittle and the hole growth stops earlier. MUDRY (Ref. 1) has determined the critical value of spherical hole growth R/R_o for different triaxialities. His results show a decrease of $(R/R_o)_c$ with the increase of $\sigma_m/\bar{\sigma}$.

Heterogeneous cementite distribution

For ST, the heterogeneous cementite microstructure was strained up to fracture. Although more scattered, the results show that the porosity at fracture is more important (5 %) than for a homogeneous cementite distribution (2 %). When heterogeneously scattered, cementite particles are gathered in clusters, there are strong interactions between adjacent holes during growth. So, the local porosity grows faster and for the same stress triaxiality, porosity at fracture is significantly higher. Table 3 shows that the hole size is also bigger for this heterogeneous microstructure : 1 μm as compared to 0.5 μm for the homogeneous state.

Fracture surface analysis

The fracture surfaces are classical ones with two kinds of dimples : bigger initiated on inclusions mainly manganese sulfides and smaller initiated on cementite particles. The mean size of the dimples has no physical meaning because it takes into account two different metallurgical features. A study of the influence of stress triaxiality has to separate these two dimple populations. We have considered that small dimples initiated on carbides were smaller than 4 μm² and large dimples initiated on inclusions were larger than 4 μm². When the triaxiality increases, the smaller remain unchanged (about

1.5 μm²) while the larger grow from 13 to 40 and 43 μm². Although the porosity at fracture decreases with stress triaxiality, dimple size increases up. This indicates that the last deformation stages are difficult to understand.

Upsetting

To study fracture conditions produced for completely different stress and strain histories, upsetting tests were performed on cylinders. We shall describe the phenomena in the classical cylindrical set of coordinates (longitudinal z, radial r and tangential θ). At the beginning of the deformation, the material is submitted to an uniaxial compression parallel to z and then, due to friction forces between specimen and die, there is a barrelling of the specimen which leads to a biaxial stress state on the equator : $\sigma_\theta > 0$ and $\sigma_z < 0$ first and near the fracture, $\sigma_\theta > 0$ and $\sigma_z > 0$. In this study, the macroscopic crack orientation indicates that the fracture appears under a biaxial tensile stress state : it takes place parallel to the z direction and penetrates in the specimen at 45° with the axis θ or r. As soon as the macrocrack is visible, the test is interrupted. The limiting relative reduction in height is around 75,5 % and the macrocrack is between .5 and 1 mm long. The specimen is cut in the "equatorial" plane perpendicular to z and containing the macrodefect. The porosity is determined parallel and perpendicular to the fracture surface. We shall consider the coordinates X and Y defined previously for the tensile fracture analysis. The area is divided in rectangular areas of 3,2 μm depth (Y direction) and 30 μm length (X direction). The analyzed area covers 120 μm in the X direction starting from the free surface of the upsetting specimen and 16 μm in the Y direction starting from the fracture surface of the specimen. The results are presented in figures 4 and 5.

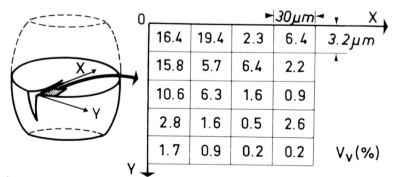

Fig. 4 - Location of the analyzed area and results.
(Porosity : %). Upsetting test.

Although a certain scatter is observed, inherent to this kind of delicate observations, two porosity gradients are clearly visible respectively in the X and Y directions. The intensity of these gradients are much higher than for tensile tests. The highest value of porosity is obtained for X and Y equal to 0, that is to say at fracture initiation on the specimen surface. The porosity at fracture reaches 15 % and is much larger than in tensile tests, which confirms the influence of the stress triaxiality. The Y gradients are very intense. Porosity goes from 10-15 % to 1-2 % over a distance of 16 µm. The Y gradients are less pronounced. The porosity is divided by 10 over 120 µm. We confirm that a critical porosity is not a good fracture criterion and that, for this kind of fractures, critical porosity decreases during crack propagation.

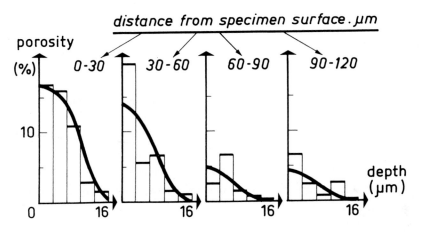

Fig. 5 - Porosity gradient perpendicular to the fracture surface in relationship with the distance from specimen surface or fracture initiation site. Histogramms from fig. 4.

Due to the important porosity gradients, the elementary representative volume or surface for our observations is very small : around 5 x 15 µm².

CONCLUSIONS

Damage defined as porosity is very localized at fracture. Its value is not a material constant but depends on the mechanical parameters : stress triaxiality evolution during deformation until fracture. It reaches 15 % for upsetting but in tensile specimens it is around a few percents. The porosity at fracture decreases strongly with an increase in stress triaxiality. The elementary representative volume which can be considered as homogeneous varies for our material between (in section) 30 x 30 µm² to 5 x 15 µm². This is very small and a consequence is that most of the experimental methods determining the critical damage value at fracture do not hold. The methodology we have developed : fractured specimen - Ion Beam Polished Sections - manual porosity cartography - quantitative metallography is tedious, but perhaps the only one today which allows the determination of fracture porosity with statistical significance on so small volumes. Mechanical models must be able to predict fracture with low localized porosity, the overall specimen porosity "far" from the fracture (more than .1 or .2 mm) being less than .5 %.

ACKNOWLEDGEMENTS

The authors gratefully acknowledge the assistance of Mr C. QUENNEVAT in the delicate experimental work and of Mrs H. DEMAGNY and F. HEINRICH for the development of the quantitative metallography analysis.

REFERENCES

1. F. MUDRY, Etude et Modélisation de la Rupture Ductile et de la Rupture par Clivage d'Aciers Faiblement Alliés. Thesis UTC. ENS Mines de Paris (1982).
2. F.M. BEREMIN, Met Trans. 12-A (1981) 723.
3. J. LEMAITRE and J.L. CHABOCHE, J. Méc. Appl. 2 (1978) 317.
4. L.E. MALVERN. Introduction to the Mechanics of a Continuous Medium. Prentice-Hall, Inc. Englewood Cliffs, New Jersey (1969) V.
5. J. LEMAITRE and J.L. CHABOCHE, Mécanique des Matériaux Solides, Dunod, Paris (1985) 72.
6. O. RICHMOND and W.A SPITZIG, Proc. Int. Cong. Theor. Appl. Mech, 15 Toronto (1980).
7. M. PREDELEANU, Ecole d'Eté Montréal 7-10 July 1986. To be published.
8. M. G. COCKCROFT and D.J. LATHAM; J. Inst. Metals, 96 (1968). 33
9. V. TVEERGARD, J. Mech. Phys. Solids. 30 (1982) 265.
10. F.A. Mc CLINTOCK in Ductility ASM (1968) 255.
11. M. OYANE, S. SHIMA and T. TABATA. J. Mech. Work. Tech. 1 (1978) 325.
12. J.W. HANCOCK and A.C. MACKENZIE. J. Mech. Phys. Solids 24, (1976) 147.
13. L. ROESCH, G. HENRY, M. EUDIER and J. PLATEAU. Mem. Sci. Rev. Met. 63 (1966) 927.
14. D. BENOIT and H. DEMAGNY. Caracterization of the Limiting Damage in Ductile Fracture by mean of Image Analyser. 7th International Congress for Stereology. 2-9 September 1987 - CAEN France.

Computational Methods for Predicting Material Processing Defects, edited by M. Predeleanu
Elsevier Science Publishers B.V., Amsterdam, 1987 — Printed in The Netherlands

METHOD OF CALCULATION OF CURVE OF DEFORMABILITY OF K18 STEEL

J. POŚPIECH

Research Institute of Ferrous Metallurgy, ul. K. Miarki 12,
44-101 Gliwice /Poland/

SUMMARY

   Description of chosen mechanical tests used for determination
of curve of deformability. Methods of calculation of boundary
strains and stresses in chosen tests. Presentation of data
obtained for K18 steel. Critical analysis of curve of deformability obtained for K18 steel.

CURVE OF DEFORMABILITY

   The capacity of materials for deformation is a function of
two variables – the material and the state of stress. For
quantitative evaluation of the dependence of deformation upon
the state of stress, the concept of the deformability curve
– also called the curve of fracture – has been introduced. This
curve is plotted in the y-x system, where y denotes a quantity
describing the deformation /effective strain will be used here/
and x a quantity characterizing the state of stress – usually
called the stress – state factor: the ratio of mean stress to
effective stress /σ_m/$\bar{\sigma}$/ will be used here. There is a new
proposal /ref. 1/ to use tensile test under different values
hydrostatic pressure and torsion test under different values
hydrostatic pressure to determine whole curve of deformability.
Deformation is a function of two variables: state – stress
factor and coefficient of Lode in this work /ref. 1/. However,
it is expensive and for practical purposes the curve of deformability can be determined by using upsetting test, torsion test
and tensile test on notched specimens. Cylindrical specimens
were used in all tests.

Upsetting test

The upsetting test is carried out using specimens with different ratios ho/do although there is a proposal to standarise a specimen /ref. 2/ with ho/do = 1.5 /ho = 21, do = 14/. The test was continued until the first cracks - visible to the naked eye - appeared on the curved surface of the specimen.

Effective strain and state of stress factor were determined by the method described in /ref. 3/.

Torsion test

The torsion test is carried out on specimens, shown in Fig. 8 /ref. 4/ although there is a tendency to decrease gauge length. Specimens have a line of about 0,1 mm depth cut on the inside, parallel to the specimen axis. State of stress factor k and effective strain ξ in this test are calculated from the following equations /ref. 1/

$$k = 0 \qquad \text{/1/}$$
$$\xi = \text{tg}\,\psi \qquad \text{/2/}$$

where tg is the mathematical function tangent; ψ is the angle of inclination between the line on the side surface of a specimen cut parallel to the specimen centre-line and the generating line, measured at the point of fracture.

Tensile test on notched specimens

The tensile test carried out on the specimen shown in Fig. 1 /ref. 5/ permits to obtain several points of the curve of reserve of plasticity by changing d/R ratio. This test does not allow to determine the start of failure /appearance of the first microcrack revealed by the naked eye/, as the fracture occurs suddenly and covers the whole cross section of the specimen. The unquestionable advantage of the tensile test is the fact that the plastic deformation at the place of failure progresses monotonically. To determine the state-of-stress factor and the effective strain, the formula given by Dawidenko and Spiridonowa /ref. 1/ have been used.

At the spot of fracture in the neck of specimen, the state of stress is three-axial tensile. The state-of-stress factor will take the form /basing on formula derived by Dawidenko and Spiridinowa/

$$k = 0.577 + /0.433 \ d_1 //R_1 \qquad\qquad /3/$$

The effective strain at the moment of fracture will be

$$\overline{\varepsilon} = 2\sqrt{3} \ \ln /d_0/d_1/ \qquad\qquad /4/$$

where d_0, d_1 - dia. of specimen at the spot of fracture before and after breakage, resp; R_1 - radius of curvature of specimen generating line at the spot of fracture after breakage.

EXPERIMENTAL RESULTS

Material tested

Investigations have been carried out on K18 steel having the following analysis /in per cents/ : 0.19 C, 0.98 Mn, 0.27 Si, 0.028 P, 0.023 S, 0.13 Cr, 0.08 Ni, and 0.08 Cu.
The steel was reheated in furnace and then forged at 1180° C to bars of 18 mm dia. After forging the bars were cooled in air. From this test material standard test specimens have been prepared by turning and polishing for mechanical tests. Tensile test specimens with the turned notch were turned and additionallly ground at the spot of notch.

Specimens for the upending test were turned. A part thereof had a grid on the surface, obtained by turning and milling. The depth of the grid was 0.2 mm.

Tensile test on notched specimens

For this test have been used seven specimens with initial dimensions $d_0 = 10$ mm and $R_0 = 2.5$ mm and seven specimens with dimensions of the notch $d_0 = 4$ mm and $R_0 = 8$ mm. Such dimensions have been adopted. In practice, there were some deviations therefrom. The lengths and outside diameters of the applied specimens were as follows : $D_0 = 15$ mm and $l_0 = 250$ mm, $D_0 = 14$ mm and $l_0 = 240$ mm.

The tensile test has been carried out on 10 tonf /9.97 kN/ tensile testing machine of Amsler manufacture. During the test the advance of the jaw was 1 mm/min. The radius of curvature was measured by the method described in /ref. 6/ under the workbench microscope.

Upsetting test

This test has been carried out on a press so as to obtain the flow rate approximately equal to that obtained in tensile testing. Because the press was controlled by a lever, this flow rate could not be established accurately. The ratio of h_o/d_o was 1 for all the specimens.

Tested were both the specimens with a grid cut on the side surface and the specimens without the grid. On the surface of some specimens there was an ortogonal grid with the mesh of 1 mm.

The upending tests have been carried out under the following conditions of lubrication:

1. without any lubricant,
2. with the use of a paste composed of drawing practice powder and palm oil, applied on both the end surfaces of specimen,
3. with the use of molybdenum disulfide powder which was rubbed into that end surface of specimen which came in contact with the movable part of the press,
4. with the use of molybdenum disulfide powder rubbed into the both end surfaces of the specimen.

CONCLUSIONS

Obtained have been six points of the curve of reserve of plasticity for K18 steel. Final results /values of state-of-stress factor and effective strain/ are given in table 1. /ref. 5/. Basing on these data the curve of reserve of plasticity for K18 steel has been plotted Fig. 1 /ref. 5/. This curve is the image of the decreasing function. As the state-of-stress factor increases, decreases the degree of deformation as necessary to cause fracture of material. Thus, there is a decrease in plasticity of the material.

Confirmed has been the well known fact that plasticity characterizes not only the material but also the state of stress.

It can be seen from Fig. 1 /ref. 5/ that with the state-of-stress factor k = 2 the plasticity of material is very low. Hence it can be concluded that there exists a certain boundary value of k, at which the degree of deformation at fracture is zero. With such a state of stress the steel of K18 grade will behave as a brittle material. Considering the data listed in table 1 it can be said that the tensile test on notched specimens and the upending test on cylindrical specimens did not allow to obtain such a state of stress for which the state-of-stress factor would be negative.

TABLE 1

Values of state of stress factor and effective strain for K18 steel

	Upseting test			Tensile test on $n.s.$		
	Case 2	Case 3	Case 4	Case 1	$d_o/R_o = 0.5$	$d_o/R_o = 4$
k	0.18	0.25	0.30	0.38	1.46	1.88
$\bar{\varepsilon}$	1.67	1.45	1.50	1.19	0.32	0.22

REFERENCES

1. A. A. Bogatow, O. I.Mizirickij, S. W. Smirnow, Riesurs Plasticznosti Mietallow pri Obrabotkie Dawlieniiem, Mietallurgija, Moskwa, 1984.

2. B. de Meester, Y. Tozawa, P.O. Strandell, Annals of the CIRP, 28/2/ /1979/ 577-580.

3. S.A. Smirnow-Alajew, Soprotiwlienije Materiałow Plasticzeskomu Dieformirowanju, Maszgiz, Leningrad, 1961.

4. J. Pośpiech, J. Mech. Work. Techn., 10 /1984/ 325-347.

5. J. Pośpiech, Prace IH, 24/4/ /1972/ 237-245.

6. Z. Affanasowicz, J. Darlewski, Z. Vogel, J. Wójcikowski and C. Tobiasz, Laboratorium Technologii Maszyn, Part 1, Gliwice, Politechnika Śląska, 1967.

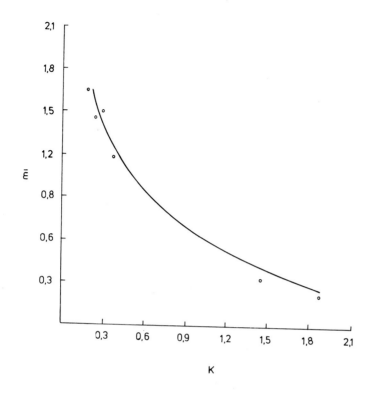

Fig. 1. Curve of deformability of K18 steel

Computational Methods for Predicting Material Processing Defects, edited by M. Predeleanu
Elsevier Science Publishers B.V., Amsterdam, 1987 — Printed in The Netherlands

FINITE STRAIN PLASTICITY ANALYSIS OF DAMAGE EFFECTS IN METAL FORMING PROCESSES

M. PREDELEANU

Laboratoire de Mécanique et Technologie

E.N.S. de Cachan - C.N.R.S. - Université Paris 6 - France.

Summary
    The most important damage effects occuring in forming processes and their modelling within the frame work of finite strain plasticity theory are reviewed.The influence of different material parameters defining the constitutive models on localization phenomena is emphasized. Applications concerning damage evaluation and forming limits are mentioned.

INTRODUCTION

    Among the microstructural rearrangements induced by extensive plastic straining during the forming processes, nucleation and growth of voids and cracks are the most important ones. This phenomenon, called damage (or plastic ductile damage) leads to a progressive deterioration of the material in the sense that it diminishes its resisting capacity to subsequent loading. An advanced evolution of damage favours the apparition of plastic instabilities and ductile fractures, internal or external, by coalescence of the microcavities.

    Practical consequences of these straining effects are unsatisfactory limitations in working operations and/or inacceptable products. Apart from the determination of forming limits, another interest equally important of the damage analysis is the knowledge of the "soundness" of the material undergoing a forming operation. As it was noted, the absence of apparent cracks on the free-stress boundary of a product does not exclude an advanced state of damage within the material itself. So in (ref. 1) is noted that metallographic examinations indicate that a sub-surface void formation occurs prior to the appearance of a surface crack during same bulk forming operations. It was also noticed that, for less ductile materials, fracture occurs through the full section and is not confined to the bulge equator area of cylindrical upset specimen.

    The internal microstructure of the material being strongly affected by damage phenomenon, the mechanical and physical properties are consequently also modified. The global (macroscopical) rheological behavior of the material will be therefore depending both on the "initial" state of damage and also on its evolution during the deformation process. Conversely, every stage of the damage evolution (nucleation, growth, coalescence) is controlled by stress and strain fields at microscopical and macroscopical levels. Therefore, the coupling of damage effects with plastic deformation must be included in constitutive modelling.

    Restricting the present analysis only to rate-independent plasticity theory, the following elements must be considered in their dependence with damage evolution : elastic properties, plastic yielding, strain hardening and strain softening and induced anisotropy. Unfortunately, there are not many experimental data in this field and what exist does not cover all the above features. Instead, theoretical models for containing-void materials provide supplementary conclusions on the behaviour of damaged materials. In the following, we shall give a non-exhaustive review of same results.

Elastic behaviour change

The effects of plastic deformation on elastic properties, although noted in (refs. 2 - 4) have been put forward for different metallic materials by many recent experimental studies (refs. 5 - 8). Most results concern Young's modulus and Poisson's ratio, obtained by loading-unloading uniaxial tests for large range of plastic deformation (usually till rupture of the specimen). These elastic characteristics diminish progressively (most of them quasi-linearly) with plastic deformation.

The degradation of elasticity under complex modes of deformation has been proved by comparing the results with those obtained for a specimen pre-strained by cold extrusion or drawing (refs. 8 - 10). The diminution in stiffness moduli increases as a function of the severity of the forming conditions (reduction ratio and die angle growing). This degradation can be imputed to the growing of damage amount such as determined by metallographic examinations. Moreover in (ref. 11) by a careful experiment in order to eliminate the influence of other effects (strain hardening and texture formation), the decrease of Young's modulus has been correlated with the increase of the porosity.

On the other hand, several studies have been devoted to the determination of the effective elastic moduli of the materials weakened by a field of planar cracks or volumic cavities by using either physical models (ref. 12), or theoretical models (refs. 13 - 15). The influence of the elastic properties of the material matrix, distribution and geometrical characteristics of the cavities on the effective elastic moduli has been quantitatively established.

Plastic yielding . Material softening.

Within the framework of the continuum plasticity theory, two elements of the constitutive model are used at macroscopic level : i) the rational stress-strain relation $\varepsilon_H \to \sigma(\varepsilon_H)$ (or Cauchy true stress σ versus logarithmic strain curve and ii) the yield surface.

(i) The first part of a typical rational stress-strain curve shows that stress increases monotonically with strain up to a maximum value $\sigma_m = \sigma(\varepsilon_H^m)$. This stage of the deformation is dominated by the material strain hardening, generated essentially by piling of dislocations and formation of textures, although an incipient damage is also developed. The last part of the curve describes a possible material strain softening due to an advanced evolution of damage ; the stress decreases with the strain up to the rupture of the volume element, at ε_H^f. It is very difficult to deduce this part of the curve rigo rously from experimental measurements. In this case, the damage effects, eventually including softening, are described by introducing the damage variables into the relation $\varepsilon_H \to \sigma(\varepsilon_H)$ established experimentally for the first stage of the deformation, or, theoretically, affecting the current yield stress for matrix material by a damage reduction factor. Some remarks are to be noted. Obviously, the presence of the softening range on the macroscopical rational stress-strain curve is depending on the choice of the scale used to represent the volume element in continuum mechanics (the one depending on the nature of the material considered) and of averaging procedures. Recent metallographic investigations supply, for some metals, new data concerning the damage pattern leading to rupture (ref. 16) and conditions to passing from microsoftening behaviour to strain-softening behaviour for macroscopic field variables (ref. 17).

Generally, the volume elements exhibiting material softening belong to zones of intenses localization of deformation into shear hands and/or localized neckings. Although for metalic materials the strain range between ε_H^m and ε_H^f is not very large, this instable phase of the deformation process has important consequences on the redistribution of the stresses and strains within the whole structure.

(ii) The investigation of the yield surfaces in the range of large plastic deformations generating damage effects has not received the proper attention, it deserves .The few experimental results, which could be cited in this field, are first those concerning the influence of a pre-straining (usually small and moderate) on the subsequent yielding behaviour and, second, those concerning porous materials (as sintered powder materials). The results from the latter, although obtained for small strain range, emphasize the influence of the amount of porosity on the yield surface (refs.18, 19). Moreover, by theoretical and numerical analysis of the solids weakened by a spherical void network, corresponding yields surfaces have been deduced (refs. 20, 21).

Anyway, the formulation of an appropriate yield surface for damaging materials must take the following into account :

1. the yield surface and its evolution is dependent on the amount of damage ;

2. the development of damage (nucleation and growth of voids) induces irreversible volumic changes. For that, the definition of the yield criterion in stress space by means of the second stress invariant and only that one could be unsatisfactory. The introduction of the first invariant of stress tensor is necessary to described the dilatational deformation and pressure sensitivity (ref. 22).

MODELLING OF DAMAGE EFFECTS

The need to evaluate quantitatively the damage induced by a forming process was pointed out in several papers (refs. 23, 24) two decades ago. In the same period, global ductile fracture criteria have been proposed (refs. 25 - 28), these being used also to appreciate the "state"of the material damage by comparison with the critical value of the criterion at rupture. These criteria do not take into account the evolutive degradating state of the material and do not allow for the prediction of the behaviour under plastic straining.

The inclusion of the degradating microstructure of the material during deformation into a continuum mechanics model needs the introduction of new field variables describing damage state. That idea was originated by Kachanov (ref. 29) for uniaxial creep behaviour and developed by Rabotnov (ref. 30) for three dimensional cases. So, continuum damage mechanics was born, the name of this new branch of continuum mechanics appeared only ten years later (ref. 31).

Essentially, two classes of damage variables can be distinguished :

a) damage variables which take into account the distribution and the morphology of the microvoids and their effects on the kinematic fields (a formidable task !). By using averaging procedures and simplified geometrical models for the void-containing material, such "kinematical" damage variables have been defined (refs. 32, 33). Most of these models are surfacic models, imagined so as to evaluate the resisting net area. These approaches lead to representation of the damage variables by scalars for isotropic damage or by vectors and tensors of different orders (second or fourth order) for anisotropic damage (cf. review paper ref. 34) ;

b) damage variables which describe the changes in mechanical and physical parameters or even can be identified with some of these parameters (elastic moduli or density for instance). These damage variables represent partial "indicators" of the degraded state of the material. This approach will be adopted in this paper (ref. 35). Damage variables, belonging to one or the other class can be considered internal variables in the sense of thermodynamics of irreversible processes (ref. 36).

Remark : Damage variables, interpreting the loss of carrying-load area, have encouraged the used of the concept of effective stress $\tilde{\sigma}$ generalized formally from one-dimensional to three-dimensional cases, by means of an operator $\mathbf{H}(\mathbf{d})$ such that $\tilde{\sigma} = \mathbf{H}(\mathbf{d})\sigma$ where, by $\tilde{\sigma}$ and σ, is noted the Cauchy stress tensor acting acting respectivelly on a cross section in damaged or undamaged material and by $\mathbf{d}$ the damage variables. The concept applied to a special mechanical response (elasticity, creep, fatigue, etc.) has permitted to define the damage variables $\mathbf{d}$ indirectly in terms of behaviour characteristics (ref. 37).

A finite strain elasto-plastic constitutive model can be developed within the framework of irreversible thermodynamics with internal variable in order to include damage effects as :

i) gradual degradation of elastic properties,

ii) strain softening effects on yielding behaviour of material volume element,

iii) plastic dilatancy due to nucleation and growth of microvoids.

The general structure of the stress-strain relations can be based on two thermodynamical potential : a free-energy potential determining the elastic damage response and a dissipation potential to define the plastic flow and damage evolution equations. Restricting the present analysis to isotropic behaviour, specific forms of the two potentials can be considered, to obtain generalization of the classical rate-independant plasticity equations (of Prandtl-Reuss type).

Elastic-damage response

The elastic component of the total deformation can be described conceptually by destressing to zero, locally, every material element. Mathematically, if the multiplicative decomposition of the deformation gradient is chosen, i.-e. : $\mathbf{F} = \mathbf{F}^e \mathbf{F}^p$, $\mathbf{F}^e$ defines the elastic part and $\mathbf{F}^p$ the plastic part of the total transformation $\mathbf{F}$. Using the elastic deformation gradient $\mathbf{F}^e$, let us consider the left Cauchy-Green tensor $\mathbf{B}^e = \mathbf{F}^e \mathbf{F}^{eT}$, defined on the current configuration as an elastic strain measure. The isotropic elastic response can be formulated in terms of the free-energy potential $\Psi(\mathbf{B}^e, \alpha)$ where α denotes a discreet set of internal variables describing the damage and plastic effects. Then, the stress strain relations are given by

$$\tau = 2\rho_0 \frac{\partial \Psi}{\partial \mathbf{B}^e} \mathbf{B}^e \tag{1}$$

where τ is the Kirchhoff stress tensor and ρ_0 the density in initial free-stress configuration.

Using the multiplicative split of the elastic deformation gradient into dilatational and volume preserving parts $\mathbf{F}^e = J^{e1/3} \overline{\mathbf{F}}^e$, $J^e = \det \mathbf{F}^e$ (ref. 38), we particularly postulate the following uncoupled form of the free-energy potential, by assuming that the elastic-damage response is decomposed into volumetric and deviatoric part over any range of deformation (linear and non-linear) :

$$\Psi(\mathbf{B}^e, \alpha) = \Psi_1 (J^e, K) + \Psi_2 (\mathbf{B}^e, G) + \Psi_3 (d_s) + \Psi_4 (p) \tag{2}$$

where $\overline{\mathbf{B}} = \overline{\mathbf{F}}^e \, \overline{\mathbf{F}}^{eT}$, K and G are the current elastic-damaged stiffness moduli characterizing respectively volumetric and deviatoric response, d_s is an internal damage variable describing the softening effects in yielding behaviour (in particular d_s can be identified with void volume fraction or with a reduction factor of the yield stress of the material matrix), p is the internal variable governing the isotropic strain hardening as in classical plasticity theory.

For simplicity, the K and G which play the role of damage variables can be written as :
$K = K_0(1 - d_K)$, $G = G_0(1 - d_G)$, with $d_K, d_G \in [0,1]$, the values $d_K, d_G = 0$ characterizing the undamaged state and $d_K = d_G = 1$ the rupture of the material volume element .

Two examples can be given :

1. The non-linear elastic-damage model of Hadamard type. The stored elastic-damage energy potential is assumed as :

$$\Psi_1 (J^e,K) + \Psi_2 (\overline{\mathbf{B}}^e,G) = K_0(1-d_K) \, U(J^e) + 1/2 \; G_0(1-d_G) \, \text{tr} \, \overline{\mathbf{B}}^e \tag{3}$$

with various choices for the function U as for instance $U(J^e) = 1/2 \, (\log J^e)$ (ref. 39). Then,

$$\tau = K_0 (1-d_K) \, J^e \, \frac{dU(J^e)}{dJ^e} + G_0 (1-d_G) \, \text{dev} \, \overline{\mathbf{B}}^e \tag{4}$$

where dev [.] = (.) - 1/3 tr(.) $\mathbf{1}$ denotes the deviator of the indicated argument in the spatial description.

2. The linear elastic-damage model. Assuming that the elastic-strains are small and linearizing about $\tau = 0$, one obtains the elastic-damage model defined by

$$\Psi_1(J_e,K) + \Psi_2(\overline{\mathbf{B}}^e,G) = 1/2 \, K_0(1-d_K) \, (\text{tr} \, \varepsilon^e)^2 + G_0(1-d_G)\text{tr}(\text{dev} \, \varepsilon^e)^2 \tag{5}$$

and, therefore, following stress-strain relations

$$\tau = K_0(1-d_K) \, \mathbf{1}\text{tr} \, \varepsilon^e + 2 \, G_0(1-d_G) \, \text{dev} \, \varepsilon^e \tag{6}$$

where $\varepsilon^e \cong 1/2 \, (\mathbf{B}^e -1)$ denotes the linearized strain tensor (refs. 8, 40).

Plastic flow and damage evolution.
Let us consider that the plastic flow is controlled by the plastic part $\mathbf{D}^p$ of the additive decomposition of the total strain rate tensor $\mathbf{D}$ defined as symmetric part of the velocity gradient $\mathbf{L} = \dot{\mathbf{F}} \mathbf{F}^{-1}$, where a superposed dot denotes the material time derivative with respect to the present time t. Such decomposition is not unique, depending on the choice of reference configuration and strain measure. Thus, in spatial description, several definitions have been proposed as for example :

i) $\mathbf{D}^p \equiv \overset{\sim}{\mathbf{D}}^p = \text{sym} \, \mathbf{F}^e \dot{\mathbf{F}}^p \, \mathbf{F}^{p-1}\mathbf{F}^{e-1}$, $\mathbf{D}^e = \text{sym} \, \dot{\mathbf{F}}^e \, \mathbf{F}^{e-1}$.

For small elastic strain, $\mathbf{F}^e \simeq 1$, approximative decomposition of total strain rate can be obtained (ref.

41) ;

ii) $\mathbf{D}^p \equiv \hat{\mathbf{D}}^p = \delta^c(\mathbf{e} - \mathbf{e}^e)$, where by δ_c is noted the Lie (convective) derivative (ref. 42). $\mathbf{e}$ is the Euler-Almansi strain tensor and $\mathbf{e}^e$ its elastic part obtained by push-forward operation of the lagrangian strain additive decomposition (ref. 43).

iii) $\mathbf{D}^p \equiv \bar{\mathbf{D}}^p = \mathbf{R}^e(\dot{\mathbf{F}}^p \mathbf{F}^{p-1})\mathbf{R}^{eT}$, obtained by multiplicative decomposition of the deformation gradient and using a fictitious rotated configuration defined by the elastic rotation tensor $\mathbf{R}^e$, arose from the polar decomposition of $\mathbf{F}^e$ (ref. 44). If the elastic strains are small, one obtains $\mathbf{D} \sim \overset{\triangledown}{\mathbf{e}}^e + \bar{\mathbf{D}}^p$, where by "$\triangledown$" is noted the Jaumann derivative with respect to total spin in current configuration.

It is worth noting that, generally, $\mathbf{d}^p \neq \hat{\mathbf{D}}^p \neq \bar{\mathbf{D}}^p$. Other definitions of $\mathbf{D}^p$ can be obtained by using lagrangian descriptions in intermediate configuration (ref. 45).

Defining the thermodynamic variables conjugated to α by $Y = \partial \Psi / \partial \alpha$, let us admit the existence of a dissipation (dual) potential $\varphi(\tau, Y, \alpha)$, scalar valued convex function, non-negative and zero for $\tau = 0$, $Y = 0$ with α as parameters. Then, for generalized standard materials (ref. 46), evolution equations are given by :

$$\mathbf{D}^p = \frac{\partial \varphi}{\partial \tau} \quad , \quad \dot{\alpha} = - \frac{\partial \varphi}{\partial Y} \tag{7}$$

By Y, we noted the discreet set of variables $Y(Y_K, Y_G, Y_s, Y_p)$ associated respectively to scalar internal variables $\alpha(d_K, d_G, d_s, p)$.

For rate independent plastic flow :

$$\mathbf{D}^p = \Lambda \frac{\partial \bar{\varphi}}{\partial \tau} \qquad \dot{\alpha} = - \Lambda \frac{\partial \bar{\varphi}}{\partial Y} \quad , \quad \Lambda \geq 0 \tag{8}$$

The multiplicative factor Λ is determined by consistence condition $\dot{\bar{\varphi}} = 0$. If the potential $\bar{\varphi}$ is not known, but contrary, a yield function $F(\tau, Y, \alpha)$ is defined, the associated plastic law gives :

$$\mathbf{D}^p = \Lambda \frac{\partial F}{\partial \tau} \tag{9}$$

The evolution laws of other internal variables may be postulated in the form
$$\dot{\alpha} = \Lambda \, r(\tau, Y, \alpha) \tag{10}$$
provided that the Clausius-Duhem inequality is verified. Some of the functions r can be defined as in rate-independent plasticity theory by an associated law using a damage convex criterion, postulated in stress or strain space.

To include plastic dilatational behaviour of the material in the model, we shall define the plastic potential F as an isotropic function of τ, in terms of the first and second invariant of τ, $F(J_1, J_2, Y, \alpha)$ where $J_1(\tau) = tr \, \tau$, $J_2(\tau) = 1/2 \, tr \, \tau^2$.

So, the plastic flow will be defined by uncoupled deviatoric/volumetric relations such as :

$$dev \, \mathbf{D}^p = \Lambda \frac{\partial F}{\partial J_2(dev \, \tau)} \frac{\partial J_2(dev \, \tau)}{\partial \tau} \qquad tr \, \mathbf{D}^p = 3 \, \Lambda \frac{\partial F}{\partial J_1(\tau)} \tag{11}$$

Special forms for F have been proposed to account for plastic dilatation and pressure sensitivity by incorporating or not the effects of the material porosity assumed to be constant or variable during deformation. Generally, these yield functions are defined adding to the second invariant $J_2(dev \, \tau)$ a function of the first invariant $J_1(\tau)$, affected by a factor, constant or depending on the porosity (refs. 20, 22, 47 - 53).

Isotropic models using a single scalar damage variable

1. Young's modulus change-based damage model.

This model can be defined by means of the effective stress introduced by Kachanov (ref. 29) as :

$$\tilde{\sigma} = \frac{\sigma}{1-d} \tag{12}$$

where d, a scalar damage variable, $d \in [0,1]$, is interpreted as describing the reduction of resisting net area S due to cavity formation : $\tilde{S} = S_0(1-d)$, independently of the orientation of surface material elements. S_0 is the initial non deformed area.

Adopting the hypothesis of the strain equivalence (ref. 54), the constitutive coupled equations of the elasto-plastic damaged material are obtained by replacing the stress tensor in constitutive relations of undamaged material by the effective stress tensor. Thus, the elastic behaviour is described by (5) - (6) with $d_K = d_G = d$, i.e.

$$\sigma = (1-d) [K_0 \, \mathbf{1} \, tr \, \varepsilon^e + 2 \, G_0 dev \, \varepsilon^e] \tag{13}$$

where the Kirchhoff stress τ has been replaced by Cauchy stress tensor σ.

The yield function is defined in the space of effective stresses i.e. $F(\tilde{\sigma}, d, Y_d, p, Y_p)$, where Y_d and Y_p are the variables associated respectively to d and p, $d_s \equiv Y_s \equiv 0$.

For small elasto-plastic strain range, a yield function F of Von Mises type has been proposed (ref. 37) in the form :

$$F(\sigma, d, Y_p) = \frac{\sigma_{eq}}{1-d} - Y_p - \sigma_Y \tag{14}$$

where $\sigma_{eq} = 3J_2(dev \, \sigma)^{1/2}$ and σ_Y the initial yield stress.

From (13), one obtains for uniaxial stress state the relation :

$$d = 1 - \frac{E}{E_0} \tag{15}$$

which permits to determine d by measurements on Young's modulus changes during plastic straining For a three-dimensional stress state, the evolution of d is defined by assuming particular forms for the dissipation damage potential φ_D as for instance :

$$\varphi_D = \frac{S_0}{s_0+1} \left(-\frac{Y_d}{S_0}\right)^{s_0+1} (1-d)^{-1} \tag{16}$$

where s_0 and d_0 are two material constants (the total dissipation potential is defined by $\bar{\varphi} = F + \varphi_D$). This model has been used for large deformation (ref. 55) by formulating the field variables in a rotated configuration defined by rotation $\mathbf{Q}$, determined by differential equation $\mathbf{Q}^T\mathbf{Q} = \mathbf{W}$ where $\mathbf{W}$ is the total spin (ref. 56).

2. Void volume fraction change-based damage model

Originally proposed by Gurson (ref. 20), this model was developped and applied for ductile rupture prediction (refs. 51-53) and plastic flow calculations for forming processes. The basic assumptions of the model are the following ones. The elasticities remain constant during plastic straining ($d_K = d_G \equiv 0$), the damage being identified with void volume fraction $d_s \equiv f$ and included only in yield criterion. The elastic behaviour is defined by a hypoelastic law as :

$$\mathbf{D}^e = \frac{1}{2\,G_0}\,[\,\overset{\triangledown}{\sigma} - \frac{v_0}{1-v}\,\mathbf{1}\ \mathrm{tr}\ \overset{\triangledown}{\sigma}\,] \tag{17}$$

where $\overset{\triangledown}{\sigma}$ is the Jaumann derivative of macroscopic Cauchy stress tensor σ. $\mathbf{D}^e$ the elastic component of the total strain rate $\mathbf{D}$, such that $\mathbf{D} = \mathbf{D}^e + \mathbf{D}^p$ and v_0 Poisson's ratio. The plastic flow is given by an associative rule $\mathbf{D}^p = \Lambda \partial F / \partial \sigma$ where F is the yield function of Gurson' type, generalized as (ref. 51).

$$F\,(\sigma, \sigma_M, f) = \frac{3\,J_2\,(\mathrm{dev}\ \sigma)}{\sigma_M^2} + 2\,fq_1\ \cosh\frac{J_1\,(\sigma)}{2\sigma_M} - [\,1 + q_1^2\ f^{*2}\,] \tag{18}$$

where σ_M is the matrix flow stress, $f^* = f$, for $f \le f_{cr}$, $f^* = f_{cr} + k\,(f - f_{cr})$, for $f > f_{cr}$, f the void volume fraction defined by the following evolution equation

$$\dot{f} = (\dot{f})_{nucleation} + (\dot{f})_{growth} = A\,\dot{\sigma}_M + \frac{B}{3}\,\mathrm{tr}\ \dot{\sigma} + (1-f)\,\mathrm{tr}\ \mathbf{D}^p \tag{19}$$

q_1, k, A, B and f_{cr} are material constants. The generalization of this model including the coalescence stage of damage process, was given in (ref. 57).

CONSTITUTIVE BRANCHING

The limitations of the forming operations are due mostly to the plastic instability phenomena which lead to localizations of the deformations in shear bands or strictions inside or at the surface of the product. The predictions of these are important to avoid not only the apparent defects but, inside a localized deformation zone, the weakening of the material by induced stress concentration favouring the growth of the damage and the ductile rupture. The onset and the development of the localized deformation zones condition the continuation of the deformation process of the solide (structure) from an equilibrium configuration in a stable or instable way. The sufficient criteria for stability have been proposed and they are based essentially on the evolution of the internal and/or external incremental work in the neighbourhood of an equilibrium configuration (refs. 58, 59, 55). These global criteria depend on the geometry of the body, the loading and the constitutive law. It is proved experimentally and also theoretically that instability phenomena are very sensitive to different aspects of the material behaviour. Thus, the faculty of a material to localize the deformation into shear bands can be analized as a bifurcation problem (constitutive branching) by using directly its constitutive equations. This analysis can be performed using the theoretical results concerning acceleration waves applied to the degenerate case of "stationary" discontinuity surfaces (considered a planar band of localization) (refs. 60 - 63). The necessary conditions for the existence of such zones across which the velocity gradient field can be discontinuous are obtained by assuming that the governing equations of equilibrium lose ellipticity (or in an equivalent way, admit real characteristics). Alternatively, equivalent conditions can be deduced assuming that during a homogeneous finite deformation field a bifurcating field can occur localizing deformation within a thin planar band of orientation $\mathbf{n}$, homogeneously deformed (refs. 64, 65). The constitutive branching conditions are given in the form :

$$\det\{\mathbf{n}.\mathbf{M}.\mathbf{n} + \mathbf{A}\} = 0 \tag{20}$$

where $\mathbf{M}$ is the spatial incremental elasto-plastic modulus tensor defined by $\sigma = \mathbf{M} : \mathbf{D}$ and
$\mathbf{A} = 1/2 \, [-\mathbf{n} \, (\mathbf{n}.\sigma) + (\mathbf{n}.\sigma.\mathbf{n}) \, \mathbf{1} + (\mathbf{n}.\sigma)\mathbf{n} - \sigma \,]$ or equivalent, as

$$\det \, \{\mathbf{n}_0 \mathbf{M}_0.\mathbf{n}_0\} = 0 \tag{21}$$

where $\mathbf{M}_0$ is the lagrangian incremental elasto-plastic modulus tensor defined by $\mathbf{s} = \mathbf{M}_0 : \mathbf{F}$. $\mathbf{s}$ is the first Piola-Kirchhoff stress tensor, $\mathbf{n}$ and $\mathbf{n}_0$ the unit normal defining the orientation of the shear band respectively in current or reference configuration. It should be noted that not all localization phenomena can be predicted by using this approach, which, on the contrary, has the advantage of appreciating easily the stabilizing or distabilizing parameters defining the constitutive model.

Same of the most important results of such analysis based on isotropic elasto-plastic constitutive relations of Prandtl-Reuss's type are the following :

1. The influence of the stress-strain curve on the onset of localization is analysed by defining from (20) the critical value of tangent hardening modulus (h_{cr}) in terms of other material parameters. For classical Prandtl-Reuss model (without damage effects and using smooth yield surface and normality rule) the localization is captured only for negative values of h_{cr} (strain softening) for axisymetric extention and compression except plain strain for which $h_{cr} \simeq 0$ (ref. 65).That is in contradiction with the experimental results (ref. 66), which show that the shear band localization occurs when the tangent hardening modulus is positive, but small.

2. The presence of the vertex points on the yield surface introduces an important destabilizing effect (ref. 65). The same conlusion appears by using other constitutive models (refs. 67, 68). In addition, the models involving smooth yield surface but presenting a high curvature (associated with kinematic hardening description) predict shear and necking type localizations more in accordance with experimental results (refs. 69, 70).

3. The introduction of damage effects in the constitutive model strongly modifies the localization predictions. Most results are based on Gurson type model. The role of an initial inhomogeneity of material properties is studied, assuming that the localized zone has material characteristics slightly larger than the outside imperfect zone (refs. 65, 71). The localization of deformation is drastically sensitive to such initial inhomogeneities. The influence of the damage on limit strains can be considered also via elasticity moduli which diminish progressively with plastic deformation as described by d_K and d_G evolutions.

4. The pressure sensitive yield criteria (dilatant materials) introduce destabilizing effects, determining more realistic critical strains at bifurcation. Also in (ref. 65) it was shown that non-normality effects arising from pressure sensitivity of yield criteria are destabilizing. Obviously, many studies, theoretical and experimental must be made and other feature of the material behaviour must be considered such as anisotropy, for instance, in order to predict the onset of the plastic instabilities, responsible for limiting ductility and ductile rupture.

APPLICATIONS

It was mentioned above, the inclusion of a damage model into description of a plastic deformation process is necessary for a new approach of formability limits and also to characterize the forming-induced effects in particular the deterioration of the material, i.e. damage state. The first studies have appeared recently once the modelling of damage, even under simplified form for isotropic behaviour, has been developed. Two kinds of applications will be mentioned : the first one concerning the iso-damage charts for forged products and the second one, the sheet forming limit curves.

Iso-damage charts. In (refs. 72, 73), a damage evaluation for extrusion and upsetting operations was performed, based on the one-scalar variable model defined by relations (13) - (15) and governed by evolution equation :

$$\dot{d} = [\frac{< \sigma^* - \sigma_d >}{S(1 - d)}]^s \frac{\sigma^*}{S}$$

where σ^* is the damage equivalent stress defined as $\sigma^* = 2E_0(1-d)^2 Y_d$ and σ_d is the damage threshold (ref. 6). S and s are damage material constants determined experimentally via Young's modulus change measurements during plastic flow ($<a> = a$ if $a > 0$, $<a> = 0$ if $a \le 0$). The numerical calculations were simplified by a sequential uncoupled numerical treatment of the field equations. The same model has been used to analyse the torsion-traction formability test (ref. 74). The metal extrusion using a coupled strain-damage model based on void volum fraction change (Gurson type model) defined by relations (17) - (19) was analysed in (refs. 75, 76). The results obtained, in agreement with experimental observations are very useful in predicting defects, such as central burst formations.

Corroborating the damage evolution with plastic instability analysis of cold upsetting in (ref. 77), important deterioration effects have been pointed out inside the product and on its free-stress boundary. It was shown that the formability analysis by means only of the "apparent" cracking observations is not always satisfactory.

Sheet forming limit curves. From a practical point of view, two types of forming limit curves (f.l.c) are equally interesting : fracture f.l.c. and necking f.l.c. To be determined, appropriate ductile fracture criteria must be added to governing equations. Many such criteria have been proposed. We can say that the criteria taking into account the damaged state of the material can lead to reliable predictions of the occurance of the defects. In (refs. 55, 78, 79), the fracture f.l.c. are determined by using the damage model defined by the relations (13) - (16) and the local fracture condition defined by a critical value of damage variable d.

The determination of the striction f.l.c. including damage state of the material has been considered in several studies, but including only an initial state of damage, non-evolutive, either as geometrical imperfection or structure imperfections (ref. 80).

The more suitable predictions of the striction f.l.c. with respect to experimental data have been obtained by using a coupled strain-damage model defined by (17) - (19) and localized striction criterion (refs. 81, 82).

REFERENCES

1. P.W. Lee and H.A. Kuhn, Fracture in cold upset forging. A criterion and model - Metallurgical Transactions, 4, April 1973, 969-974.
2. M. Feigen, Inelastic behaviour under combined torsion, in Proc. 2nd U.S. National Congress of Applied Mechanics, 1954.
3. A. Jukov, Elastic properties of plastically deformed metals and combined loading, Inj. Sbornik 30, (1960), 3-16.

4. J. Bell, The physics of large deformation of crystalline solids, in Springer Tracts in Natural Phylosophy, Vol. 14, 1968, Springer-Verlag, Berlin.
5. J. Lemaitre et J. Dufailly, Modélisation et identification de l'endommagement plastique des métaux, in 3ème Congrès Français de Mécanique, Grenoble, France, 1977.
6. J. Dufailly, Modélisation mécanique et identification de l'endommagement plastique des matériaux, Thèse de Docteur de 3ème Cycle, Université Paris 6, 1980.
7. D. Nouailhas, Etude expérimentale de l'endommagement plastique ductile anisotrope, Thèse de Docteur de 3ème Cycle, Université Paris 6, 1980.
8. B. Gattoufi, Effets de la prédéformation due au filage sur le comportememt des métaux, Thèse de Docteur de 3ème Cycle, Université Paris 6, 1984.
9. B. Gattoufi, P. Le Nevez and M. Predeleanu, Effects of the cold forming on subsequent elasto-plastic behaviour of metals, International Symposium on current trends and results in plasticity, Udine, Italie, June 1983.
10. L. Chevalier, Simulation de l'opération de tréfilage. Répartition de l'endommagement, Rapport GRECO GDE - G.I.S. Mise en Forme, 6ème année, 1986.
11. B. Gattoufi et F. Moussy, Influence propre à la porosité sur la valeur du module d'Young, in Compte rendu du Groupe de Réflexion "Endommagement", RCP-CNRS, I.R.S.I.D., septembre 1983.
12. J.P. Cordebois, Comportement et résistance des milieux métalliques multi-perforés, J.M.A., Vol. 3., (1979), pp. 119-142.
13. Z. Hashin, Mechanics of composite materials, in Proc. 5th Symp. Naval Struct. Mech., Pergamon Press, New York, 1967, 201-242.
14. M. Hlavacek, Effective elastic properties of materials with high concentration of aligned spheroidal pores, Int. J. Solids Structures, Vol. 22, (1986), n°3, 315-332.
15. B. Budiansky and R.J. O'Connell, Elastic moduli of a cracked solid, Int. J. Solids Structures, 12 (1976), 81-97.
16. A. Pineau, Review of fracture micromechanisms and a local approach to predicting crack resistance in lowstrengh steels, in D. François (Ed.), Advances in Fracture Research, Vol. 2, Pergamon Press, 1981, pp. 553-577.
17. F. Moussy, Les différentes échelles du développement de l'endommagement dans les aciers, Influence sur la localisation de la déformation à l'échelle microscopique, in Proc. Int. Symposium on Plastic Instability, Paris, France, septembre 9-13 1985, pp. 263-272.
18. H.A. Kuhn and C.L. Downey, Deformation characteristics and plasticity theory of sintered powder materials, Int. J. Powder Met., 7, 1971, 15.
19. J. Mielniczuk, Plasticity of porous metals, in Proc. Colloque Intern., CNRS, Villars de Lans, Juin 1983.
20. A.L. Gurson, Continuum theory of ductile rupture by void-nucleation and growth : Part I : Yield criteria and flow rules for porous ductile media, Journal of Engineering Materials and Technology, 2-15, January, (1977), pp. 2-15.
21. C. Richard, Comportement macroscopique d'un matériau poreux en écoulement plastique, Thèse de Docteur de 3ème Cycle, Université Paris 6, 1986.
22. L.A. Berg, Plastic dilatation and void interaction, in Proceedings of the Batelle Memorial Institute Symposium on Inelastic Processes in Solids, 1969, pp. 171-209.
23. H.C. Rogers and J.L.F. Coffin, Structural damage in metal working, in CIRP Int. Conference on manufacturing technology, Ann. Arbor, Michigan, U.S.A., 1967, pp. 1137-1145.
24. H.C. Rogers, Prediction and effects of materials damage during deformation processing, in ed. A.L. Hoffmaner (ed.),"Metal Forming. Interaction between theory and practice" Plenum Press, New York, London, 1971, pp. 453-474.
25. M.G. Cockcroft and D.J. Latham, A simple criterion of fracture for ductile metals, J. Inst. metals 96, (1968), pp. 33-39,.
26. M.G. Cockcroft, Ductile Fracture in cold working operations in "Ductility", A.S.M. Metals Park, Ohio, 1968, pp. 199-225.
27. M. Oyane, Criteria of ductile fracture strain, Bull of the J. S. M. E. 15, (1972), pp. 1507-1513.
28. D.M. Norris Jr. and al, A Plastic-Strain, Mean-Stress Criterion for Ductile Fracture, J. Eng. Mat. Tech., 100, (1978), pp. 279-286.
29. L.M.Kachanov, Time of the rupture process under creep conditions, Izv. Akad Nauk.SSR, Otd. Tekh. Nauk, (1958), n° 8, pp. 26-31 (in russian).
30. Y.N. Rabotnov, Creep rupture, in Proceedings of the 12th Int. Crongress Appl. Mech., Stanford, 1968, Springer-Verlag, 1969, pp. 342-369.
31. J. Janson and J. Hult, Fracture mechanics and damage mechanics a combined approach, J.

306

Méc. Appl., 1, (1977), pp. 69-84.

32. A.A. Vakulenko and M.L. Kachanov Jr., Continual theory of a medium with cracks, Mechanics of Solids, 6, (1971), p. 145.

33. A. Dragon and A. Chihab, On finite damage : ductile fracture-damage evolution, Mechanics of materials, 4, (1985), pp. 95-106.

34. D. Krajcinovic, Continuum damage mechanics, Applied Mechanics Reviews, 37, n° 1, (1984), pp. 1-6.

35. M. Predeleanu, Modèles d'endommagment plastique ductile appliqués à la mise en forme, Séminaire franco-canadien sur les nouveaux matériaux métalliques et nouveaux procédés de fabrication, Mc Gill University, 7-10 juillet 1986, Montréal, Canada.

36. P.Germain et al, Continuum thermodynamics, J. Applied Mechanics, 50, (1983), pp. 1010-1020.

37. J. Lemaitre and J.L. Chaboche, Mécanique des matériaux solides, Dunod, Paris, 1985.

38. R.J. Flory, Thermodynamic relations for high elastic material, Trans, Faraday Soc. 57, (1961), pp. 829-838.

39. J.C. Simo et al., Variational and projection methods for the volume constraint in finite deformation elasto-plasticity, Comp. Meth. Appl. Mech. Engngr, 51, (1958), pp. 177-208.

40. P. Ladevèze, Sur une théorie de l'endommagement anisotrope, Rapport interne, mars 1984, n° 34, L.M.T. Cachan.

41. M. Kleiber, J.A. König and A. Sawczuk, Studies on plastic structures : stability, anisotropic hardening, cyclic loads, Comput. Meth. Appl. Mech. Engng., 33, (1982), pp. 487-556.

42. J.C. Simo and M. Ortiz, A unified approach to finite deformation elastoplastic analysis based on the use of hyperelastic constitutive equations, Comput. Meth. Appl. Mech. Engng. 49, (1985), pp. 221-245.

43. A.E. Green and P.M. Naghdi, A general theory of an elastic plastic continuum, Arch. Rat. Mech. Anal. 18, (1965), p. 251.

44. F. Sidoroff, Incremental constitutive equation for large strain elastoplasticity, Int. J. Engng. Sci., Vol. 20, (1982), n° 1, pp. 19-26.

45. J. Mandel, Thermodynamics and plasticity, In. J.J. Delgado et al (Eds), Macmillan, London, (1974), pp. 283-304.

46. B. Halphen et Q.S. Nguyen, Sur les matériaux standards généralisés, J. de Mécanique, 14, (1975), pp. 39-63.

47. M. Oyane, S. Shima and Y. Kono, Theory of plasticity for porous metals, Bull. J. S.M. E. 16 (1973), pp. 1254-1262.

48. S. Shima, K. Mori and K. Osakada, Analysis of metal forming by the rigid plastic finite element method based on plasticity theory for porous metal, , in H. Lippman (ed.) in Proc. of the I.U.T.A.M. Symposium on Metal Forming Plasticity, Tutzing, pp. 306-317, 1978, Germany.

49. S.M. Doraivelu et al., A new yield function for compressible P/M materials, Int. J. Mech. Sci., Vol. 26, (1984), pp. 527-535.

50. P. Perzyna, Constitutive modeling of dissipative solids for postcritical behavior and fracture, Journal of Engineering Materials and Technology, (1984), pp. 1-10.

51. V. Tvergaard and A. Needleman, Analysis of the cup-cone fracture in a round tensile bar, Acta Metallurgica 32, (1984), pp. 157-169.

52. V. Tvergaard, Influence of voids on shear band instabilities under plane strain conditions, International Journal of Fracture, 17, 1981, pp. 389-407.

53. G. Rousselier, Three-dimensional constitutive relations and ductile fracture, in S. Nemat-Nasser (Ed.), Proceedings of the I.U.T.A.M. Symposium on three-dimensional constitutive relations and ductile fraction, Dourdan, France, 2-5 June 1980, pp. 331-355.

54. J. Lemaitre, Evolution of dissipation and damage in metals submitted to dynamic loading, in Proc. I.C.M. 1, Kyoto Japan, 1971.

55. J.P. Cordebois, Critères d'instabilité plastique et endommagement ductile en grandes déformations. Application à l'emboutissage, Thèse de Docteur ès Sciences Physiques, Université Paris 6, 1983.

56. P. Ladevèze, Sur la théorie de la plasticité en grandes déformations, Rapport interne 1981, n° 9, L.M.T. Cachan.

57. J.C. Gélin, Modèle numériques et expérimentaux en grande déformations plastiques et endommagement de rupture ductile, Thèse de Docteur ès Sciences Physiques, Univrsité PARIS 6, 1985.

58. R. Hill, A general theory of uniqueness and stability inelastic plastic solids, J. Mech. Phys. Solids 6, (1958), pp. 236-249.

59. M.E. Gurtin and S.J. Spector, On stability and uniqueness infinite elasticity, Arch. Rat. Mech. Anal. 70, (1979), pp. 153-165.

60. J. Hadamard, Leçons sur la propagation des ondes et les equations de l'hydrodynamique, Paris, 1903.

61. R. Hill, Acceleration waves in solids, J. Mech. Phys. Solids, 10, (1962), p. 1.

62. T.Y. Thomas, Plastic flow and fracture in solids, Academic Press, 1961.

63. J. Mandel, Conditions de stabilité et postulat de Drucker, in J. Kravtchenko and P. Sineys (eds), in proc. I.U.T.A.M. Symposium Rheology and Soil mechanics, Grenoble, France, Springer Verlag, 1966, 58-68.

64. J.R. Rice, The localization of plastic deformation in W.T. Koiter (Ed.), Proc. 14th, I.U.T.A.M. Congress north-Holland, (1976), pp. 207-220.

65. A. Needelman and J. R. Rice, Limits to ductility set by plastic flow localization, in D.P. Koistinen and N.M. Wang (eds), "Mechanics of sheet metal forming",, Plenum Press, New-York, London, (1978), pp. 237-267.

66. L. Anand and W.A. Spitzig, Initiation of localized shear bands in plane strain, J. Mech. Phys. Solids, Vol. 28, (1980), pp. 113-1128.

67. S. Stören and J.R. Rice, Localized necking in thin sheets, J. Mech. Phys. Solids, Vol. 23, (1975), pp. 421-441.

68. J.W. Rudnicki and J.R. Rice, Conditions for the localization of deformation in pressure-sensitive dilatant materials, J. Mech. Phys. Solids, Vol. 23, 1975, pp. 371-394.

69. T. Tvergaard, Effects of kinematic hardening on localized necking in biaxially stretched sheets, Int. J. Mech. Sci. 20, 1978, pp. 651-658.

70. J.W. Hutchinson and V. Tvergaard, Shear band formation in plane strain, Int. J. Solids Struct. Vol. 17, 1981), pp. 451-470.

71. H. Yamamoto, Conditions for shear localization in the ductile fracture of void-containing materials, Int. J. of Fracture 14, (1978), pp. 347-364.

72. N. Dahan et P. Le Nevez, Méthode de calcul des grandes déformations plastiques et endommagement dans les pièces extrudées, Revue de Métallurgie, (1983), pp. 557-566.

73. L. Belkhiri, Etude de la déformabilité des métaux lors de l'écrasement à froid, Thèse de Docteur de 3ème cycle, Université Paris 6, 1985.

74. J.F. Fontaine, Contribution à l'étude de l'aptitude au formage à froid des métaux, Thèse de Docteur de 3ème Cycle, Université Paris 6, 1985.

75. J.C. Gélin, J. Oudin, and Y. Ravalard, An improved finite element method for the analysis of damage and ductile fracture in cold forming processes, Annals of the CIRP, Vol. 34/1, (1985), pp. 209-213..

76. N. Aravas, The analysis of void growth that leads to central bursts during extrusion, J. Mech. Phys. Solids, Vol. 34, n° 1, (1986), pp. 55-79.

77. M. Predeleanu, J.P. Cordebois, and L. Belkhiri, Failure analysis of cold upsetting by computer and experimental simulation, in K. Mattiasson et al. (eds.) Proceedings of the Numiform'86 Conference, Gothenburg, 25-29 August 1986, pp. 277-282.

78. J. Lemaitre, How to use damage mechanics, Nuclear Engineering and Design, 80, (1984), pp. 233-245.

79. J. Lemaitre, A three-dimensional ductile damage model applied to deep-drawing forming limits, in J. Carlsson and N.G. Ohlson (eds.), "Mechanical behavior of materials", Vol. 2, Pergamon Press, 1984, pp. 1047-1053.

80. J.M. Jalinier, Mise en forme et endommagement, Thèse de Docteur ès Sciences Physiques, Université de Metz, 1981.

81. A. Needleman and N. Trandafyllidis, Void growth and local necking in biaxially stretched sheets, J. Eng. Mat. Tech., 100, (1978), pp. 164-169.

82. A. Needleman and V. Tvergaard, Limits to formability in rate-sensitive metal sheets, in J. Carlsson and N.G. Ohlson (eds.), "Mechanical behaviour of materials, Vol. 1, (1982), Pergamon Press, pp. 51-65.

Computational Methods for Predicting Material Processing Defects, edited by M. Predeleanu
Elsevier Science Publishers B.V., Amsterdam, 1987 — Printed in The Netherlands

PLASTIC INSTABILITY OF PRESTRAINED MATERIALS

J.H. SCHMITT, J. RAPHANEL, E. RAUCH and P. MARTIN

Institut National Polytechnique de Grenoble, Ecole Nationale Supérieure de Physique, Laboratoire Génie Physique et Mécanique des Matériaux, Unité Associée au CNRS n° 793, B.P. 46, 38402 Saint Martin d'Hères Cedex (France)

SUMMARY

The prediction of plastic instability of prestrained materials has often been based on damage and its evolution with strain history. It has been shown that the results, such as they appear on forming limit diagrams for instance, are very sensitive to the models of plastic behavior that are considered. A knowledge of microstructural evolutions at the intragranular level is necessary since these evolutions may be very drastic in the case of a change of loading path and thus influence greatly the macroscopic response, giving rise to transient stages with decreased strain rates or to macroscopical heterogeneities of deformation.

INTRODUCTION

The accurate prediction of plastic behavior of metals requires the knowledge of its thermomechanical history. The analysis of subsequent deformation of a prestrained material is then a step in that way. In the case of sheet metals, various studies have already been performed (see for example Refs. 1-7) which show for low carbon steel an increase of the reloading yield point (latent hardening effect) and a decrease of strain hardening in most cases of sequential strain paths with the exception of Bauschinger tests.

From a practical point of view, such studies are of importance as, for instance, each part of metal is deformed in a complex way during deep-drawing. It is then necessary to estimate the forming limit strain along complex strain paths. Previous studies have shown that the shape and the level of the forming limit diagrams (FLD) strongly depend on the strain history. For example, in the case of low carbon steel, any FLD lies between two extreme curves (Ref. 8) : the higher is obtained along sequences of uniaxial tension followed by equibiaxial stretching while the lower is for the reverse sequence.

A former explanation of this effect has been proposed (Ref. 9) based on the concept of equivalent strain. However, this does not account for the whole influence of prestrain. Beyond the purely mechanical analysis, microstructural evolutions during prestrain need to be taken into account. For a single-phase polycrystalline metal, the main structural parameters are : crystallographic textures, morphological textures, dislocation substructures and volume damage.

They play different roles in instability calculations. In a first approximation, one may consider that the first three parameters are influent on the plastic behavior and the fourth represents an initial and evolutive defect.

This paper aims at synthesizing previous works which have analyzed the influence of constitutive law and geometrical defect (through damage concept) on the FLD prediction. Moreover, one shows the effect on the FLD of particular macroscopical behavior of prestrained materials during reloading. A careful analysis of microscopical behavior of prestrained low carbon steel sheets reveals the frequent development of strain heterogeneities. The phenomenon is described and some ways for introducing it in instability calculations are proposed.

FORMING LIMIT DIAGRAMS

Damage effect on plastic instability

For a long time, the effects of particles and hard second phases on the ductile fracture of metallic alloys have been established. Cavities form mainly at the particle-matrix interface, grow and then coalesce leading to rupture. More recently, the concept of evolutive damage (Ref. 10) has been introduced for accounting of cavities which are generated during plastic deformation. The influence of damage may then be evaluated on the macroscopical plastic instability. It is characterized through the calculation of FLD in the case of thin sheet metals.

The instability calculation may be performed with an hypothesis of two-zone material in a manner similar to the one proposed by Marciniak and Kuczynski (Ref. 11). The geometrical defect represents the diminution of sheet thickness due to the presence of voids in the sample section. Through the proposed modelling, the growth of the geometrical defect arises from the two following causes : (i) an increase of the actual geometrical defect owing to the strain localization as in the classical approach by Marciniak and Kuczynski ; (ii) an evolution of damage — void growth and initiation of new cavities during deformation — which may be determined by direct physical measurements or estimated through void growth models.

This approach has led to FLD calculations for various materials (Ref. 12). The calculated curves are in better agreement with experimental results than are those which are obtained with the classical assumption of initial defect. The effect of the initial thickness of the sheet on the level of the FLD is also well predicted and the influence of damage is established in this case (Refs. 12,13). As usually observed, the thinner sheets have less formability than the thicker ones. This may be related to a dependence of the damage on the initial sheet thickness. In fact, the equivalent geometrical defect is more important for the thinnest sheets for a given void size and void distribution.

Moreover, one has proved that the effect of initial thickness with respect to the strain path depends on the mechanism of damage. In the case of particle failure, this effect is almost independent of loading conditions, while, for decohesion process, an important effect is noticed only for biaxial stretching. This analysis has been extended in the case of sequential loadings such as uniaxial tension followed by equibiaxial stretching or the reverse sequence (Ref. 14). For each sequence the specific damage evolution gives rise to different FLD levels, in agreement with experimental tendencies.

Influence of constitutive law

A careful analysis of the two-zone method points out the importance of the constitutive law on the plastic instability calculations. The influence of the shape of the yield surface and of its evolutions along the deformation is specially noticed (Ref. 15). When one assumes the existence of a geometrical defect, two different points are imposed on the yield surface ; each one describes the stress/strain state of a zone. The major principal stress is greater in the defect area than in the bulk from the beginning. Owing to the deformation and to the damage evolution, the stress state thus evolves to accommodate plane strain in the neck. As a consequence, the shape of the yield surface, more specifically the difference in major principal stresses between the imposed stress state in the bulk and plane strain conditions, is one of the main parameters with the strain hardening effect which governs the plastic instability development (Ref. 16).

An account of different anisotropic conditions through the classical formulation by Hill has been made for the calculation of FLD along single and sequential strain paths (Ref. 17). These models do not however describe accurately the physical behavior of materials. This is mainly due to the fact that rheological parameters are evaluated using tension tests which are inadequate to describe multiaxial experiments. For instance, crystallographic texture which bears a large part of plastic anisotropy, evolves differently according to the imposed strain path.

Different authors (Refs. 16,18-21) attempt to account for the crystallographic nature of intragranular plastic behavior. The yield surface is calculated by the Taylor or Bishop-Hill model. The differences between such yield surfaces and those classically obtained by continuum mechanics are small, but some variations of curvature appear and lead to large deviations for determining the strain increment orientations. The use of such yield surfaces in the Marciniak calculations induces large modifications of the shape and position of the FLD in the biaxial region (Ref. 18). The results are then in very good agreement with experimental data for realistic values of initial defect. Moreover, this approach allows the estimation of predeformation effects on FLD

312

level. Mainly in the case of large prestrain (equivalent strain larger than 1.), the evolution of crystallographic texture induces some rounded vertices on the yield surface which may have a drastic effect on the stability calculations (Ref. 20). It is noteworthy that such models lead to an effect on the FLD which is comparable with the damage influence in the equibiaxial region.

Transient work hardening effects

Effects of work hardening are well known on sheet formability. An important hardening favors a large stretchability. The Marciniak approach allows such influence to be measured. First calculations have then been performed during sequential loadings with a work hardening which does not depend on strain path (Ref. 22). Nevertheless, in most cases, stress-strain curves of prestrained metals exhibit a transient part after the macroscopical reloading yield stress. This part characterize a very low strain hardening region. The amplitude of this effect, i.e. the value of strain hardening and the extent of the transient region, depends of the prestrain amount and on the difference between prestrain and reloading conditions. As an example, one presents in Fig. 1 the schematic

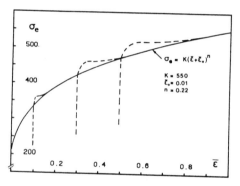

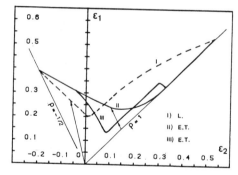

Fig. 1. Schematic behavior during tension test of low carbon steel sheets prestrained at different strain amount along equibiaxial stretching.

Fig. 2. Calculated FLD through a two zone model : I-linear strain path, II-along a sequence equibiaxial stretching-tension without transient, III-the same as II., but with a strain hardening transient.

behavior of steel sheets during uniaxial tension tests after different amount of equibiaxial prestretching. Such a mean law has been introduced in a two-zone model calculation (Refs. 23,24) in order to measure the influence of the transient on the FLD. A large effect is predicted for a prestrain value larger than 0.1 along equibiaxial stretching (Fig. 2). An account of a transient behavior with low work hardening tends to decrease the limit strain. The theoretical FLD is in better agreement with experimental results.

A complete calculation has moreover to take into account the damage. For equibiaxial stretching followed by uniaxial tension, damage and hardening transient effects play in the same direction ; it is then difficult to weight their respective influence. More generally, instability calculations are so sensitive that they cannot be used to justify a model of plastic behavior. This sensitivity implies that FLD predictions require a precise knowledge of the actual behavior of the material, accounting for its history and based on physical reality.

MICROSTRUCTURAL APPROACH OF PLASTIC BEHAVIOR OF PRESTRAINED METALS

Intragranular substructural evolutions

In the following, more special attention will be paid to dislocation substructures which develop in low carbon steel sheets during simple and sequential strain paths.

After about 5% of plastic deformation along monotonic loading paths, dislocation tangles appear in grains and divide them in cells. Previous studies (see for example Ref. 25) have shown that dislocation walls, i.e. areas of large dislocation density, lie on the slip planes which are the most active in a given grain. Owing to the number of active slip systems in a grain, dislocation substructures present two main aspects : long straight parallel walls when one slip plane is mainly active ; equiaxial cell structure when multislip occurs. So, dislocation substructures are closely linked to the actual deformation in grain and influence the anisotropy and the work hardening of the intragranular response.

During sequential strain paths, after a small amount of deformation under the new loading conditions, the substructure evolves into one which is more typical of this new strain mode. This evolution occurs gradually and starts at different stages depending on the grain. This tendency exists whatever the sequence ; nevertheless, the more important the prestrain, the more delayed the "new" substructure. After 10% deformation along the second part of the strain path, the memory of the prestrain seems to have vanished inside most grains. At this point, a careful analysis shows that dislocation cells are only correlated to the subsequent loading mode for a given grain orientation (Ref. 25). These results point out the existence of a transient period of substructural evolution after a change of loading conditions.

During prestraining, dislocation walls develop in close connection with the slip systems which are active in grains. Since other slip systems are activated during reloading, the substructures may then be considered as "alien dislocation substructures" with respect to the new loading mode in the same way as the description of latent hardening tests of single crystals. As a consequence, the relationships which exist between active slip systems and dislocation walls induce different behaviors according to whether the substructures developed during prestrain are reinforced during subsequent loading or are "swept" leading to a new structure. After a reversal of loading conditions which represents an extremal case, the microscopical yield stress is very low owing to the internal stresses while work hardening is higher than along monotonic loading.

Our observations on low carbon steel (Ref. 25) also reveal the influence of cell shapes as they are created during prestrain. The transition from one family of parallel dislocation walls to equiaxial cells is rather continuous through a progressive "dissolution" of the previous structure while the reverse sequence is quite unstable. In the latter case, microbands appear inside grains (Fig. 3) which manifest a quite heterogeneous intragranular deformation. This accounts for a local rearrangement of the dislocation substructure which induces a decrease of the obstacle strength on the active slip plane. Slip line observations confirm the localized deformation by their coarse aspect.

Fig. 3. TEM observation of microbands (noted by arrows) developed in a grain of low carbon steel during a tension test following a rolling prestrain (equivalent strain of 0.1 by rolling, then of 0.01 in tension along the transverse direction).

Two causes may then induce an abrupt change of substructure with the appearence of microbands : i) a large interaction that exists between walls and mobile dislocations, mainly in the case of a dramatic change of strain mode ; ii) second loading conditions which produce preferentially single slip in grains where closed cells have been generated by prestrain. This event is essentially connected to anisotropic hardening of slip systems.

In other cases, substructural transitions are more progressive and lead to a more homogeneous deformation inside grains. A large number of dislocations which are present in walls are annihilated by dynamical recovery which is then important during the first stage of reloading. This phenomenon may increase the stability in some circumstances.

Micro-heterogeneities and plastic instabilities

The intragranular microbands would then be characteristic of unstable intragranular deformation which develops as shear bands in single crystals for complex loading experiments. The localized deformation requires collective movements of a large number of dislocations on particular planes. It induces instabilities on the single crystal stress-strain curve. The amplitude of the phenomenon — i.e. yield drop or not, low work hardening or softening,... — is mainly related to the relative orientations of the active slip systems during prestrain and subsequent deformation.

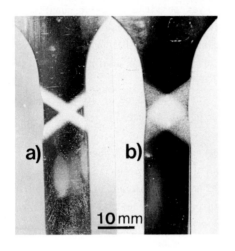

Fig. 4. Surface observation of tensile specimens prestrained by rolling (up to an equivalent strain of 0.2) showing bands of strain localization :
a) tension along the rolling direction ;
b) tension along the transverse direction.

The substructure instability leads to polycrystalline instability and to the development of deformation bands when microbands propagate through the entire section of the sample. This has been observed for some experiments of sequences of uniaxial tensions along different axes (Ref. 6). It has been specially described and analyzed for low carbon steel sheets submitted to uniaxial tension after a prestrain by rolling (Ref. 26). After about 1% of homogeneous elasto-plastic deformation, a sudden decrease of load is measured which corresponds to the development through the sample of one or two localized bands of deformation (Fig. 4). The orientation of the band seems to be independent of the prestrain amount and of the prestrain mode (either parallel or transverse subsequent tension). It might be governed by some purely mechanical reasons owing to the heterogeneity of the sample. As the imposed deformation increases, bands tend to spread out and seem to carry the whole macroscopical strain. A careful surface analysis shows that bands are mainly composed of grains showing coarse slip lines oriented about the macroscopical deformation band direction. The macroscopical bands then appear as the result of a cooperative effect of localized deformation in some grains (microbands), though the "propagation" process is not yet understood.

The propagation must be easier when crystallographic texture accounts for a neighbourhood of "favorably" oriented grains, namely grains in which microbands form first in a crystallographic manner. This would explain a part of the relationships between texture and shear banding phenomenon as it has been commonly described. The existence of local vertices on the macroscopical yield surface makes possible the existence of shear components which favour macroscopical instability development. From a purely physical point of view, Korbel (Ref. 27) has proposed a "plaston" model. It is based on dynamic collective movements of dislocations that begin in a crystallographic manner in some grains then propagate by localized shearing in neighbouring grains (Ref. 28). From our knowledge, it is not yet possible to solve quantitatively the problem of the development of instability from the microscopical level to the macroscopical one. It seems nevertheless well established that microbands which develop in some grains, specially after a change of loading conditions, conduce to a macroscopical localization of deformation.

To conclude, three main cases may occur during a change of loading mode :

i) Substructural evolution takes place and induces a decrease of work hardening rate in some grains without any development of localized deformation. The polycrystalline media may be considered as an heterogeneous media with different volume fraction of grains having different mechanical behavior, for example, a soft phase in a hard matrix. The volume amount of grains which are concerned with the evolution of work hardening and the "matrix" behavior are the main parameters for the analysis of the strain localization and the necking development. The problem of stability may then be investigated by a two-zone model as it has been done for introducing a damage parameter. In this case, a mechanical defect has to be taken into account. The results may be greatly dependent on the volume fraction of grain with a "softening" behavior which can induce macroscopical effects and allow a band of defect to be assumed. The grain distribution may also have an influence in the sense that the macroscopical response depends certainly on the existence of homogeneous distribution of grains having a similar orientation or not.

ii) Microbands appear in some grains but they cannot propagate through grain boundaries or induce particular shearing in neighbouring grains. This event is similar to the previous one except that particular attention must be paid to the local internal stress state induced by microbands.

iii) Microbands propagate and induce macroscopical deformation bands through a large number of grains. The bands may then propagate at a macroscopical level in a similar way as Lüders bands and/or lead to an early rupture. The model would have to take into account heterogeneities at different scales. A two-zone model seems no longer adequate. The knowledge of local parameters are then needed such as : the number of shear bands, their size and the deformation rate that is carried, the macroscopical propagation of the process either by scattering of few bands or by appearence of new ones, the possible initiation of a geometrical defect owing to the decrease of effective section induced by plastic deformation inside the bands.

The development of one or another event depends on the strength of the dislocation walls considered as obstacles in each grain, on the crystallographic texture, on the characteristics of the microscopical behavior (cross-slip and dynamical recovery, for example, do not allow pile-ups, microbands developments and energy concentration large enough to produce localized propagation), that is the main structural parameters which characterize the successive deformation modes and the prestrain amount.

CONCLUSION

One has recalled the large influence of a geometrical defect on the plastic stability of materials. This defect originates very often from volume damage associated with hard particles or inclusions. Instability calculations are also, and to the same extent, sensitive to constitutive law formulations. The effects of plastic behavior, on the one hand, and of defect, on the other, seem to be so comparable that they cannot be weighed separately.

A physical analysis of plastic behavior of prestrained materials points out the importance of intragranular substructure and more specially of its rapid evolution during reloading. It frequently leads to the development of microscopical and macroscopical heterogeneities and to strain localizations. It seems very influent on the plastic stability though no model can yet be proposed owing to the complexity of the phenomenon.

ACKNOWLEDGEMENTS

The authors whish to express their grateful thanks to Professor B. Baudelet for fruitful discussions.

REFERENCES

1 W.B. Hutchinson, R. Arthey and P. Malmström, On anomalously low work hardening in prestrained metals, Scripta Metall., 10 (1976) 673-675.
2 J.V. Laukonis and A.K. Ghosh, Effects of strain path changes on the formability of sheet metals, Metall. Trans., 9A (1978) 1849-1856.
3 D.J. Lloyd and H. Sang, The influence of strain path on subsequent mechanical properties - Orthogonal tensile paths, Metall. Trans., 10A (1979) 1767-1772.
4 R.H. Wagoner and J.V. Laukonis, Plastic behavior of aluminum-killed steel following plane strain deformation, Metall. Trans., 14A (1983) 1487-1495.
5 P.L. Charpentier and H.R. Piehler, Yielding anisotropy from the Bauschinger effect and crystallographic texture in drawn HSLA steel sheet, Metall. Trans., 15A (1984) 1699-1710.
6 J.L. Raphanel, J.H. Schmitt and B. Baudelet, Effect of a prestrain on the subsequent yielding of low carbon steel sheets : experiments and simulations, Int. J. Plasticity, 2 (1986) 371-378.
7 J.H. Schmitt, Contribution à l'étude de la micro- macroplasticité des aciers, Doctorate Thesis, INP-Grenoble, France, March 1986.
8 T. Kikuma and K. Nakazima, Effects of deforming conditions and mechanical properties on the stretch-forming limits of steel sheets, Trans. Iron Steel Inst. Japan, 11 (1971) 827-831.

318

9 M. Grumbach and G. Sanz, Effet des chemins de déformation sur les courbes limites d'emboutissage à striction et à rupture, Mém. Sci. Revue Métall., 71 (1974) 659-671.

10 J.J. Jonas and B. Baudelet, Effect of crack and cavity generation on tensile stability, Acta Metall., 25 (1977) 43-50.

11 Z. Marciniak and K. Kuczynski, Limit strains in the process of stretch-forming sheet metal, Int. J. Mech. Sci., 9 (1967) 609-620.

12 J.M. Jalinier and J.H. Schmitt, Damage in sheet metal forming - II. Plastic instability, Acta Metall., 30 (1982) 1799-1809.

13 J.H. Schmitt, J.M. Jalinier and B. Baudelet, Effect of initial thickness of sheets on the FLD's at necking, Comm. at Deep Drawing Research Working Group, Tokyo, Japan, May 1981, DDR/WG2/81/4.

14 F. Barlat, A. Barata da Rocha and J.M. Jalinier, Influence of damage on the plastic instability of sheet metals under complex strain paths, J. Mater. Sci., 19 (1984) 4133-4137.

15 R. Sowerby and J.L. Duncan, Failure in sheet metal in biaxial tension, Int. J. Mech. Sci., 13 (1971) 217-229.

16 F. Barlat, Crystallographic texture, anisotropic yield surfaces and forming limits of sheet metals, Mater. Sci. Eng. (1987) in press.

17 A. Barata da Rocha, F. Barlat and J.M. Jalinier, Predictions of the forming limit diagrams of anisotropic sheets along linear and non-linear strain paths, Mater. Sci. Eng., 68 (1984-1985) 151-164.

18 P. Bate, The prediction of limit strains in steel sheet using a discrete slip plasticity model, Int. J. Mech. Sci., 26 (1984) 373-384.

19 P.M.B. Rodrigues, P.S. Bate and D.V. Wilson, Structure and strain localisation fracture in aluminum sheet stretching, in : J.D. McQueen et al. (Eds.), Proc. 7th Int. Conf. on Strength of Metals and Alloys, Montréal, Canada, August 1985, Pergamon Press, 1985, Vol.1, pp. 323-328.

20 R.J. Asaro and A. Needleman, Texture development and strain hardening in rate dependent polycrystals, Acta Metall., 33 (1985) 923-954.

21 F. Barlat and O. Richmond, Prediction of tricomponent plane stress yield surfaces and associated flow and failure behavior of strongly textured fcc polycrystalline sheets, submitted for publication.

22 A. Barata da Rocha and J.M. Jalinier, Plastic instability of sheet metals under simple and complex strain paths, Trans. Iron Steel Inst. Japan, 24 (1984) 132-138.

23 J.H. Schmitt, F. Barlat, A. Barata da Rocha and J.M. Jalinier, Influence des phénomènes transitoires d'écrouissage lors d'un changement de chemin de déformation, Comm. Journées Métall. d'Automne, Soc. Franç. Métall., Paris, France, October 1984.

24 K. Chung and R.H. Wagoner, Effect of stress-strain-law transients on formability, Metall. Trans., 17A (1986) 1001-1009.

25 J.V. Fernandes and J.H. Schmitt, Dislocation microstructures in steel during deep-drawing, Phil. Mag., 48 (1983) 841-870.

26 P. Martin and A. Korbel, Microstructural events leading to macroscopic strain localization in prestrained tensile specimen, submitted for publication.

27 A. Korbel, The real nature of shear bands - plastons?, in : Plastic instability, Proc. Int. Symp. on Plastic Instability - Considère Memorial, September 1985, Ecole Nale Ponts et Chaussées, Paris, France, 1985, 325-335.

28 A. Korbel and P. Martin, Microscopic versus macroscopic aspect of shear band deformation, Acta Metall., 34 (1986) 1905-1909.

Computational Methods for Predicting Material Processing Defects, edited by M. Predeleanu   319
Elsevier Science Publishers B.V., Amsterdam, 1987 — Printed in The Netherlands

TAKING INTO ACCOUNT THE FORGING PROCESS IN COMPUTER AIDED DESIGN TO INCREASE THE LIFE OF THE DIES

S. TICHKIEWITCH

Laboratoire de Mécanique et Technologie

E.N.S. Cachan/C.N.R.S./Université Paris 6/ G.R.E.C.O. "Calcul de Structures"

61, Avenue du Président Wilson, 94230 Cachan (France)

ABSTRACT

   Computer aided design of mechanical parts has to take into account the manufactured process to minimize the cost of the part. In forging process, the life of the die is one of the factors increasing this cost, and is a consequence of the forged part design. The rules of such a forged part design are given in order to minimize the fatigue of the die. These rules are introduced in COPEST, expert system in design and optimization of parts.

INTRODUCTION

   COPEST is an expert system for the design of mechanical parts mainly used in power transmission. Its main topic is an optimization loop to minimize the manufacturing and using cost of a part, for which it knows the technological use and the forcast loads. Taking into account the real cost of the manufacturing in design process is a far-away objective in which we must have a look on stamping, tooling, completion, setting up... HEINRITZ in [1], McROBERT and WATSON in [2] affirm that it is only after those conditions that we effectively could obtain an optimal product. Today, we have introduced into our system the stamping step.

   This system was previously described in [3], [4], [5] in terms of expert system and stresses computations. So this paper will present the part of automatic cost determination, including the forged design features and machining allowance used to obtain the stamped part. This work was previously done by a stamper with interactive process [6].

OPTIMIZED FORGING DESIGN FEATURES :

   The cost determination of the necessary stamped part is a double shoot problem. We have first to choose the stamped part design, and in a second time to evaluate the best machine to be used in the process. BADAWY et al. say in [7] : "Given a final finished geometry, the forging engineer is faced with the problem of determining the optimum forming sequence to forge the part. The selection of the optimum forming sequence is an art, and so far it has been mainly achieved using past experience." So, in COPEST we use the sense of the forger, his knowledge and experience built through study of thousands of cases. A part of this knowledge is given in CHAMOUARD [8].

Stamped modes :

The variety of the stamped processes induces the matching of the undressed part among a lot of classes, which is useful to choose the best process, and, consequently, to define the needed features. "A.D.E.T.I.E.F." wrote in [9] such a maping. For rounded parts, parameters we have to take into account are :
- the greatest diameter   D,
- the length (or height)   H,
- the projected surface on a meridian plan   s,
- the rectangular and circumscribe surface   S.

Four families (among seven) are proposed for the revolution parts, and are function of the ratios : D/H and S/s. Copest works to day on the first and second family. Examples of these families are shown in figure 1.

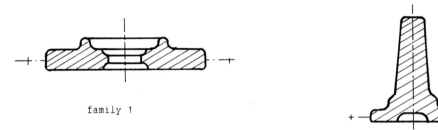

family 1

Fig 1 : Morphological classification of stamped parts   family 2

Tooling allowance :

After being stamped, the billet must be machined on a turning lathe, and so we have to reserve tooling allowances. These out-depths are dependent on the dimension of the concerned surfaces, and on the type of these surfaces, cylinder, cone or plan. COPEST questions the user to have the starting surface for tooling, and so is able to determine the length to take into account in searching allowances. The standard values are given on figure 2.

diameter or length in mm	<30	30-60	60-120	120-250	250-500	>500
allowance in mm	1	1.5	2	2.5	3.5	6

Fig 2 : Tooling allowance

Specific surfaces for stamped parts :

Those surfaces must be only the choice of stampers, as they are the most important factors causing defects to occur [10].'

We only are concerned here by the surfaces used on the two families dressed by COPEST (1 and 2).

The choice of the process to stamp the part gives the separate plan for the dies, and so, gives also the extracting direction. To make this extraction easier, it is useful to forecast surfaces in demi-intaglio, or to give an inclination on all the parallel surfaces to the extracting direction. The inclination can be inner or outer.

When cooling the part due to the contact with the dies, inner faces have a natural disposition to band on the die, and this part of the die, called nucleus, has to have an angle of 4°. We usually take 2° for all the other angles, normal values for stamping on vertical forging-press.

Such a part with inclination surfaces is shown on figure 3a.

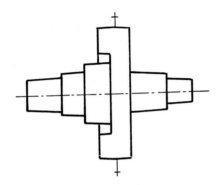

Fig 3a: Part with surfaces in demi-intaglio

In case of two nuclei move forward during the strike, we obtain a thin depth of metal named the web. Thinner is the web and higher is the strength to stamp the part, as we will show below, so the dimensions of the web are fixed on account of the resistance of the nucleus. Thinness of the web is the ratio between its depth and its diameter. The forger's experience gives this depth e in function of the diameter D and the height of the nucleus H as: (in mm) (Figure 3b).

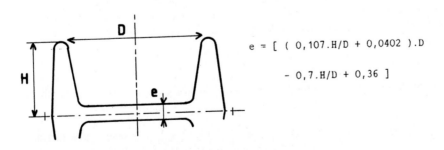

$$e = [(0,107.H/D + 0,0402).D$$
$$- 0,7.H/D + 0,36]$$

Fig 3b : Dimension of the web.

The corners of a dressed part are not indifferent. We have to distinguish three kinds of possible corners:

- salient corners obtained by flattening,
- salient corners obtained by spining,
- fillet. First and second are the fact of pressure level at the end of the process, and we will have a look at them later on. Fillet do not need the final pressure to be obtained, as metal naturally assumes it when flowing. Two different fillets are recognized during stamping, those with material slipping on it, placed on the beginning of a spining part, and those with no slipping, placed on the way of the oppression. Figure 4a displays such corners. When slipping exists, we have an increase of the radius due to erosion during the process, and all long the die life.

In absence of complete simulation of the metal flow with the thermic effects, we use mean conditions. In COPEST, we consider that the temperature of the part is of 1100°C at the beginning, to finish at 1000°C, and so, with dies in H12 steel , we use for the ratio of the fillet radius over concerned diameter the rule:

$r/D = 0,002.P_0$ where P_0 is the pressure on the fillet in daN/mm^2.

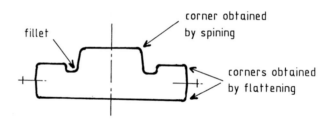

Fig 4a: Different corners in stamping

Flow and closing strength :

The easiest problem we can study at first is the problem of forging a round billet between swage-blocks. We can solve such a problem by a slab method, in order to obtaine the pressure distribution on the faces of the cylinder :

$$p = - \sigma_e.e^{2.(f/h).(R-r)}$$

and so, we have a mean compressive stress, the plasticity threshold, such as:

$$\sigma_0 = - F/S = 2.\sigma_e .K/f.[(K/f).(e^{f/K} - 1) - 1]   \text{avec}   K = h/2.R$$

The curve a), with an horizontal level, is sufficient to explain why we have barelling effect during compressive test. We add for this the curve b) which gives the mean compressive stress for an effort F and different diameter D. So we can now, still with the force F, do a third curve c) : height h for each diameter D. A D_0, maximum diameter is obtained in correlation with the level of the plasticity threshold. The h_0 which agrees with D_0 is the minimal height of the cylinder. Such a curve is a boundary curve, as we have not taken in the slab problem the strain induced by shear stresses. Another boundary curve is given when the friction factor f is taken equal to 0, because there is no shear stress in this case, the cylinder remaining a cylinder. The real strain is in fact between these two boundaries, and we can define an effective radius r for this case. (Figure 4b).

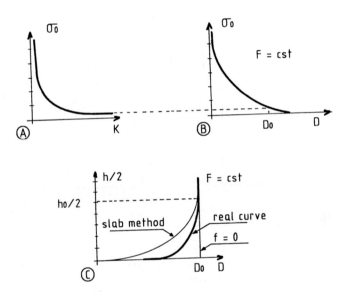

Fig 4b : Definition of radius

We propose to modify the starting equation for the plasticity threshold, in order to match with experimental curves. The new term is only active with small values of K, where the shear stresses are efficient. So we take :

$$\sigma_0 = 2.\sigma_e .K/f.(\ (K/f).(\ e^{f/K} - 1\ ) - 1\ ) + a.(\ K-b\ ).e^{-c.K^n}$$

The same study is possible for the spinning mode and gives the same kind of equation, with a ratio $K' = 1/K = D/h$.

The values of the mean pressures are established with static hypothesis and free lateral surface. We have to analyse what appends when the billet is placed in closed dies, as shown in figure 5a.

In a first time, only plane surfaces are in contact, and the process is as previously discribed, with barel effect.The s surface undergoes the global displacement. When the billet arrives in touch with the die on lateral surfaces, the load is in accordance with the ratio H/D. The increase of the load cannot create a bigger diameter, so the salient corners decrease their radius. M' is the point where the plasticity treshold for ratio H'/D is equal to the driving pressure.

As soon as M' point exists, we can consider a mean distibuted pressure on lateral surface whose effect is to spin the billet. Without draft surface and friction effect, the necessary load to spin the section s is constant until this queue meets the bottom of the cavity. There, if r is the radius, we must increase the driving pressure to ensure a reaction pressure m on the bottom of

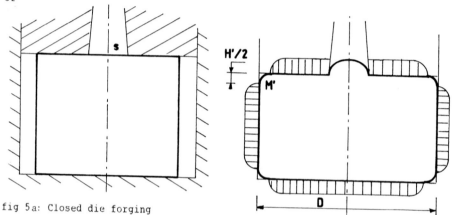

fig 5a: Closed die forging

the cavity given by the pasticity threshold for ratio K = 2r/d. In fact, the queue with height h needs a minimal pressure α, with ratio K'=1/K=d/h to be spinned, and we find the add pressure $(m-\alpha)$ on the latteral surface of the queue. The effect of draft, associated with the friction factor is such that the lateral pressure has to be decreasing along the queue. At entrance of spinning, that pressure is given by : (Fig 5b)

$$t_0 = (m - \alpha).(1 + 2.\ln(d_2/d_1)).e^{8.f.h/(d_1+d_2)}$$

The driving pressure in front of the queue is so :

$$P_0 = t_0 + \alpha \qquad \qquad \text{Fig 5b}$$

COPEST analyses the geometry of the part to determine the strain mode, spinning or flattening, of each elementary volume. For this, it searches where the first surfaces in contact with the billet are, which is an equivallent volume of the part with a ratio h/D of 3. Then, it separates the spinning regions of the flatting regions and is able to compute the different ratios K.

The radius are chosen in function of these terms, step by step, and also in respect to the computed pressure. The choice of the first radius is necessary to start the process. This choice is made in accordance with the trade wonts, given below.

If the radius concerns a flattening corner, we take :

      $r = 1.5$ if the corresponding diameter is less than 100

      $r = 1.5.(D/100)$ in other cases

If the radius concerns a spinning corner, we take :

      $r = .10 D$ if H/D is less than 1.5

      $r = .17 D$ if H/D is less than 2.5 and greater than 1.5

      $r = .29 D$ if H/D is less than 3.5 and greater than 2.5

      $r = .40 D$ if H/D is greater than 3.5

This fisrt corner chosen, COPEST evaluates the useful pressure in accordance with its radius and integrates this pressure all along the directrix, taking into account the friction if necessary, in order to obtain the pressure in front of the flash and the closing strength. Then, it starts with this pressure to define the radius of the corners on the other side of the die line. With such determination of radius, we assume that the part is filling the die cavity at the same final time, and so, the contact of corners with the die is minimal, the temperature of the die does not increase more than necessary at the same time that the temperature of the stamped part decreases, and the local pressures are also just adapted.

This is an essential choice for the life of the dies and thus, for the manufacturing cost. The variation of the static failure load with temperature (figure 6), read in [11], shows the brutal degradation of this characteristic between 500°C and 600°C, for three different die steels.

The compressive force useful to obtain a part with diameter D is only conditionned by the minimal allowable height h_0 in accordance with its h_0/D ratio. For many radius values, and for a diameter D, it is so possible to evaluate F. We can see on figure 7 the consequence of choosing any radius on a part, and why it is useful to give the designer system the knowledge of such a behaviour.

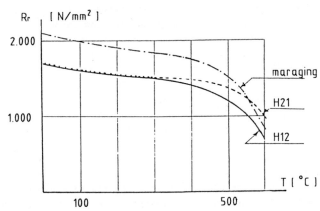

fig 6 : Static failure load for die steel

Anisothermal adjustment :

All the choices of radius are made with a constitutive equation for plasticity threshold whose factors are evaluated at a 1000°C. This temperature is reasonable for 95 % of the part but is over-estimated for small corners. There the real pressure is greater and we take into account this fact by introducing an anisothermic factor to obtain a radius used in computations instead of the real radius.

$$r_{calculus} = r_{real} / C_a$$

with

$$C_a = max(0,822.e^{1/(1.276*R)} ; 1)$$

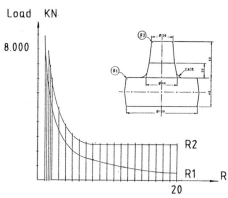

fig 7 : Useful closing load as a radius function

Flash design :

On a die, the gutter is a necessary space to recover the overweight we have to add to the normal weight of a part in order to assume the complete filling of the die whatever the wear. We place between the die cavity and the gutter a flash, or supplementary thin ring. To flow through the flash, the pressure must be sufficient and in relation with the ratio $\epsilon/(D+2\lambda)$, where ϵ is the thickness of the flash and λ its width. So, the design of the flash is done in order to assume some per cent more than the requested pressure.

GIVING A COST FOR A PART :
Fixing a hammer :

As we have dressed the finished part to obtaine a stamped part, it is now possible to evaluate the cost of stamping. Therefore, we have to fix which hammer or press we use, in order to get the stamping rate, and so the cost of the process.

We first determine the best hammer, by fixing a useful energy to obtain the part. With the machine efficiencies, we can infer from the hammer an equivallent press.

Choosing a hammer needs two steps: first to fix the falling mass, second the falling height. Instead of running an optimization loop to chose these factors, we use in COPEST the rules established by trade and based on experience. We consider in a first time than the part is an equivalent cylindric part (with the same volume as the final part and the same maximum diameter), and that the mass falls from a 1.4 meter height. The compressive problem is in accordance with the viscoplastic formulation of the constitutive equation :

$$\sigma = K.[\ 1 + \varepsilon^m\ ].\ \varepsilon^n$$

the theorem on kinetic energy :

$$F = M.(\ g + \dot{h}\ )$$

the law of volume conservation :

$$r = r_0 .\sqrt{h_0\ /\ h}$$

and the starting height, H, which gives the boundary condition :

$$\dot{h}_0 = \sqrt{2.g.H}$$

So, each strike is the result of the dynamic equation :

$$M.(\ g + \ddot{h}\ ) = \frac{\pi.h}{f}.K.[\ 1 + \dot{h}^m/h\ ].(\ \log h_0/h\ )^n.[\ \frac{h}{2.f}.(e^{2.f.R/h} - 1) - R\ ]$$

Study of such a problem can give the optimized mass in accordance with the minimal energy with :
- determination of a mass factor :

$$a = 21,17.(\ S - 15\ )^{1.08} + 286,7$$

where S is the surface of the part for maximum diameter,
- then the falling mass :

$$M_p = a\ /\ E^{0.4}$$

where E is the mean depth,
- and the number of strikes :

$$N_b = P_m^{0.353}/\ 15,53$$

where P_m is the weight of the part. The given number of strikes is for a revolution part, which accepts a flash of about 5 per cent of the weight.

In a second part, we modify the falling height of the mass and the number of strikes in accordance with the useful energy for the real part, with its particular complexity. Therefore, the closing strength previously got is used and is compared with the strength we had to get to stamp the equivalent cylinder. The added strength grades the part in complexity classes. Each of them gives the falling height H_p and a corrective factor for the final number of strikes N_p.

Equivalent engine :

COPEST has determined the best hammer for the stamped part. It is able now to consider an equivalent press. The total energy spent by the hammer has not been used to manufacture the part, in account of the lack efficiency of the hammer. The knowledge of compared efficiency between hammer and press can give the energy spent with a press. As the energy is dependent on the velocity of the engine, stampers have introduced in their determination a ratio giving the effective energy spent with falling velocity V as a function of the minimal useful energy with nule velocity.

Example of COPEST work :

Starting with the finished part geometry given below, COPEST proposes the stamped shape and its manufactured cost.

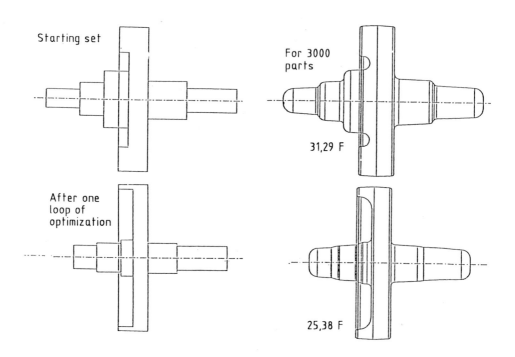

Fig 8 : A best solution proposed by COPEST.

Surface with maxi diameter: 10
Diameter: 120.29 mm
Position of the maxi diameter: 62.16 mm
Total volume: 602387.60 mm^3
Mini diameter of the billet: 63.46 mm

Pressure in front of the flash: 75.96 hB
Width of the flash: 10.00 mm
Thickness of the flash: 2.00 mm.

Total projected surface: 25044.80 mm^2
Mean thickness with flash: 24.47 mm
Mass factor: 8002.14
Falling mass: 1781.57 kg
Basic number of strikes: 7.5
Falling height: 1.40 m
Corrected number of strikes
  used for used energy: 7.5
Number of strikes
  used for spent energy: 8.

Used power: 10868.46 kg.m
Spent power with hammer: 19953.58 kg.m
Spent power with press
  (V=0.8 m/s):   6728.09 kg.m
Number of parts with hammer
  / hour: 75.00

Hammer cost / hour: 855.85 F
Labour cost / hour: 130.19 F

Metal cost / part: 11.95 F
Stamped cost / part: 13.14 F
Preparing cost / part: 0.28 F

FINAL COST / PART: 25.37 F

REFERENCES

1 M. HEINRITZ, Optimization of stamped parts, Proc. of the 12th International Forging Congress, ORLANDO, Florida, November 1986 p-p IV-C-1-20.

2 S.C. McROBERT, P. WATSON, Research of optimal characteristics of the stamped parts, Proc. of the 11th International Forging Congress, KOLN, June 1986, p-p 257-290

3 S.TICHKIEWITCH, Copest : système expert pour la conception optimisée de pièces estampées, Les systèmes experts et leurs applications, AVIGNON, May 1984

4 S.TICHKIEWITCH, Prise en compte de contraintes mécaniques et technologiques dans l'optimisation de structures. Application à l'estampage, Tendances actuelles en Calcul de Structures, BASTIA, November 1985, Pluralis editor p-p 801-816

5 S.TICHKIEWITCH, L.PROSLIER, Optimization of structures with technological constraints. Application of the 'R.d.M. Tridmensionnelle', Internal repport, Laboratoire de Mécanique et Technologie, May 1987

6 T. ALTAN, Computer-Aided Design and Manufacturing (CAD/CAM) of Hot Forging Dies, Proc. of the 11th International Forging Congress, KOLN, June 1983, p-p 295-332

7 BADAWY, A.A.., P.S.. RAGHUPATHI, D.J.. KUHLMANN, T. ALTAN, Computer-aided design of multistage forging operations for round parts, Journal of Mechanical Working Technology, Vol. 11, 1985, p-p 259-274

8 A. CHAMOUARD, Estampage et forge, Tomes 1 à 4, Dunod editor, PARIS, 1964

9 Classification morphologique des pièces estampées, A.D.E.T.I.E.F., PARIS, March 1974

10 W.T. WU, S.I. OH, T. ALTAN, Investigation of defect formation in rib-web type forging by 'ALPID', Proc. of NAMRC XII Michigan Technological University, HOUGHTON, Michigan, May 30-June 1, 1984.

11 V. MOLS, Tenue des outillages d'estampage, Proc. of the 124,th International Forging Congress, ORLANDO, Florida, November 1986 p-p IV-B-1-22

Computational Methods for Predicting Material Processing Defects, edited by M. Predeleanu
Elsevier Science Publishers B.V., Amsterdam, 1987 — Printed in The Netherlands

DAMAGE ARREST BY CLOSURE OF VOIDS IN DILATANT SOLIDS SUBJECTED TO COMPRESSIVE LOAD

J. TIROSH

Faculty of Mechanical Engineering, Technion, Haifa, Israel

SUMMARY

A continuum mechanics solution is offered to explain the role of remoted confining pressure (encountered often in metal forming processes) in closing voids in dilatant visco-plastic materials. The solution accounts for the simultaneous large scale rotation, contraction and shortening of an ellipsoidal void subjected to non-symmetric far field load, either unidirectionally or bi-directionally. The interaction of the void with its surrounding voids is achieved in a 'self consistant' sense (i.e. the bulk solid behaves like the unit cell with its single void). The volume fraction of the voids represents a scalar measure of the damage in the solid. The geometrical and material nonlinearities associated with the damage arrest by voids closure is simulated by piece-wise linear time-integration of the equivalent elastic solution. It is applied to the visco-plastic situation via the Rayleigh analogy. The major outcome is the ability to simulate with some generality the evolution of the closure strain and the inhibition of the material dilatancy. The advantage of operating the forming process under confining pressure environment and the usefulness of having materials with relatively high rate sensitivity index (of around 0.5, thus characterizing the superplastic condition) are assessed quantitatively with regard to damage arrest.

INTRODUCTION

Experimental observations on the effect of hydrostatic pressure on material behavior indicate an apparent increase of ductility [1]. This property has been beneficially utilized in metal forming processes where compressive environment, either by confining fluid pressure [2,3] or by improved die design [4,5] improved the performance of the process. Otherwise internal defects (mainly voids of various kinds) may grow beyond an acceptable size and defect the product. Such studies were considered primarily from the stress analysis view point via Finite element numerical approach [6,7] or from an upper bound consideration [7]. Analytical-oriented investigations [8,9,10] amplified more sharply the role of hydrostatic tension in enhancing void growth. A peculiar feature concerning the rotation of a void due to nonsymmetric remoted load was noted by Berg [12] for linear viscous solid. It became later more evident in the study of localized shear [13]. By allowing the voids to interact and to undergo simultaneously large scale rotation, expansion and elongation, a damage evolution model has been recently developed [14]. It was based on voids coalescence in power-law creep of porous solids. This new model matches very well other damage-evolution theories [15, 16], but it has some extra features. One of them is being extendible into biaxial compressive loading. It is the intention of this work to employ this extension in order to study the closure evolution of internal voids as caused by external compression. By doing so the predictive capability of this mathematical model can be used to assess the time and the strain required for arresting the damage growth of a given porous solid subjected to remoted compressive load. It may also enable to predict the material whose properties lead to faster closure-rate. The clear advantage of operating the forming process in superplastic conditions (rate sensitivity index of around 0.5) for arresting damage growth is emerged.

CONSTITUTIVE EQUATION

We consider an unbounded 2D solid with equally spaced (and also equally oriented) ellipsoidal voids as shown in Fig. 1. The volume fraction of the voids, f, represents a measure of the damage and defined as

$$f = \frac{\sum V_{voids}}{V_{solid}} \tag{1}$$

where V_{solid} is the volume sum of all the unit cells, each of which includes a single void. The matrix volume $V^{(m)} = V_{solid} - \sum V_{voids}$ is considered to be incompressible, $\varepsilon_{ii}^{(m)} = 0$, though the bulk solid is inherently dilatant. The behavior of the matrix (with superscript m) follows the power-law constitutive relation as

$$\frac{\sigma^{(m)}}{\sigma_0^{(m)}} = \left(\frac{\dot{\varepsilon}^{(m)}}{\dot{\varepsilon}_0^{(m)}}\right)^n \qquad 0 \leq n \leq 1 \tag{2}$$

where $\sigma_0^{(m)}$ and $\dot{\varepsilon}_0^{(m)}$ are predetermined reference values of nominal stress and strain-rate, and n is the rate sensitivity index of the matrix. This index spans from n = 0 (representing rigid-perfectly plastic solids) to n = 1 (representing linear viscous solids). Thus, eq. (2) comprises a very wide family of materials behaviour, at which $\sigma^{(m)}$ and $\dot{\varepsilon}^{(m)}$ (of the damaged material) are evaluated from their respective tensorial components according to

$$\left.\begin{aligned}
\sigma^{(m)} &= \left(3/2 S_{ij}^{(m)} S_{ij}^{(m)}\right)^{1/2}, & s_{ij}^{(m)} &= \sigma_{ij}^{(m)} - 1/3\, \delta_{ij}\, \sigma_{kk}^{(m)} \\
\dot{\varepsilon}^{(m)} &= \left(2/3\, \dot{\varepsilon}_{ij}^{(m)} \dot{\varepsilon}_{ij}^{(m)}\right)^{1/2} & \dot{\varepsilon}_{ij}^{(m)} &= 1/2\left(u_{i,j}^{(m)} + u_{j,i}^{(m)}\right)
\end{aligned}\right\} \tag{3}$$

and $u_i^{(m)}$ is the velocity field at the considered point.

The tensorial representation of (2) is equivalent to

$$\dot{\varepsilon}_{ij}^{(m)} = \frac{1}{2\eta^{(m)}} \cdot S_{ij}^{(m)} \tag{4}$$

where the nonlinearity is embedded in the 'viscosity', $\eta^{(m)}$, defined as

$$\eta^{(m)} = \left(\frac{1}{3} \, \bar{\sigma}_o \, \dot{\varepsilon}_o^{-n} \right) \dot{\varepsilon}^{\,n-1} \tag{5}$$

By noting that Hooke's law can be written as

$$\varepsilon_{ij} = \frac{1}{2\mu} S_{ij} - \frac{\nu}{1-2\nu} \varepsilon_{\kappa\kappa} \delta_{ij} \tag{6}$$

the so called 'Rayleigh analogy' between visco-plastic behaviour of (4) and elastic behaviour (6) is readily established, provided the elastic solid is incompressible ($\varepsilon_{kk} = 0$)

THE VOID MOTION AND VOIDS INTERACTION

The analogy cited above between solutions of incompressible elastic body and linear viscous solid, having the same void configuration and subjected to the same far-field load (P_1 and P_2) enables to express the void surface motion in a precise form as [17]

$$u(s) + iv(s) = \frac{\varkappa + 1}{2\eta} R \left[\left(\frac{P_1 + P_2}{4} \right) \left(\xi - \frac{m}{\xi} \right) + \left(\frac{P_1 - P_2}{4} \right) \frac{2 e^{2i\alpha}}{\xi} \right] \tag{7}$$

for $\xi = e^{i\theta}$ and $0 \leqslant \theta \leqslant 2\pi$ $\qquad \varkappa = \begin{cases} 3 - 4\nu & \cdots\cdots \text{ for plane strain} \\ (3-\nu)/(1+\nu) & \cdots\cdot \text{ for plane stress} \end{cases}$

where R and m are the two independent geometrical variables of the void, defined in Fig. 2. The angle α is the orientation of the major axis to the 'vertical' traction P_1, P_2 is the transverse traction and η is defined by (5). After a short time interval, Δt, (short enough not to affect significantly the solution (7) by variations in α, m or R) one should update the shape of the void by

$$\Delta z(s) = [u(s) + iv(s)] \Delta t \tag{8}$$

Hence, the new shape $\hat{z}(s) = z(s) + \Delta z(s)$ is, by (7) and (8),

$$\hat{z}(s) = R(1+\delta)(e^{i\theta} + e^{-i\theta} \{ m \cdot \frac{1 - \delta + \delta(\frac{P_1 - P_2}{P_1 + P_2}) \frac{2 e^{2i\alpha}}{m}}{1 + \delta} \}) \qquad 0 \leqslant \theta \leqslant 2\pi \tag{9}$$

where

$$\delta = \frac{\varkappa + 1}{8\eta} \cdot (P_1 + P_2) \Delta t$$

Assuming that the induced new configuration $\hat{z}(s)$ retains an ellipsoidal shape (though with updated variables $\hat{R}$, $\hat{m}$, $\hat{\alpha}$), the expression for this new ellipse is written most generally as

$$\hat{z}(s) = \hat{R}\left(e^{i\theta} + e^{-i\theta}\,\hat{m}\,e^{2i\hat{\alpha}}\right) \tag{10}$$

By comparing the structure of (9) to that of (10), one recognizes that the updated variables are related to their previous values by

$$\hat{R} = R(1+\delta) \tag{11}$$

$$\hat{m} = \text{complex absolute value of } \{\ \} \text{ in (9)} \tag{12}$$

$$\hat{\alpha} = \tfrac{1}{2} \text{ argument of } \{\ \} \text{ in (9)} \tag{13}$$

Equations (11), (12) and (13) constitute the basic algorithm for the numerical tracking of the ellipsoidal damage evolution. As a typical example, a time history simulation of void rotation and contraction due to unidirectional compression is shown in Fig. 3. In view of the contraction process, the porosity is continuously diminished which calls for updating of the related field variables (as ϵ, η etc.). In the 'self consistent' approach the interaction between the voids is gained by considering the bulk material response as if it behaves homogeneously with the same properties as the representative cell with the void. This bulk strain rate $\bar{\epsilon}$, is defined in terms of the current porosity, f, the average matrix strain rate, $\bar{\epsilon}^{(m)}$ and the void strain-rate, $\bar{\epsilon}^{(c)}$, in a weighted-sum form as

$$\bar{\epsilon} = \bar{\epsilon}^{(m)}(1-f) + \bar{\epsilon}^{(c)}\cdot f \tag{14}$$

where,

$$\bar{\epsilon}^{(m)} = \left(\frac{\bar{\sigma}^{(m)}}{C}\right)^{1/n}, \qquad C = \sigma_0\,\dot{\epsilon}_0^{-n} \tag{15}$$

and $\bar{\sigma}^{(m)}$ is the equivalent matrix stress (eq. 3) based on the stress components at the ligament between voids. $\bar{\epsilon}^{(c)}$ is calculated from the current geometry and boundary motion of the void via

$$\bar{\epsilon}_{ij}^{(c)} = \frac{1}{V_c}\int_{V_c} \tfrac{1}{2}\left(u_{i,j} + u_{j,i}\right)dV_c = \frac{1}{S_c}\oint_{S_c}\tfrac{1}{2}\left(u_i n_j + u_j n_i\right)ds \tag{16}$$

The accumulated plastic strain components in the matrix are computed by the following

$$\bar{\epsilon}_{ij}^{(m)} = \frac{1}{2\eta^{(m)}}\,S_{ij}^{(m)} \tag{17}$$

where

$$\eta^{(m)} = (1/3)\,C\left(\bar{\epsilon}^{(m)}\right)^{n-1} \tag{18}$$

and

$$\bar{\varepsilon}_{ij}^{(m)} = \int_{o}^{t} \bar{\dot{\varepsilon}}_{ij}^{(m)} \, dt \qquad (19)$$

With the matrix strain components of (19) the new dimensions of the cell is re-computed with associated modification of the stresses within the matrix. By a proper account for the current growth of the void, the porosity is updated which affects the bulk strain rate $\bar{\dot{\varepsilon}}$ via (14), and so forth at any time step increment. This description of dilatant materials is totally different from others [7, 18, 19].

RESULTS AND DISCUSSION

The equations (11), (12) and (13) simulate the closure process at constant remoted compressive load, and plotted in Fig. 3. Each configuration is slightly changed from its previous configuration (described at constant time intervals) showing the rotation of the void in conjunction with closure and dimensional contraction. After a certain time the void is transformed into a closed slit (m = 1) and traction can be transmitted through the contact area. This condition terminates the closure process. The poroisty of the material disappears and the damage is hence considered as being arrested. Consider, for example, various material sheets distinguished one from the other by having different rate sensitivity index n with initial porosity composed of circular voids (m = 0). The solution of equations (9) to (19) simulates the evolution of the damage arrest as shown in Figures 4 to 6. It is seen from the figures that the materials with higher rate sensitivity index leads to faster damage arrest often desired in metal forming application. Thus, deep drawing processes for instance in superplastic conditions (n = 0.5) have a clear advantage in comparison to common sheet forming materials ($n \ll 0.1$) by their quicker time to arrest internal damage. It is also noteworthy that any engineering effort to increase n further on is rarely payoff, since close to n = 0.5 there is only slight sensitivity to changes in closure time. The reason for this is that the strain rate at such high values of n is almost independent of the plastic strain accumulation whereas for small values of n the creeping rate is dropped as shown in Fig. 7.

Various initial damages(f_o = 3.5%, 5%, 6.5 %) at a given n are arrested at the same time and at the same strain. By superimposing additional bi-axial pressure of 20%, the arresting time is decreased (by 17%) as shown in Figure 8. This result 'legitimates' the   endeavor by practical designers to operate the forming process under confining pressure.

By reversing the load from compression to tension, one can predict by this model the damage growth till rupture [14]. Because of the nonlinearity of the process in both the geometry and the material properties, it is impossible to infer results of one loading system from its reciprocal. It is however concei-

336

vable that the continuity of the solution from tension load (validated by experiments from [20] in Fig. 9) to compressive load may give a physical ground to the present study. If so, the outcome cast light on a major issue in the technology of metal forming processes.

Acknowledgement

The work was funded partially by SIMA (Stanford Institute for Manufacturing and Automation) and partially by the Center for Robotics and Manufacturing at the Faculy of Mech. Eng. at Technion, Haifa, ISRAEL.

REFERENCES

1 P.W. Bridgeman, Physics of High Pressure. Int. Textbook of Exact Sciences, G. Bell & Sons, (1949).
2 B. Avitzur, Hydrostatic Extrusion, Trans. ASME, 87, J. Engng Ind. (1965), pp. 487-494
3 J. Tirosh and P. Konvalina, On the Hydrodynamic Deep Drawing Process, Int. J. of Mech. Sci., 27 (1985) pp. 595-608.
4 B. Avitzur, Trans. ASME, 90, J. Engng. Ind. (1968), pp. 79-85.
5 Z. Zimerman, H. Darlington and E.H. Kottcamp, Proceeding of a Symposium on the realtion between Theory and Practice of Metal Forming. Cleveland, Ohio, October (ed. A.L. Hoffmanner) (1970) pp. 47-62.
6 E.H. Lee, R.L. Mallett and R.M. McMecking. Numerical modelling of Manufacturing processes (Ed. R.F. Jones, Jr. H. Armen and J.T. Fond) ASME, PVP-PB-025, 1977, pp. 19-32.
7 N. Aravas, The Analysis of void growth that leads to central burst during extrusion, J. Mech. Phys. Solids, 34, No.1 (1986) pp. 55-79.
8 W. Johnson and H. Kudo. The Mechanics of metal extrusion. Manchester Univ. Press, Manchester (1962).
9 F.A. McClintock "A Criterion for Ductile Fracture by Growth of Holes", Trans. ASME, Series E, J. Appl. Mech., 35, (1986) 303-371.
10 J.R. Rice and D.M. Tracey, "On the Ductile Enlargement of Voids in Triaxial Stress Field," J. Mech. Phys. Solids, 17, (1969), pp. 201-217.
11 B. Budiansky, J.W. Hutchinson and S. Slutky, "Void Growth and Collapse in Viscous Solids," Mechanics of Solids (ed. H.G. Hopkins and M.J. Sewell, Pergamon Press (1982), pp. 13-45.
12 C. Berg, "Deformation of Fine Cracks Under High Pressure and Shear", J. of Geophysical Research, 70, (1965), 3447-3452.
13 F.A. McClintock, S.M. Kaplan and C.A. Berg, "Ductile Fracture by Hole Growth in Shear Bands," Int. J. Fracture Mech., 2, (1966) pp. 614-627.
14 J. Tirosh and A. Miller, SIMA report, Stanford Univ. Feb. 1987 (submitted for publication).
15 F.A. Leckie and D.R. Hayhurst, "Constitutive Equations for Creep Rupture", Acta Net., 25, (1977), pp. 1059-1070.
16 B.F. Dyson and D. McLean, "Creep of Nimonic 80A in Torsion and Tension", Metal Sci., 11, (1977), pp. 37-45.
17 N.I. Muskhlishvili, "Some Basic Problems in Mathematical Theory of Elasticity," P. Noordhof Ltd. Groningen, The Netherlands (1953).
18 A.L. Gurson, "Continuum Theory of Ductile Rupture by Void Nucleation and Growth", Trans. ASME 99, Series H, J. Engng. Mater. Tech. (1977), pp. 2-15.
19 C.A. Berg, Inelastic Behavior of Solids (ed. M.F. Kanninen,W.F. Adler, A.R. Rosenfield and R.I. Jaffee) McGraw-Hill, N.Y. (1970) pp. 171-209.
20 D.A. Woodford, "Strain-Rage Sensitivity as a Measure of Ductility". Quarterly Trans. ASM, 62, (1969), pp. 291-293.

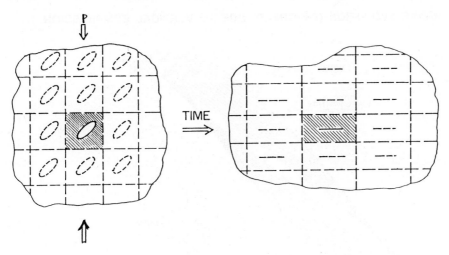

Fig. 1   The dilatant material is composed of periodic array of cells. Each
cell encloses a single void. Due to remoted compressive load (held
constant for simplicity) the material 'creeps' into its final state
while transforming the ellipsoidal voids into narrow slits.

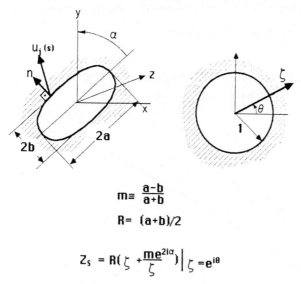

$$m \equiv \frac{a-b}{a+b}$$

$$R = (a+b)/2$$

$$Z_s = R\left(\zeta + \frac{me^{2i\alpha}}{\zeta}\right)\Big|_{\zeta = e^{i\theta}}$$

Fig. 2   The shape of the void is expresses as an ellipse in a parametric form
$m(a,b)$ and $R(a,b)$. The solution of the void motion is performed on a
unit circle in an auxiliary plane (ξ) and expressed via the shown
transformation $z(\xi,r,m,\alpha)$ in the physical plane (z) by equation (7).

SHAPE EVOLUTION (BY CREEP) DUE TO VERTICAL COMPRESSION

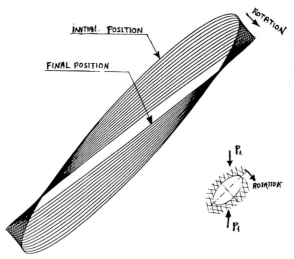

Fig. 3   Numerical simulation of the void closure process solved by eq. 7 for
          equal intervals of time. Note the evolution of rotation and contraction
          of the ellipsoidal void. The evolution rate is inferred from the
          density of the consecutive shapes.

THE SHAPE OF THE VOID VS. NORMALIZED CREEPING TIME

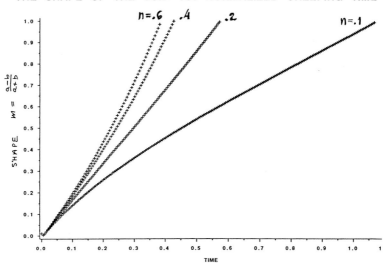

Fig. 4   The voids, starting from circular shapes (m = 0) contract in time
          (while rotating) towards a slit-like shape (m = 1) at closure.

THE S T R A I N STATE VS. NORMALIZED CREEPING TIME

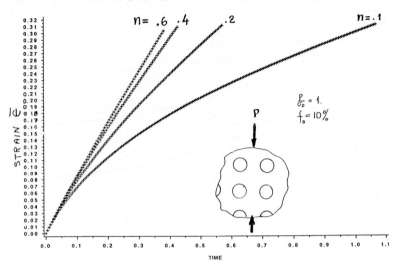

Fig. 5   The global strain accumulated during the closure process is reached at
different rates depending on the rate sensitivity indexes. The higher
the index the faster the closure-time.

THE   DAMAGE STATE VS. NORMALIZED CREEPING TIME

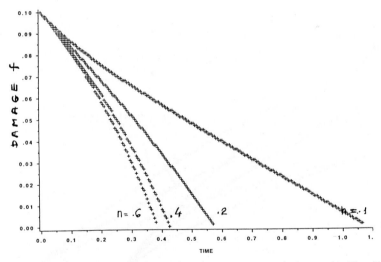

Fig. 6   The damage (the volume fraction of the voids) is gradually diminished
till closure, the rate of which is a function of the rate sensitivity
index (n) of the material. Superplastic condition (n = 0.5) exhibits
a fast damage arrest.

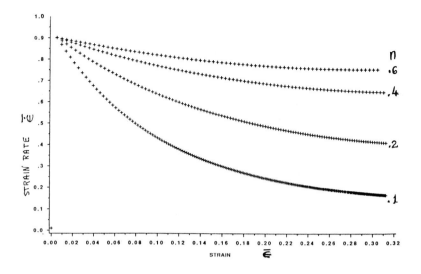

Fig. 7   The accumulated strain during the closure process is associated with
a decrease in the strain rate. Evidently it is more pronounced in
lower rate-sensitivy index.

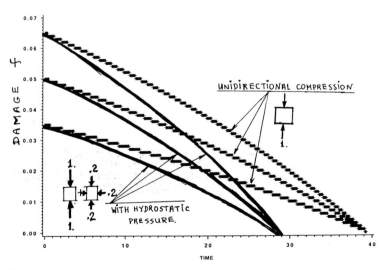

Fig. 8   The effect of hydrostatic pressure on various initial damage
is to enhance the closure rate of the voids.

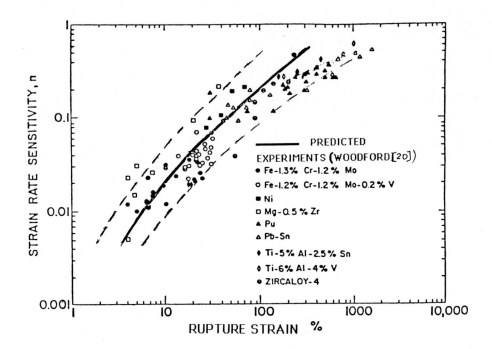

The legend in the figure reads:

— PREDICTED
EXPERIMENTS (WOODFORD[20])
• Fe-1.3% Cr-1.2% Mo
○ Fe-1.2% Cr-1.2% Mo-0.2% V
■ Ni
□ Mg-0.5% Zr
▲ Pu
△ Pb-Sn
♦ Ti-5% Al-2.5% Sn
◊ Ti-6% Al-4% V
● ZIRCALOY-4

Axes: STRAIN RATE SENSITIVITY, n (vertical, 0.001, 0.01, 0.1, 1) — RUPTURE STRAIN % (horizontal, 1, 10, 100, 1000, 10,000)

Fig. 9   The application of the present solution to predict rupture strain
under unidirectional tensil load [14] exhibits the role of the
rate-sensitivity index in controlling the total elongation of
tensile specimen before rupture intercepts.

TEMPERATURE DISTRIBUTION IN STRIP ROLLING OF RATE SENSITIVE MATERIAL

Z. K. Yao and Y. Chen*
Dept. of Mechanics and Materials Science
Rutgers, The State University of New Jersey
P.O. Box 909, Piscataway, NJ 08855-0909

SUMMARY
    In the paper by Frick and Chen [3] the stress field in strip rolling of a
rate sensitive material was solved. There would be a temperature rise in the
material due to the plastic work during deformation. Added to the original
steady state computational algorithm in [3] was the discretized heat conduction
equation and its associated boundary condations. The temperature solutions for
rolling speed at 2500 mm/sec at two defferent times were obtained. Plots of the
temperature distributions were presented.

INTRODUCTION

Investigations on the temperature rise during metal forming have shown that

nearly all the energy dissipated in the plastic deformation process is converted

into heat. Farren and Taylor [2] reported that for steel, copper and aluminium

the heat due to dissipation represents 86.5, 90.5 and 95 per cent, respectively,

of the plastic work. The remainder of the plastic work is stored as internal

energy associated with the deformations.

Zienkiewicz et al. [4] treated the forming of metals as a flow problem.

The constitutive equations used were rate independent. In [3] Bodner-Partom

constitutive equation was used, which is a strain rate dependent equation.

Briefly speaking, Bodner-Partom equation assumes the following exponential

from, i.e.,

$$D_2^P = D_0^2 exp\left[-[\frac{Z^2}{3J_2}]^n\right]$$

where

$$D_2^P = \frac{1}{2} d_{ij} d_{ij}$$

d_{ij} is the rate of deformation tensor D_0, a scalar factor related to the strain

rate, n, a strain rate dependent material constant, J_2, the second invariant of

the stress deviator, Z, an internal state variable. For details of Bodner-

Partom equation the readers are referred to [1].

---

*Currently visiting professor at the Institute of Applied Mechanics, National
Taiwan University, Taiwan, ROC.

MATHEMATICAL ANALYSIS

The steady state heat conduction can be written as follows,

$$K\nabla^2 T + Q = 0 \tag{1}$$

where T is the temperature, ∇^2, the Laplacian operator in two dimensions, K, the thermal conductivity and Q, the heat due to plastic work.

The rate of plastic work $\sigma_{ij}\epsilon_{ij}$ can be shown, using Bodner-Partom equation, to be

$$\sigma_{ij}\overset{\cdot}{\epsilon}_{ij} = \frac{2}{\sqrt{3}}D_0{}^2 exp\left[-\left[\frac{Z}{\tau^E}\right]^n\left(\frac{1}{2}\right)\right]\tau^E \tag{2}$$

where $\tau_j{}^E$ is the effective stress given by

$$\tau^E = \left[\frac{3}{2}S_{ij}S_{ij}\right]^{\frac{1}{2}} \tag{3}$$

The boundary conditions consist of two parts. For the free surface, the convective heat transfer applies, i.e.,

$$K\underline{n}^T\underline{\nabla}T = -h_c(T-T_c) \tag{4}$$

where $\underline{n}$ is the outward normal, h_c, the convective heat transfer coefficient and T_c, the ambient temperature. Over the contact area between the roller and the billet, Eq.(4) is changed by the roller temperature and it heat transfer coefficient as follows,

$$T_c = T_R \quad \text{and} \quad h_c = h_R \tag{5}$$

At both ends and center the adiabatic boundary conditions are imposed. Since the scale of the thickness of the billet is much smaller than the length, therefore the above assumptions at both ends are reasonable.

The heat conduction equation must be solved simultaneously with the equilibrium equation. Since the original algorithm [3] contains all the elements necessary for the solution of the stress field, the only additional analysis is to discretize the heat conduction system of equations, and implement the discretized system and incorporate it with the stress field through the plastic work term.

The discretization can be done in the standard fashion by assuming the following,

$$T = \sum_{j=1}^{n} N_j T_j \tag{6}$$

$$\tau^E = \sum_{j=1}^{n} \overline{N}_j \tau_j{}^E \tag{7}$$

where N_j and $\overline{N}_j$ are the shape function, T_j and $\tau_j{}^E$ are the nodal temperature and the nodal effective stress at node j. Applying the wellknown Galerkin method and using weighting functions W and $\overline{W}$ for equations 1, 4 and 5, we have

$$[\int_{\Omega}(K\nabla W^T\nabla N)d\Omega + \int_{\Gamma}(h_c WN)d\Gamma]T - \frac{2}{\sqrt{3}}D_o\int_{\Omega}\left[exp[-(\frac{1}{2})(Z/\bar{N}\tau^E)^{2n}]\bar{W}\bar{N}\tau^E\right]d\Omega$$

$$-\int_{\Gamma}h_c WT_c d\Gamma = 0 \tag{8}$$

Note that the superscript 'T' denotes the transpose, $W_j = -N_j$ and $\bar{W}_j = N_j$.
Eq. (8) can be abbreviated as

$$\underline{E}\ \underline{T} + \underline{q} = 0 \tag{9}$$

where

$$E_{ij} = \int_{\Omega}K\nabla W_i^T\nabla N_j d\Omega + \int_{\Gamma}h_c W_i N_j d\Gamma \tag{10}$$

and

$$q_i = -\frac{2}{\sqrt{3}}D_o\int_{\Omega}\left[exp[-(\frac{1}{2})(Z/\bar{N}_j\tau^E)^{2n}]\bar{W}_j\bar{N}_j\tau^E\right]d\Omega - \int_{\Gamma}h_c W_i T_c d\Gamma \tag{11}$$

The original computer program in [3] was run until a steady state was achieved.
At that time the plastic work was introduced into the new system of equations
implemented in this work. The dimension of the billet and all constant used in
calculation are displayed in Figure 1.

For roll speed of 2500 mm/sec the temperature distributions were shown in
Figures 2 and 3.

CONCLUDING REMARKS

A complementary computational program for temperature determination in strip
rolling has been developed and made operational in conjunction with the original
program in [3]. The temperature distribution for roll speed of 2500 mm/sec at
two different times were obtained.

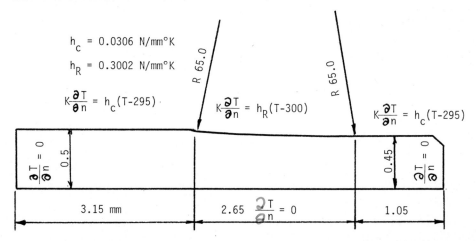

Figure 1: Billet Geometry and Boundary Conditions

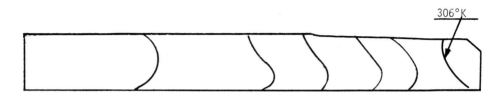

Figure 2:   Temperature Contours at 72% of Full Contact

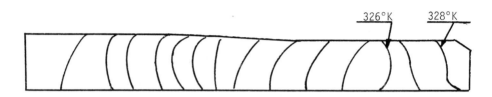

Figure 3: Temperature Contours at Full Contact (Entry T = 295° K)

REFERENCES

1 S. R. Bodner, and Y. Partom, Constitutive Equations for Elastic-Viscoplastic
   Strain Hardening Materials, ASME J. Appl. MECH 42 (1975) 385-389.
2 W. S. Farren, and T. I. Taylor, The Heat Developed during Plastic Extension
   of metal Proc. Roy. Soc. A, 107 (1925) 422.
3 R. Frick, and Y. Chen, Strip Rolling of Rate Sensitive Material, ASME
   Manufacturing Simulation and Processing, PEP-Vol. 20, 63-76, Dec. 1986.
4 O. C. Zienkiewicz, E. Onate, and J. C. Heinrich, A General Formulation for
   Coupled Thermal Flow of Metals using Finite Elements, International Journal
   for Numerical Methods in Engineering 17 (1981) 1497-1514.

Computational Methods for Predicting Material Processing Defects, edited by M. Predeleanu
Elsevier Science Publishers B.V., Amsterdam, 1987 — Printed in The Netherlands

DAMAGE DURING FORMING OF COATED SHEET STEEL

J.Z.Gronostajski

Technical University of Wrocław, ul.Łukasiewicza 3/5
50-371 Wrocław, Poland

SUMMARY

   The effect of deformation of zinc coated steel sheet on the
nucleation and propagation of voids and cracks in the zinc layer
and base steel was investigated. It has been found, that the density and size of voids and cracks were effected by strain paths.

INTRODUCTION

   The increasing demands for corrosion protection of very wide
range of structural components have increased the tonnage of coated
steel used in industry. The coated steel sheets play a very important role in the automotive industry. During forming of coated
steel sheet, besides the phenomena which limit the formability of
uncoated steel sheets, different types of coating damage often
occur, as cracking, flaking and galling. The damages decrease corrosion protection of steel sheets (refs.1-3).

   The cracking can be formed even if the elastoplastic properties
of coatings are better than those of the base steel, coatings could
break first as a results of the adverse thickness ratio. The lattice of crack could occur over the whole surface of deformed area.
The flaking is caused by different mechanical properties of base
steel and coatings. During deformation the shear stress at the contact surface of the two layers could be higher than the adhesion
force. The galling is connected with surface roughness of tools
and adhesion force between materials of tools and coatings. These
phenomena accelerate the corrosion damage of steel.

   The main aim of the paper is to determine the effect of deformation mode of the steel sheets on the above mentioned phenomena and
on the nucleation and propagation of voids inside the steel core
and the zinc coating.

EXPERIMENTAL PROCEDURE

   The material was a commercially produced zinc hot dip galvanized
steel. The coating was applied in a continuous hot-dip galvanized

line. The thickness of zinc layers was approximately 25 µm. Total thickness of sheet was about 0.8 mm. The steel contains 0.09% C, 0.44% Mn, 0.035% P, 0.018% S, 0.1% Ni and 0.1% Cu. The zinc layer contains 0.19% Al, 0.025% Fe, 0.01% Mg and 0.1% Pb. The mechanical properties of the as-received materials are listed in Table 1.

TABLE 1
Mechanical properties of investigated coated steel sheet

Inclination to rolling direction deg	Yield stress MPa	Ultimate tensile strength MPa	Work hardening coefficient	Normal anisotropy coefficient
0	304	341	0.134	1.06
45	317	354	0.155	0.80
90	324	355	0.182	1.28

Plane anisotropy coefficient calculated from $(r_{max} - r_{min})/\bar{r}$ is equal to 0.49. The method of determination of forming limit diagram was described elsewhere (ref.4). The forming limit diagrams were constructed using hemispherical and flat punches.

The cracking limit curve was evaluated on the specimens deformed to necking. From each specimens several samples of 10x10 mm were examined in a scanning electron microscope. Since the state of deformation of each specimen was known this could be related to the presence of zinc cracks. The results were used to draw cracking limit curves of the coated steel sheet.

The same specimens were also used to determine the coating flaking by adhesive tape tests. Where flaking was present, the zinc coating was ripped off by the tape, leaving the steel sheet completly clean.

The nucleation and growth of voids was observed on the fracture surface of sheet and on the cross-section of specimens in longitudinal direction by using optical and scanning electron microscopes. Such observations were performed after different stages of deformation and at different places on each specimen. The size and density of voids were measured only on the specimens deformed by uniaxial tensile, plane strain and equibiaxial stretching.

RESULTS AND DISCUSSION

Forming limit curve (FLC), flaking limit curve (FKLC) and cracking limit curve (CLC) of zinc hot-dip galvanized steel when ε_1

is parallel to rolling direction, for linear strain paths is shown
in Fig.1.

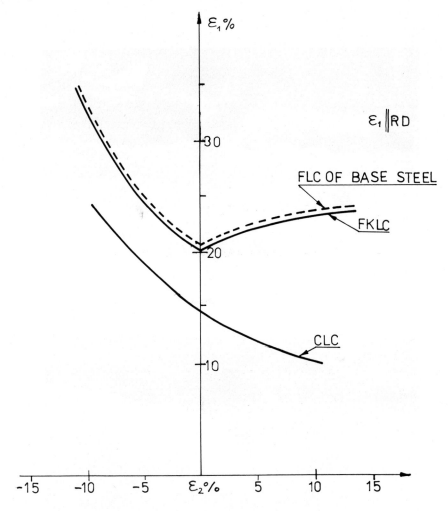

Fig.1. Forminmg limit curve, flaking limit curve and cracking limit
curve of the zinc hot dip-galvanized steel sheet.

From that figure one can notice that the cracking limit curve
lies at lower level than the forming limit curve and the flaking
limit curve. The discrepancy between these curves is rising in the
direction of positive value of ε_2. This is caused by a brittle
iron-zinc alloy layer between the zinc coating and the base steel.
The nucleation of the cracks takes place in the layer and propaga-

tes into the zinc coating and base steel (Figs. 2 and 3).

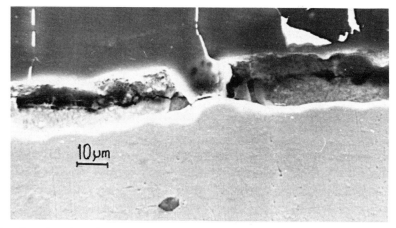

Fig.2. Cracks in zinc layer. Cross section.

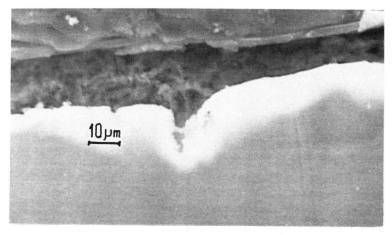

Fig.3. Cracks in zinc layer and base steel. Cross section.

In early forming stages the intercrystalline cracks were formed
for all forming strain paths (Fig.4). At higher straining also intracrystalline cracks were formed for plane strain deformation
(Fig.5) and sometimes for equibiaxial deformation (Fig.6). In the
Fig.6 the cracks were formed as a set of main parallel cracks and
a set of short perpendicular cracks. For uniaxial tension such

cracks were rarely observed.

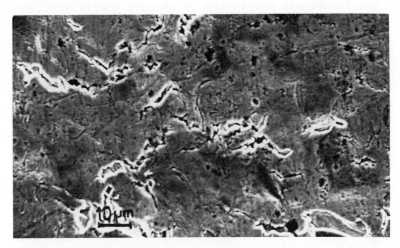

Fig.4. The coating studied in the normal direction. The sample is formed in equibiaxial stretching. The rolling direction is horizontal.

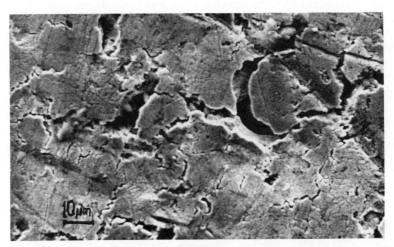

Fig.5. The coating studied in the normal direction. The sample is formed in plane strain. The rolling direction and major principal strain are horizontal.

At the bottom of the cracks the base steel is usually exposed (fig.7) and sometimes the intermetallic compound can be visible. The quantitative analysis of the area fraction of cracks f_s in

352

the deformed surface of specimens was determined by microcomputer
analyser. The area fractions of cracks determined for the plane
strain and equibiaxial deformation are shown in Fig.8. The solid
line on the figure was calculated by using a simple model suggested
by Mäkimattila and Ranta Escola (ref.5).

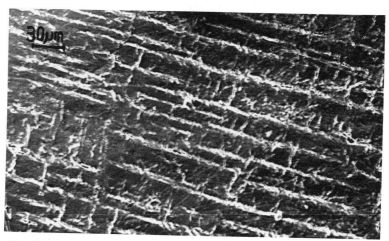

Fig.6. The coating studied in the normal direction. The sample is
formed in equibiaxial stretching. The rolling direction is horizontal.

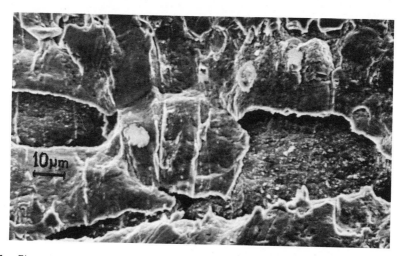

Fig.7. The coating studied in the normal direction. The sample is
formed in plane strain. The rolling direction and major principal
strain are horizontal.

In the model the coating is assumed to be stiff during plastic
straining. The area fraction of cracks is then determined by the
area increase due to deformation. Experimental curves fall below
the theoretical line, it means that the zinc layer was plastically
deformed.

Coefficient of zinc layer deformation k was defined by equation

$$f_s = 1 - exp \left[(1 - k) \right] \varepsilon_t \qquad (1)$$

where ε_t = thickness strain.

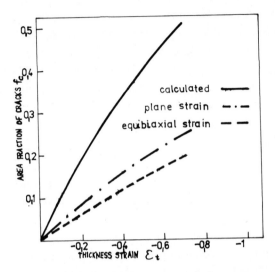

Fig.8. Area fractions of cracks as a function of effective strain.

If k=0, the coating is not deformed, if k=1 the coating deforms as
the base steel without cracks. Values of k calculated by means of
the least square method for plane strain is 0.7 and for equibiaxial
stretching is 0.8. It means that the formability of zinc layer is
lowest in plane strain. Similar value of k=0.8 was obtained by
Schedin et al. (ref.6) for the zinc hot-dip galvanized steel as a
average value for uniaxial tension, plane strain and equibiaxial
stretching.

Voids nucleated by decohesion at the interface between the inclusion and the matrix. Nucleation of voids by cracking of the second-
phase particles or inclusions has not been observed. Void growth
follows, and this may be enhanced due to strain concentration in
the vicinity of inclusion, caused by the interaction of the voids.
Void growth can be described by various equations each relating to
circumstances associated with void nucleation, but each of which

depends on the strain concentration near the particles being directly proportional to the frontal radius of curvature. At the start
of voids nucleation the frontal radius of curvature is equated with
the radius of inclusion.

If it is assumed that all of the inclusions behave in an identical manner, then it can be shown that the total volume fraction of
voids, f, can be expressed as (ref.7)

$$f = f_o \left\{ (1 + k/r^2) \exp[2 (\varepsilon - \varepsilon_o)] - k/r^2 \right\}^{1/2} \tag{2}$$

where f_o = initial volume fraction of inclusion; r = length to
width ratio of the inclusion; ε = true strain; ε_o = constant of
integration; and k = strain concentration constant.

Melander (ref.8) assumes that inclusions are stiff during metal
forming and voids have already occured during straining. This means
that voids exist in the as-received sheet and can grow from the beginning of metal forming, accordingly to equation

$$f = f_o \exp (\varepsilon + \varepsilon_v) - 1 \tag{3}$$

where ε_v = fraction of voids in the as-received sheet.

Based on experimental results of Gladman et al (ref.7) the
strain concentration constant, k, was 2. The ratio of length to
width of inclusion, r, determined by using micro-computer analizer
Robotron was 0.2. Metallographic examination of cross section revealed that voids were formed at low strains, i. e. ε_o = o, and
that initial void fraction is not higher than 1%. The theoretical
calculation gives lower values than experimental results; this should be due to coalescence of voids. The coalescence of voids was
associated with unusually closed spaced particles. The theory assumes a uniformity of particle distribution which is not usually observed in practice. This effect could explain the increasing positive deviation from predicted void fractions with the strain.

The voids have a softening effect on the material which accele-
rates necking if a large density of inclusions is nonuniformly distributed in the sheet. Voids generation and growth at inclusion
might initiate the necking because the increasing void fractions
decrease the stress and work hardening ability of steel sheet. The
effect of voids growth usually reduces FLD, but more complicated
situation might arise if voids nucleation occur during forming.

The equibiaxial stretching causes more intensive growth of voids
than uniaxial tensile deformation.

The softening proceeds more rapidly in the inclusion rich region
of the sheet than in the surrounding material even if the strain

levels are identical. Softening promotes strain localization and leads to sharp necking in the inclusion rich region of the sheet. From that point of view, the higher value of voids density at plane strain deformation (Fig.9) that at other strain path (Fig.10) might be the main reason of lowest formability in plane deformation (Fig.1).

CONCLUSION

During deformation of the zinc hot-dip galvanized steel the voids were nucleated by decohesion between matrix and inclusion. The density and size of voids are dependent on the strain path, the highest values were obtained for plane deformation. Determined void density as a function of effective strain can be used for calculation of FLC in the manner proposed by Melander (ref.8).

Cracks in the layer of zinc nucleate and grow during deformation as intercrystalline or intracrystalline. The kind of cracks is dependent on the mode of deformation. The coefficient k which defines the formability of coating varies with strain path.

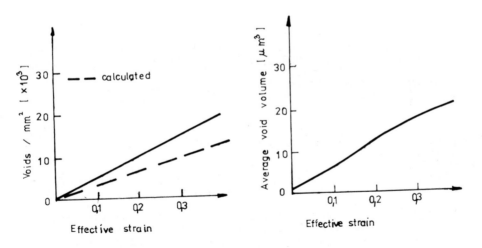

Fig.9. Void density and average void volume as a function of effective strain during plane deformation.

356

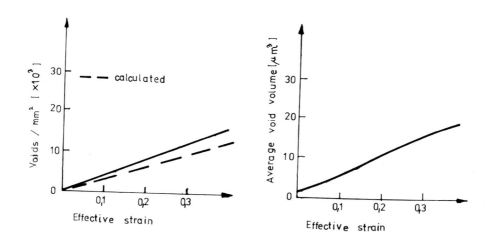

Fig.10. Void density and average void volume as a function of effective strain during equibiaxial stretching.

REFERENCES

1. G.Arrgoni, M.Sarracino, Effect of press forming deformation on the corrosion resistance of precoated steel sheets for automoti- ve industry, Proc.of IDDRG, Working Meeting, Amsterdam, May 23- 24, 1985, paper 11, pp.1-17.
2. G.Blanchard, Comparative study using simulative tests on the press formability of coated high strength steel sheets, Proc.of IDDRG Working Meeting, Amsterdam, May 23-24, 1985, paper 18, pp.1-8.
3. J.Z.Gronostajski, M.S.Ghattas, W.J.Ali, The effect of zinc and plastic coating on the formability of steel sheet, Proc.of 14th Biennial Congress of IDDRG, Köln, April 21-23, 1986, pp.454-456.
4. J.Z.Gronostajski, W,J.Ali, M.S.Ghattas, The effect of strain path on the FLD of coated steel sheet, Proc. of 2nd Internatio- nal Conference of Plasticity, Stuttgart, August 24-28, 1987, in press.
5. S,Mäkimattila, A.Ranta-Escola, Proc. of 3th Biennial Congress of IDDRG, Melbourne, February 18-20, 1984, pp.293-304.
6. E.Schedin, S.Karlsson, A.Melander, Damage development in a hot- dip galvanized coating during forming. Proc.of 14th Biennial Congress of IDDRG, Köln. April 21-23, 1986, pp.460-461.
7. T.Gladman, B.Holmes, I.D.McIvor, Effects of second-phase partic- les on strength, toughness and ductility, JISI, 1976, pp.68-78.
8. A.Melander, Comparison between experimental and theoretical forming limit diagrams, Proc.of ICM4, Stockholm 1983, pp.33-49.

Computational Methods for Predicting Material Processing Defects, edited by M. Predeleanu
Elsevier Science Publishers B.V., Amsterdam, 1987 — Printed in The Netherlands

DEFECTS AT INTERFACES IN CO-EXTRUDED METALS

C. S. HARTLEY[1], M. DEHGHANI[2], N. IYER[3], A. T. MALE[3], and W. R. LOVIC[3]

[1]Program Director, Metallurgy, National Science Foundation, Washington, D. C. 20550 U.S.A.

[2]Ph.D. Candidate, Mechanical Engineering Department, Louisiana State University, Baton Rouge, LA 70803 U.S.A.

[3]Westinghouse Research & Development Center, Pittsburgh, PA U.S.A.

SUMMARY

The strength of the interface between core and clad in hydrostatically co-extruded copper-clad aluminum has been measured as a function of the total extrusion ratio. It is found that the interface strength exhibits a minimum at a particular extrusion ratio. This phenomenon is investigated analytically by modeling the co-extrusion process using a finite element code, ABAQUS, to determine the stresses at the interface in the deforming region of the die and in the extruded product. These calculations show that the state of stress at this interface depends on the die geometry and on the extrusion ratio. The implications of this observation on the design of dies for co-extrusion are discussed.

INTRODUCTION

Composite materials of two or more different material layers are used for engineering and economic reasons; consequently, the processing of such materials has been of interest for several years. Among the simplest forms of such composite materials are bi-metal rods; these are rods of two dissimilar metals wherein the core of one metal is surrounded by a concentric sleeve of another metal. The two metals are intimately interlocked to function in unison. The usefulness of bi-metal rods stems from the possibilities presented for combination of qualities of dissimilar metals. Typical commercial uses for these materials include clad fuel elements for nuclear applications, copper-clad aluminum wires for the electrical industry, complex multi-filamentary wires having Al or Cu matrices for superconducting applications, etc.

Various techniques have been used for producing bi-metal configurations. Of these, the co-extrusion process offers the most versatility, and a number of researchers have investigated the technique with objectives ranging from determination of the forming loads (ref. 1) to the analysis of residual stress patterns developed in the product (ref. 2), and finally to the investigation of the failure modes and crack formation (ref. 3). These analyses, however, are not sufficient to evaluate the state of stress at the interface of bi-metallic products, and little systematic effort has been expended in developing the relationship between the processing variables and the interfacial shear strength of such products.

In this study, an index Δ, defined as the ratio of the mean deformation zone diameter, d, to the contact length between tool and work metal, L is used to relate the process variables (die angle and extrusion ratio) and the inter-facial bond strength of copper-covered aluminum rods. Low values of Δ repre-sent a combination of low die angles and high reduction ratios, while high values of Δ represent a combination of high die angles and low reduction ratios.

The interfacial shear strength of hydrostatically extruded copper-clad aluminum rods having 75 percent and 85 percent aluminum cores were determined experimentally by the interfacial strength tests (ref. 4). In a parallel theoretical investigation, the co-extrusion process was modeled numerically using a large deformation finite element code, ABAQUS, and calculated values of the bond strength are compared with those measured experimentally.

THEORETICAL BACKGROUND

Finite Element Program

The governing equation which serves as the fundamental formulation of the elasto-plastic program, ABAQUS, is the well-known formulation of McMeeking and Rice (ref. 5), which was further improved for incremental procedures by Lee and Mallett (ref. 6), and is given as

$$\int_V (\rho/\rho^o)(\overset{*}{\tau}_{ij}\delta\dot{\varepsilon}_{ij} + \tau_{ij}\delta(\dot{\varepsilon}_{ik}\dot{\varepsilon}_{kj} - v_{k,i}v_{k,j}/2))dv = \int_S \dot{f}_i \delta v_i ds + \int_V \dot{b}_i \delta v_i dv \tag{1}$$

The constitutive equation for an elastic-plastic material is given gene-rally as a combination of a linear elastic behavior, usually Hooke's Law, a yield criterion, usually that due to von Mises or Tresca, and the Prandtl-Reuss equations relating stress components to plastic strain rates. The mathematical description of this combination defines the stress-strain relationships for a given material, and they can be inverted and generalized to express the stress rate in terms of the rate of deformation. This was carried out by Yanada, et. al (ref. 7), who obtained the expression

$$\overset{*}{\tau}_{ij} = 2G(\varepsilon_{ij} + \frac{\upsilon}{1-2\upsilon}\delta_{ij}\dot{\varepsilon}_{kk}) + \frac{1}{1+(h/3G)} \cdot \frac{3}{2} \cdot \frac{\tau'_{ij}\tau'_{kl}}{\bar{\tau}^2}\dot{\varepsilon}_{kl}, \tag{2}$$

where $\overset{*}{\tau}_{ij}$ is the Jaumann stress rate tensor, and the primes indicate deviatoric components. An overview of the above equations is given in (ref. 8).

INTERFACIAL STRENGTH EVALUATION

Residual and Interfacial Stress Evaluations

Let σ_i and σ_o be the effective flow stress of the inner and outer materials, respectively, both measured at the value of R (extrusion ratio) by

$$\bar{\sigma} = 1/\varepsilon_f \int_o^{\varepsilon_f} \sigma d\varepsilon \qquad (3)$$

During plastic deformation, both core and clad are constrained to experience the same total deformation; therefore, the total axial strain is given as

$$\varepsilon_T = \ln R_i - (2\bar{\sigma}_i/3B) \qquad (4)$$

or

$$\varepsilon_T = \ln R_o - (2\bar{\sigma}_o/3B) \qquad (5)$$

At the end of the forming process and before removal of the force, the elastic extensions of core and clad are:

$$\delta_i = \bar{\sigma}_i L/E_i \qquad \delta_o = \bar{\sigma}_o L/E_o \qquad (6)$$

where E is the Young's modulus and L is the length of the work piece. Upon removal of the external forces, however, each component contracts elastically and the elastic extensions after relaxation are

$$\delta_i' = P_{Ri} L/A_i E_i \qquad \delta_o' = P_{Ro} L/A_o E_o \qquad (7)$$

where A_i and A_o are the cross-sectional areas of the core and clad, respectively. With the constraint that $P_{Ri} + P_{Ro} = 0$, the residual stresses in the core and the clad are given, respectively, as

$$\sigma_{Ri} = P_{Ri}/A_i \qquad \sigma_{Ro} = P_{Ro}/A_o \qquad (8)$$

Furthermore, if the product is to be sound and if there is not to be a failure at the interface, the elastic relaxation must occur such that each component contracts by the same amount, i.e.,

$$\delta_i - \delta_i' = \delta_o - \delta_o' \qquad (9)$$

Therefore, given equations (6) and (7) and the fact that $P_{Ri} = -P_{Ro}$, equation

(9) becomes

$$\bar{\sigma}_i/E_i - P_{Ri}/A_iE_i = \bar{\sigma}_o/E_o + P_{Ri}/A_oE_o$$

or $\qquad\qquad\qquad\qquad\qquad\qquad\qquad\qquad\qquad\qquad\qquad\qquad$ (10)

$$P_{Ri} = [\bar{\sigma}_i/E_i - \bar{\sigma}_o/E_o] [(A_iE_iA_oE_o) / (A_iE_i + A_oE_o)] \qquad\qquad (11)$$

and

$$\sigma_{Ri} = [(\bar{\sigma}_iE_o - \bar{\sigma}_oE_i)(1-f)] / [fE_i + (1-f)E_o] \qquad\qquad (12)$$

where f is defined as

$$f = A_i/A_T \qquad\qquad\qquad\qquad\qquad\qquad\qquad\qquad\qquad\qquad (13)$$

and A_T is the total cross-sectional area. Similarly

$$P_{Ro} = -P_{Ri} = [(\bar{\sigma}_o/E_o) - (\bar{\sigma}_i/E_i)] [(A_iE_iA_oE_o) / (A_iE_i + A_oE_o)] \qquad (14)$$

and

$$\sigma_{Ro} = [(\bar{\sigma}_oE_i - \bar{\sigma}_iE_o)f] / [fE_i + (1-f)E_o] \qquad\qquad (15)$$

In the above equations, $\bar{\sigma}_o$ and $\bar{\sigma}_i$ are the average flow stresses of the core and the clad at a plastic strain corresponding to an extrusion ratio R.

Finally, the interfacial shear strength between the core and the clad is calculated according to the equation

$$\tau_{rz} = F/2\pi rl \qquad\qquad\qquad\qquad\qquad\qquad\qquad\qquad\qquad (16)$$

where F is the higher value of the axial compressive or tensile forces in the core and clad, respectively; r is the radius of the interface; and l is the length equal to that of the test specimen.

EXPERIMENTAL PROCEDURES

Extrusion experiments

The extrusion experiments were carried out using a vertical press of the Bridgman-Birch type, equipped with a nominal 38mm diameter bore container which allowed a maximum extrusion pressure of 760 Mpa, Figure 1a. The press is also equipped with a versatile hydraulic system which facilitates the use of constant ram velocities up to 14.8 mm/sec. A pressure-compensated system was preset to control the ram velocity to assure "stick-slip"-free extrusions under most conditions. The instrumentation of this system, which includes a pressure and a linear transducer input module, transducer-amplifier, and limit control output module, was calibrated prior to each experiment.

The die design used in this study is shown schematically in Figure 1b. As it can be seen from the figure, the dies had a geometry consisting of an approach angle, a bearing length, a 10° included-relief angle, and various die orifice diameters. Four dies of included angles 30°, 45°, 90°, and 120°, and die orifice diameter of 5 mm were used. All the dies had been fabricated from A2 tool steel and hardened to R_c 60-62. The approach angles and bearing lengths were polished to a 200 μmm surface finish for smoother contact with the work piece. The dies were positioned at the same location in the high-pressure chamber for all experiments in order to maintain a near-constant volume of pressure-transmitting fluid.

DIE	DIE DESIGN			
#	A	B	C	D
1	3.8	22.4	30°	10°
2	5.0	22.4	60°	10°
3	3.2	22.4	90°	10°
4	3.2	22.4	120°	10°
DIMENSIONS IN mm				

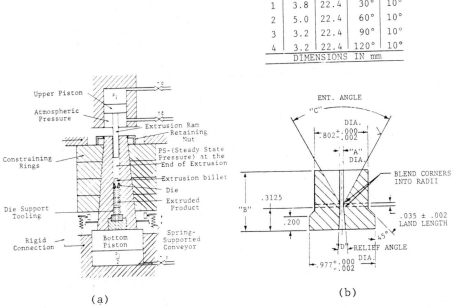

Fig. 1. (a) Apparatus for hydrostatic extrusion. (b) Extrusion die design.

The materials used in this program were annealed OFHC copper and grade 1350(EC) aluminum. The aluminum was annealed at 350°C for three hours in an argon atmosphere, while the copper was annealed at 450°C in a hydrogen atmosphere. Composite billets consisting of copper-clad/aluminum cores were fabricated for the investigation, in which the volume fractions of the cores were 0.75 and 0.85. The cores were machined for a sliding fit to facilitate ease in assembly. Steel end plugs were attached to the billets with epoxy adhesive to prevent unrestrained extrusion. Total billet lengths were 15.25 cm with a 30° to 60° included conical nose for corresponding die angles in the 30° to 120° range. The mismatch of cone angles beneficially influenced the breakthrough pressure. The billets' outer diameters were machined to an 800 μmm finish. Composite billets were extruded at six extrusion ratios ranging from 2 to 25, through dies of included angles 30°, 45°, 90°, and 120°. Every attempt was made to set the ram velocity at a value so as to achieve the same average strain rate during all the extrusion experiments. Extrusion was stopped in time to preserve approximately 50 mm of the unextruded billet for interface strength studies.

INTERFACIAL STRENGTH MEASUREMENTS

The core-clad interfacial bond strength was measured by the interfacial strength tests (ref. 4). Interfacial shear strength specimens were machined from both unextruded and extruded portions of the clad composites as detailed in Figure 2. The cross-sectional area of the core was measured prior to

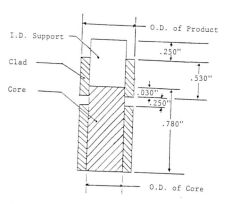

Fig. 2. Schematic diagram of an interfacial strength test specimen.

testing. The interface strength tests were conducted on an Instron Universal testing machine at a crosshead speed of 1.27 mm/min. The load was monitored as a function of crosshead travel. The shear area was measured following the test and interface strength calculated, both as a function of extrusion ratio (at a constant die angle) and as a function of the die angle.

FINITE ELEMENT SIMULATION

A large deformation finite element program, ABAQUS, was used to simulate the hydrostatic extrusion problem. Several extrusion ratios ranging from 2 to 14, in conjunction with die angles in the 30° to 120° range, were simulated. Composite billets consisting of 75 and 85 volume percent aluminum cores clad with copper were modeled. The geometries of the simulations were congruent with those of the experimental part of the investigation.

The problem was treated axisymmetrically and, therefore, half of the meridian planes of the billets were discretized as the domain of the problem. Depending upon the extrusion ratio, the domain was divided into a 42 X 6, 42 X 8, or 42 X 10 grid. Four-node axisymmetric hybrid elements with four integration stations were used, and the dies were simulated as properly-shaped rigid surfaces. A coefficient of friction of 0.05 was used at the die-billet interface, and uniform pressure boundary conditions were imposed on the exposed surfaces of the billets to simulate the hydrostatic fluid pressure.

The flow curve--that is, the equivalent true stress-equivalent true plastic strain curve--was given by a saturation type equation of the form

$$\bar{\sigma} = \sigma_s - (\sigma_s - \sigma_m) \exp \left(\frac{-\sigma_m \varepsilon_m}{\sigma_s \sigma_m} \left(\frac{\bar{\varepsilon}}{\varepsilon_m} - 1 \right) \right) \tag{17}$$

where σ_s is the saturation stress, σ_m and ε_m are the stress and strain at maximum load, and $\bar{\sigma}$ is the equivalent stress. Table 1 is the list of the material properties used in this study (ref. 9).

TABLE 1
Material properties for aluminum and copper.

	Aluminum	Copper
Saturation stress, σ_s (Mpa)	141.18	433.35
Stress at maximum load, σ_m (Mpa)	115.19	322.10
Yield stress, σ_0 (Mpa)	47.90	62.30
Modulus of elasticity, E (Gpa)	68.00	128.00
Shear modulus, G (Gpa)	26.16	48.30
Strain at maximum load ε_m	0.2884	0.4152

RESULTS

The experimental and numerical results for the interfacial strength between the core and the clad of various billet compositions are obtained and presented in Figures 3 and 4. It appears that good-quality product was obtained even when the interfacial strength between the components in the billet was as low as a few Mpa.

Figure 3a shows the relationship between change in the interfacial shear strength and the extrusion ratio. Although there is a significant difference between the numerical and experimental results, both techniques demonstrate the same behavior. In the experimental results, the increase in the interfacial strength exhibits a minimum at an extrusion ratio of 10 to 11 for both the Cu-clad/75% Al and Cu-clad/85% Al products. In the numerical results, however,

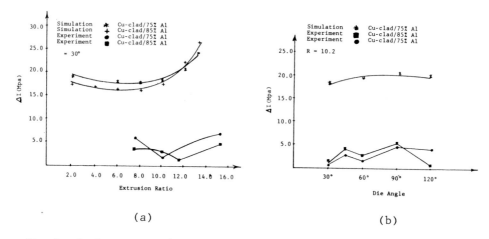

(a)   (b)

Fig. 3. Variation of the increase in interfacial strength in co-extruded bi-metallic compounds. (a) Change in interfacial strength as a function of extrusion ratio. (b) Change in interfacial strength as a function of die angle.

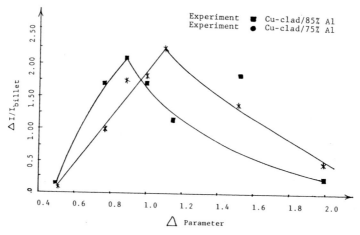

Fig. 4. Change in interfacial strength as a function of deformation zone geometry Δ.

this minimum occurs at an extrusion ratio of approximately 8.

Figure 3b shows the increase in the interfacial strength as a function of die angle. Both the numerical and experimental results correspond to a single value of extrusion ratio, namely 10.2. Again, the variation of the increase in interfacial strength obtained numerically is similar to that of the experiments. The two results, however, show a noticeable difference in magnitude. Although the change in die angle alone does not reveal any significance in the variation of the interfacial strength, its importance is demonstrated when this variation is plotted as a function of Δ of the deformation zone geometry, Figure 4.

DISCUSSION AND CONCLUSIONS

The results of previous sections indicate the importance of process variables such as extrusion ratio and die angle on the bond strength between the core and the clad of the extruded products.

Comparison of the numerical and experimental results of Figure 3 indicates that although the two results are qualitatively similar, they are quantitatively different. The explanation could lie in the fact that the interfacial strength experiments were conducted on extruded samples which had been heated considerably by deformation during the extrusion process and, thus, the temperature rise caused recovery to occur, altering the state of stress. The finite element simulation, on the other hand, completely ignores the temperature effect and, consequently, no consideration of heat is incorporated into the numerical results. More significantly, the circumferential machining of the clad during the sample preparation also caused significant alteration in the residual stresses presented in the extruded samples. These facts suggest that the actual results must lie between the experimental and numerical results, and future improvements on both techniques could further converge the two results.

In Figure 4, it is observed that there is an optimal value of the die angle-reduction ratio combination when the increase in interfacial strength reaches a maximum. This maximum occurs when Δ approaches 1.00. This observation can be rationalized on the basis of the normal stresses at the interface and the degree of proportional flow at the various extrusion conditions. At low values of Δ, the die length is large; consequently, the normal stresses at the die-billet interface are small, resulting in low values of normal stresses at the core-clad interface and in little increase in the interfacial strength. Alternatively, at high values of Δ, a high degree of inhomogeneous deformation results in a low degree of proportional flow, thus resulting in poor strength in the product. Although high interfacial strength in the billet would suggest a high degree of proportional flow during extrusion, and consequently high interfacial strength in the product, the results here show that improvement in interfacial strength is directly dependent on the combination of die angle and extrusion ratio used.

In conclusion, however, the results of this work show that finite element analysis in conjunction with experimental measurements can be applied successfully to predict and verify the qualitative interfacial strength developed in hydrostatically extruded products.

ACKNOWLEDGMENTS

This paper is based on research supported by the National Science Foundation Grant No. DMR-8400991, "Deformation Processing by Hydrostatic Coextrusion." The authors wish to thank the Louisiana Department of Transportation and Development for the use of their computer facilities. This research forms part of a dissertation submitted by one of the authors (MD) to the Graduate School of Louisiana State University, in partial fulfillment of the requirements for the Ph.D. in Mechanical Engineering.

REFERENCES

1   J.M. Alexander and C.S. Hartley, On the hydrostatic extrusion of copper-covered aluminum rods, Hydrostatic Extrusion, (1974) 72-78.
2   C.S. Hartley and M. Dehghani, Residual stresses in axisymmetrically formed products, 2nd International Conference on The Technology of Plasticity, Stuttgart, Germany, Aug. 24-28, 1987.
3   B. Avitzur, Metal Forming: Processes and Analysis, Teta-McGraw-Hill, New Delhi (1977).
4   R. Lugosi, C.S. Hartley and A.T. Male, The influence of interfacial shear yield strength on the deformation mechanics of an axi-symmetric two component system, Westinghouse Research Lab. Tech. Document No. 77-1D4-PROEN-P1, (1977).
5   R.M. McMeeking and J.R. Rice, Finite-element formulation for problems of large elastic-plastic deformation, Int. J. Solids Structures, 11 (1975) 601-616.
6   E.H. Lee, R.L. Mallett, and W.H. Yang, Stress and deformation analysis of the metal extrusion process, Comp. Meth. Appl. Mech. Engg. 10 (1977) 339-353.
7   Y. Yanada, N. Yoshimura, and T. Sakurai, Plastic stress-strain matrix and its application for the solution of elastic-plastic problems by the finite element method, Int. J. Mech. Sci. 10 (1968) 343-354.
8   M. Dehghani, Simulation of hydrostatic co-extrusion of bimetallic composites. Ph.D. thesis in preparation, Louisiana State University (1987).
9   R. Srinivasan, Study of the hydrostatic co-extrusion of aluminum and copper, Ph.D. thesis, State University of New York at Stony Brook (1983).